Compendium of Plant Genomes

Series Editor

Chittaranjan Kole, Raja Ramanna Fellow, Government of India,
ICAR-National Research Center on Plant Biotechnology, Pusa,
New Delhi, India

Whole-genome sequencing is at the cutting edge of life sciences in the new millennium. Since the first genome sequencing of the model plant Arabidopsis thaliana in 2000, whole genomes of about 100 plant species have been sequenced and genome sequences of several other plants are in the pipeline. Research publications on these genome initiatives are scattered on dedicated web sites and in journals with all too brief descriptions. The individual volumes elucidate the background history of the national and international genome initiatives; public and private partners involved; strategies and genomic resources and tools utilized; enumeration on the sequences and their assembly; repetitive sequences; gene annotation and genome duplication. In addition, synteny with other sequences, comparison of gene families and most importantly potential of the genome sequence information for gene pool characterization and genetic improvement of crop plants are described.

Interested in editing a volume on a crop or model plant? Please contact Prof. C. Kole, Series Editor, at ckoleorg@gmail.com

More information about this series at http://www.springer.com/series/11805

Minami Matsui • Keng-See Chow
Editors

The Rubber Tree Genome

Springer

Editors
Minami Matsui
RIKEN Center for Sustainable Resource Science
Yokohama, Kanagawa, Japan

Keng-See Chow
Academy of Sciences Malaysia
Subang Jaya, Selangor, Malaysia

ISSN 2199-4781 ISSN 2199-479X (electronic)
Compendium of Plant Genomes
ISBN 978-3-030-42257-8 ISBN 978-3-030-42258-5 (eBook)
https://doi.org/10.1007/978-3-030-42258-5

This Springer imprint is published by the registered company Springer Nature Switzerland AG
The registered company address is: Gewerbestrasse 11, 6330 Cham, Switzerland

This book series is dedicated to my wife Phullara and our children Sourav and Devleena

Chittaranjan Kole

Preface to the Series

Genome sequencing has emerged as the leading discipline in the plant sciences coinciding with the start of the new century. For much of the twentieth century, plant geneticists were only successful in delineating putative chromosomal location, function, and changes in genes indirectly through the use of a number of "markers" physically linked to them. These included visible or morphological, cytological, protein, and molecular or DNA markers. Among them, the first DNA marker, the RFLPs, introduced a revolutionary change in plant genetics and breeding in the mid-1980s, mainly because of their infinite number and thus potential to cover maximum chromosomal regions, phenotypic neutrality, absence of epistasis, and codominant nature. An array of other hybridization-based markers, PCR-based markers, and markers based on both facilitated construction of genetic linkage maps, mapping of genes controlling simply inherited traits, and even gene clusters (QTLs) controlling polygenic traits in a large number of model and crop plants. During this period, a number of new mapping populations beyond F_2 were utilized and a number of computer programs were developed for map construction, mapping of genes, and for mapping of polygenic clusters or QTLs. Molecular markers were also used in the studies of evolution and phylogenetic relationship, genetic diversity, DNA fingerprinting, and map-based cloning. Markers tightly linked to the genes were used in crop improvement employing the so-called marker-assisted selection. These strategies of molecular genetic mapping and molecular breeding made a spectacular impact during the last one and a half decades of the twentieth century. But still they remained "indirect" approaches for elucidation and utilization of plant genomes since much of the chromosomes remained unknown and the complete chemical depiction of them was yet to be unraveled.

Physical mapping of genomes was the obvious consequence that facilitated the development of the "genomic resources" including BAC and YAC libraries to develop physical maps in some plant genomes. Subsequently, integrated genetic–physical maps were also developed in many plants. This led to the concept of structural genomics. Later on, emphasis was laid on EST and transcriptome analysis to decipher the function of the active gene sequences leading to another concept defined as functional genomics. The advent of techniques of bacteriophage gene and DNA sequencing in the 1970s was extended to facilitate sequencing of these genomic resources in the last decade of the twentieth century.

As expected, sequencing of chromosomal regions would have led to too much data to store, characterize, and utilize with the-then available computer software could handle. But the development of information technology made the life of biologists easier by leading to a swift and sweet marriage of biology and informatics, and a new subject was born—bioinformatics.

Thus, the evolution of the concepts, strategies, and tools of sequencing and bioinformatics reinforced the subject of genomics—structural and functional. Today, genome sequencing has traveled much beyond biology and involves biophysics, biochemistry, and bioinformatics!

Thanks to the efforts of both public and private agencies, genome sequencing strategies are evolving very fast, leading to cheaper, quicker, and automated techniques right from clone-by-clone and whole-genome shotgun approaches to a succession of second-generation sequencing methods. The development of software of different generations facilitated this genome sequencing. At the same time, newer concepts and strategies were emerging to handle sequencing of the complex genomes, particularly the polyploids.

It became a reality to chemically—and so directly—define plant genomes, popularly called whole-genome sequencing or simply genome sequencing.

The history of plant genome sequencing will always cite the sequencing of the genome of the model plant Arabidopsis thaliana in 2000 that was followed by sequencing the genome of the crop and model plant rice in 2002. Since then, the number of sequenced genomes of higher plants has been increasing exponentially, mainly due to the development of cheaper and quicker genomic techniques and, most importantly, the development of collaborative platforms such as national and international consortia involving partners from public and/or private agencies.

As I write this preface for the first volume of the new series "Compendium of Plant Genomes," a net search tells me that complete or nearly complete whole-genome sequencing of 45 crop plants, eight crop and model plants, eight model plants, 15 crop progenitors and relatives, and three basal plants is accomplished, the majority of which are in the public domain. This means that we nowadays know many of our model and crop plants chemically, i.e., directly, and we may depict them and utilize them precisely better than ever. Genome sequencing has covered all groups of crop plants. Hence, information on the precise depiction of plant genomes and the scope of their utilization are growing rapidly every day. However, the information is scattered in research articles and review papers in journals and dedicated Web pages of the consortia and databases. There is no compilation of plant genomes and the opportunity of using the information in sequence-assisted breeding or further genomic studies. This is the underlying rationale for starting this book series, with each volume dedicated to a particular plant.

Plant genome science has emerged as an important subject in academia, and the present compendium of plant genomes will be highly useful to both students and teaching faculties. Most importantly, research scientists involved in genomics research will have access to systematic deliberations on the plant genomes of their interest. Elucidation of plant genomes is of interest not only for the geneticists and breeders, but also for practitioners of an array of plant science disciplines, such as taxonomy, evolution, cytology, physiology, pathology, entomology, nematology, crop production, biochemistry, and obviously bioinformatics. It must be mentioned that information regarding each plant genome is ever-growing. The contents of the volumes of this compendium are, therefore, focusing on the basic aspects of the genomes and their utility. They include information on the academic and/or economic importance of the plants, description of their genomes from a molecular genetic and cytogenetic point of view, and the genomic resources developed. Detailed deliberations focus on the background history of the national and international genome initiatives, public and private partners involved, strategies and genomic resources and tools utilized, enumeration on the sequences and their assembly, repetitive sequences, gene annotation, and genome duplication. In addition, synteny with other sequences, comparison of gene families, and, most importantly, the potential of the genome sequence information for gene pool characterization through genotyping by sequencing (GBS) and genetic improvement of crop plants have been described. As expected, there is a lot of variation of these topics in the volumes based on the information available on the crop, model, or reference plants.

I must confess that as the series editor, it has been a daunting task for me to work on such a huge and broad knowledge base that spans so many diverse plant species. However, pioneering scientists with lifetime experience and expertise on the particular crops did excellent jobs editing the respective volumes. I myself have been a small science worker on plant genomes since the mid-1980s and that provided me the opportunity to personally know several stalwarts of plant genomics from all over the globe. Most, if not all, of the volume editors are my longtime friends and colleagues. It has been highly comfortable and enriching for me to work with them on this book series. To be honest, while working on this series I have been and will remain a student first, a science worker second, and a series editor last. And I must express my gratitude to the volume editors and the chapter authors for providing me the opportunity to work with them on this compendium.

I also wish to mention here my thanks and gratitude to the Springer staff, particularly Dr. Christina Eckey and Dr. Jutta Lindenborn for the earlier set of volumes and presently Ing. Zuzana Bernhart for all their timely help and support.

I always had to set aside additional hours to edit books beside my professional and personal commitments—hours I could and should have given to my wife, Phullara, and our kids, Sourav and Devleena. I must mention that they not only allowed me the freedom to take away those hours from them but also offered their support in the editing job itself. I am really not sure whether my dedication of this compendium to them will suffice to do justice to their sacrifices for the interest of science and the science community.

Kalyani, India Chittaranjan Kole

Preface

Hevea brasiliensis Müll. Arg., also known as the para rubber tree, is the sole species that is cultivated for the commercial production of natural rubber. Natural rubber (or *cis*-polyisoprene) is synthesized in the latex of this tree and harvested by systematic tapping of the tree trunk. The productive life of the rubber tree may extend beyond 20 years from first bark tapping in plantations managed under optimal agronomic practices. The genus *Hevea* (2n = 36) comprises 10 species, however only the *brasiliensis* species has been commercialized on a wide scale, primarily in Southeast Asian countries. The success of rubber production facilitated the growth of the automobile, medical, aviation, electronics, and building and construction industries, not to mention the manufacturing of a substantial range of rubber products essential to daily human life. Southeast Asian countries generate more than 95% of the global natural rubber production to meet increasing world demand. Through the effort of breeding and selection, the production capacity of modern rubber clones has reached more than four times that of the original parental trees. Nonetheless, the plantation industry is still facing the threat of diseases and the limitations of a narrow gene pool for breeding.

The significance of the rubber tree as a world crop warrants investment into genome-based technologies to deepen the current understanding of tree physiologies, particularly in latex, which affect production capacity. The rubber tree has accumulated a wealth of fundamental biology based on conventional scientific approaches spanning most of the twentieth century. Genomics and biotechnology approaches complement the existing knowledgebase and have great potential in discovering genes, proteins, and other regulatory components that could be engineered to enhance tree productivity. The period 2013–2017 was pivotal for rubber genome sequencing: the first draft genome sequence was published in 2013, followed by the release of three higher quality genome sequences of rubber tree clones using advanced sequencing technologies. Concurrently, genome-based investigations flourished as evidenced by the large number of publications by members of the rubber tree research community, particularly among the molecular biologists, physiologists, geneticists, and breeders.

In this book, we have compiled recent genome-related research, beginning with the experiences of the first three research groups in sequencing the genomes of rubber clones Reyan 7-33-97, RRIM 600, and BPM 24. The importance of harnessing benefits from genome information is reflected in chapters discussing the development of markers, genetic linkage maps, marker-trait association, and databases to facilitate data sharing and utilization. Applications of transcriptome and proteome analyses of major metabolic pathways involved in rubber biosynthesis and latex yield are also included. This book has drawn on the experience of contributors with a broad range of expertise in rubber tree research and/or background in genomic technologies. Nonetheless, we also recognize the value of parallel developments in other plants which have similarly accumulated sequence resources, namely, the Euphorbiaceae species such as *Manihot esculenta* (cassava), *Ricinus communis* (castor bean) and *Jatropha curcas* (physic nut), and alternative rubber-producing species such as *Taraxacum kok-saghyz*. We hope that this book will serve to record genome sequencing as a milestone in rubber tree research, and also the scientific investigations made during the initial phase of the research community's engagement with genome data.

Genomics is rapidly evolving and so will the directions of rubber tree genomics as more and more cutting-edge technologies become available. In the continuing quest for high-resolution genomes, we can expect to see publications of chromosome-level genome assemblies, high-density genetic maps, and tools for genome selection in the near future. Given the potential of inter- and intra-species genetic diversity for molecular breeding, we may also expect additional genomes of different *Hevea* species and cultivated genotypes to be sequenced. Insights from comparative genomics are likely to promote closer cooperation between genome scientists in the fields of *Hevea* and non-*Hevea* rubber research.

The rubber tree research community is relatively small compared to many major agricultural crops. The International Rubber Research and Development Board (IRRDB) has played an important role in promoting links between genomics and biotechnology researchers from different natural rubber-related organizations. In the course of preparing this book, we have benefited from some of the annual activities and wish to thank the IRRDB for facilitating such useful meetings and discussions. Finally, we also thank Prof. Chittaranjan Kole and the Springer team for their valuable advice and kind assistance.

Members from three rubber genome groups and the IRRDB Molecular Biology and Physiology Specialist Group gathered at the Sungai Buloh Research Station, Malaysian Rubber Board, Selangor on April 9, 2019.
From left: Azlina Bahari (MRB), Yuko Makita (RIKEN), Han Cheng (CATAS), Minami Matsui (RIKEN), Roslinda Sajari (MRB), Thitaporn Phumichai (RAOT), Keng-See Chow (MRB), Sithichoke Tangphatsornruang (BIOTEC).
MRB: Malaysian Rubber Board, Malaysia.
RIKEN: RIkagaku KENkyusho, Japan.
CATAS: Chinese Academy of Tropical Agricultural Sciences, China.
RAOT: Rubber Authority of Thailand, Thailand.
BIOTEC: National Center for Genetic Engineering and Biotechnology, Thailand.

Yokohama, Japan
Subang Jaya, Malaysia
January 2020

Minami Matsui, D.Sci
Keng-See Chow, Ph.D.

Contents

Contributors

Azlina Bahari Genetic Resources and Improvement Unit, Production Development Division, Malaysian Rubber Board, Sungai Buloh, Selangor, Malaysia

Han Cheng Key Laboratory of Rubber Biology, Ministry of Agriculture, Rubber Research Institute, Chinese Academy of Tropical Agricultural Sciences, Haikou, Hainan, People's Republic of China

Keng-See Chow Academy of Sciences Malaysia, Selangor, Malaysia

Katrina Cornish Departments of Horticulture and Crop Science, and Food, Agricultural and Biological Engineering, Ohio Agricultural Research and Development Center, College of Food, Agricultural and Environmental Science, The Ohio State University, Wooster, OH, USA

Longjun Dai Key Laboratory of Biology and Genetic Resources of Rubber Tree, Ministry of Agriculture and Rural Affairs, Rubber Research Institute, Chinese Academy of Tropical Agricultural Sciences, Danzhou, People's Republic of China

Muhammad Akbar Abdul Ghaffar Latex Harvesting Technologies and Physiology Unit, Production Development Division, Malaysian Rubber Board, Sungai Buloh, Selangor, Malaysia

Huasun Huang Key Laboratory of Rubber Biology, Ministry of Agriculture, Rubber Research Institute, Chinese Academy of Tropical Agricultural Sciences, Haikou, Hainan, People's Republic of China

Mika Kawashima Synthetic Genomics Research Group, RIKEN Center for Sustainable Resource Science (CSRS), Yokohama, Kanagawa, Japan

Yukio Kurihara Synthetic Genomics Research Group, RIKEN Center for Sustainable Resource Science (CSRS), Yokohama, Kanagawa, Japan

Nyok-Sean Lau Centre for Chemical Biology, Universiti Sains Malaysia, Bayan Lepas, Penang, Malaysia;
Synthetic Genomics Research Group, RIKEN Center for Sustainable Resource Science (CSRS), Yokohama, Kanagawa, Japan

Dejun Li Key Laboratory of Biology and Genetic Resources of Rubber Tree, Ministry of Agriculture and Rural Affairs, Rubber Research Institute, Chinese Academy of Tropical Agricultural Sciences, Danzhou, People's Republic of China

Yuko Makita Synthetic Genomics Research Group, RIKEN Center for Sustainable Resource Science (CSRS), Yokohama, Kanagawa, Japan

David F. Marshall Scotland's Rural College (SRUC), Edinburgh, Scotland, UK

Minami Matsui Synthetic Genomics Research Group, RIKEN Center for Sustainable Resource Science (CSRS), Yokohama, Kanagawa, Japan

Ahmad Sofiman Othman Centre for Chemical Biology, Universiti Sains Malaysia, Bayan Lepas, Penang, Malaysia;
School of Biological Sciences, Universiti Sains Malaysia, Minden, Penang, Malaysia

Wirulda Pootakham National Omics Center, National Center for Genetic Engineering and Biotechnology (BIOTEC), National Science and Technology Development Agency (NSTDA), Science Park, Thailand, Khlong Nueng, Pathumthani, Thailand

Jeremy R. Shearman National Omics Center, National Center for Genetic Engineering and Biotechnology (BIOTEC), National Science and Technology Development Agency (NSTDA), Science Park, Thailand, Khlong Nueng, Pathumthani, Thailand

Chaorong Tang School of Agricultural Science, Hainan University, Haikou, Hainan, People's Republic of China

Sithichoke Tangphatsornruang National Omics Center, National Center for Genetic Engineering and Biotechnology (BIOTEC), National Science and Technology Development Agency (NSTDA), Science Park, Thailand, Khlong Nueng, Pathumthani, Thailand

Mark A. Taylor The James Hutton Institute, Invergowrie, Dundee, Scotland, UK

Shaohua Wu Key Laboratory of Biology and Genetic Resources of Rubber Tree, Ministry of Agriculture and Rural Affairs, Rubber Research Institute, Chinese Academy of Tropical Agricultural Sciences, Danzhou, People's Republic of China

Hoong-Yeet Yeang Academy of Sciences Malaysia, Selangor, Malaysia

Abbreviations

4DTv	Transversion rate at fourfold degenerate sites
ABA	Abscisic acid
ACAT	Acetyl-CoA acetyltransferase
ACO	ACC oxidase
ACS	1-aminocyclopropane-1-carboxylic acid synthase
AFLP	Amplified fragment length polymorphism
AFSM	Amplified fragment SNP and methylation
ATAC-seq	Assay for transposase-accessible chromatin using sequencing
atm	Atmosphere
BAC	Bacterial artificial chromosome
BLASTP	Basic local alignment search tool (protein)
bp	Base pairs
bZIP	Basic leucine zipper transcription family
CAGE	Cap analysis gene expression
CDS	Coding sequence
ChIP-seq	Chromatin immunoprecipitation sequencing
cM	Centimorgan
COP	Coat protein complex
COR	Coronatine
CPT	*cis*-prenyltransferase
CPTL/CPTBP	CPT-like/CPT-binding protein
CTC	CAGE tag cluster
DC	Donor clone
DEG	Differentially expressed gene
DEP	Differentially expressed protein
DNase-seq	DNase I hypersensitive sites sequencing
dpi	Days post-infection
DXR	1-deoxy-D-xylulose 5-phosphate reductoisomerase
DXS	1-deoxy-D-xylulose 5-phosphate synthase
EIN2	Ethylene insensitive 2
ER	Endoplasmic reticulum
EST	Expressed sequence tag
ET	Ethylene
ETR	Ethylene resistance 1

EU	Extension unit
FDE	Filtered differentially expressed
FL-cDNA	Full-length cDNA
FPKM	Fragments per kilobase of transcript per million fragments mapped
FPP	Farnesyl pyrophosphate
FPPS	Farnesyl diphosphate synthase
FPS	Farnesyl diphosphate synthase
Gb	Giga base pairs
GBS	Genotyping-by-sequencing
GGPS	Geranylgeranyl pyrophosphate synthase
GO	Gene Ontology
GPPS	Geranyl diphosphate synthase
GPS	Geranyl pyrophosphate synthase
GS	Genome selection
GWAS	Genome-wide association study
HiC	High-throughput chromosome conformation capture
HMGR	3-hydroxy-3-methylglutaryl-CoA reductase
HMGS	3-hydroxy-3-methylglutaryl-CoA synthase
HMM	Hidden Markov model
hpi	Hours post-infection
HRBP	HRT1-REF bridging protein
IDP	Isopentenyl diphosphate
Indel	Insertion and deletion
Inr	Initiator element
IPP	Isopentenyl pyrophosphate
IPPI	Isopentenyl diphosphate isomerase
IRRDB	International Rubber Research and Development Board
Iso-Seq	Isoform sequencing
JA	Jasmonate
JAZ	Jasmonate ZIM-domain
JC	Juvenile clones
Kb	Kilobase pairs
KEGG	Kyoto Encyclopedia of Genes and Genomes
kg/ha/yr	Kilograms/hectare/year
$K_m^{IPP\text{-}Mg}$	Binding constant for the IPP-magnesium substrate
LA	Linolenic acid
LC-MS/MS	Liquid chromatography with tandem mass spectrometry
LG	Linkage group
LRP	Large rubber particle
LTR	Long-terminal repeat
LTR-RT	Long-terminal repeat retrotransposon
MALDI-TOF	Matrix-assisted laser desorption/ionization-time of flight
MAPK	Mitogen-activated protein kinase
MAS	Marker-assisted selection
Mb	Megabase pairs
MEP	2-C-methyl-D-erythritol 4-phosphate

miRNA	Micro ribonucleic acid
MNase-seq	Micrococcal nuclease-sequencing
mRNA	Messenger ribonucleic acid
MVA	Mevalonate
MWD	Molecular weight distribution
mya	Million years ago
NCBI	The National Center for Biotechnology Information
ncRNA	Non-coding RNA
NgBR	Nogo-B receptor
NGS	Next-generation sequencing
NMD	Nonsense-mediated decay
Nr	Non-redundant
NRS	Non-redundant sequence
ORF	Open reading frame
PAML	Phylogenetic analysis by maximum likelihood
PCD	Programmed cell death
PDC	Pyruvate dehydrogenase complex
PE	Paired-end
QTL	Quantitative trait locus
RACE	Rapid amplification of cDNA ends
RAPD	Random amplified polymorphic DNA
RBIP	Rubber biosynthesis inhibitor protein
RBSP	Rubber biosynthesis stimulator protein
REF	Rubber elongation factor
RFLP	Restriction fragment length polymorphism
RNA-seq	RNA sequencing
ROS	Reactive oxygen species
RPKM	Reads per kilobase of transcript per million mapped reads
rRNA	Ribosomal RNA
RT	Rubber transferase
RT-ase	Rubber transferase
SALB	South American Leaf Blight
SAMS	S-adenosyl-L-methionine synthase
SDR	Short-chain dehydrogenase/reductase
SDS-PAGE	Sodium dodecyl sulfate polyacrylamide gel electrophoresis
SEM	Scanning electron microscope
SMRT	Single-molecule real time
snoRNA	Small nucleolar ribonucleic acid
SNP	Single nucleotide polymorphism
snRNA	Small nuclear RNA
SOD	Superoxide dismutase
SRA	Sequence read archive
SRPP	Small rubber particle protein
SRPRNA	Signal recognition particle RNA
SSH	Suppression subtractive hybridization
SSR	Simple sequence repeat

TE	Transposable element
TEM	Transmission electron microscopy
TF	Transcription factor
TFBS	Transcription factor-binding site
TGN	Trans-Golgi network
tmRNA	Transfer-messenger RNA
TPD	Tapping panel dryness
TPM	Tags per million
tRNA	Transfer RNA
TSA	Trichostatin A
TSS	Transcription start site
uORF	Upstream ORF
UPP	Ubiquitin proteasome pathway
WGCNA	Weighted gene co-expression network analysis
WGS	Whole-genome shotgun
WRP	Detergent-washed rubber particle

Cornucopia that Brazil Gifted the World

1

Hoong-Yeet Yeang

Abstract

As an engineering material, natural rubber combines elasticity with other useful properties of vibration absorption, abrasion resistance, malleability, heat resistance and dispersion, electrical insulation, gas impermeability, and water resistance. It is a unique industrial material that is not mined, but grown. High-yield rubber from the rubber tree *Hevea brasiliensis* is realized through the extraction of rubber-bearing latex by tapping, which is a non-destructive method of harvesting. High turgor pressure in the laticifer network expels latex from the tapped tree, and its flow ceases when plugs form at the cut ends of the laticifers. At the termination of its rubber producing life, the tree is felled for cultivated timber, yet another renewable industrial material. Events in the year 1876, when the rubber tree was transferred from its native South America to the Orient, were pivotal in its transformation from a jungle tree to a plantation crop plant. After almost a century and a half, rubber productivity has increased sixfold through breeding and selection and the adoption of various agronomic practices (Kadir, Rubber Chem Technol 67:537–548, 1994). Recent advances in genomics have introduced new opportunities for further crop improvement. Breeders can now select genotypes that harbor not just individual desirable genes, but use microarrays to pick sets of genes that control polygenic traits.

Hoong-Yeet Yeang—Formerly Rubber Research Institute of Malaysia

H.-Y. Yeang (✉)
Academy of Sciences Malaysia, Selangor, Malaysia
e-mail: yeanghy@gmail.com

1.1 Renewable Source of an Industrial Material

As a crop plant, the rubber tree *Hevea brasiliensis* stands out. Unlike most crops that provide food, fiber, or fuel, the rubber tree offers none of these, nor any of the niche market products such as medicines, flowers, or perfumes. Instead, its harvested product is an engineering material that is grown, rather than mined from the ground. Most other crop-based industrial products such as gums and resins have limited markets or are crop by-products (e.g., roof thatching material). Timber from purpose-planted forests is perhaps the only other "crop" produce besides rubber that is a significant industrial material. In this regard, the rubber tree is as much a timber tree, considering that "rubberwood" is its second harvested produce. As in the case of natural rubber, such

M. Matsui and K.-S. Chow (eds.), *The Rubber Tree Genome*, Compendium of Plant Genomes,
https://doi.org/10.1007/978-3-030-42258-5_1

cultivated timber too is a renewable and sustainable resource. That notwithstanding, the product the rubber tree is primarily known for is, of course, natural rubber, *cis*-1,4 polyisoprene.

The fact that the rubber tree yields latex does not in itself render it as exceptional in the plant kingdom since some 10% of flowering plant species have this ability (Agrawal and Konno 2009). Besides *Hevea*, at least two other tree species have had their latex and rubber commercialized on a substantial scale in the past. Chicle rubber, a mixture of *cis* and *trans* rubber (Tanaka et al. 1988) from *Manilkara zapota* used to feature in chewing gum, while gutta-percha, a *trans* rubber from *Palaquium gutta* found application especially in insulating under-sea telegraph cables (the forerunner of present-day Internet communication cables) (Goodman et al. 1974). Consumption of both types of rubber has since been largely superseded by synthetics today, although the latter continues to be used in endodontics. *Hevea* rubber is not exempt from competition from petroleum-based synthetic elastomers. Nevertheless, various unique characteristics of natural rubber in the *cis* configuration have enabled it to hold its place side by side with its synthetic counterparts, each having advantages in terms of specific product requirements and comparative cost. Natural rubber has vast applications across diverse industries, from heavy engineering to delicate medical devices that take advantage of the combination of its elasticity, vibration absorption, abrasion resistance, malleability, heat resistance and dispersion, electrical insulation, gas impermeability, and water resistance. The automotive industry, where a large proportion of rubber is consumed, serves as an example. While car and truck tires come immediately to mind, rubber is extensively used also in a myriad of other less obvious automotive components that include tubing, belts, seals, gaskets, and suspension mountings.

The value of natural rubber as a product to humankind is indisputable, but what's in it for the rubber tree itself? Of what value is rubber to the tree, and how does it add to its "fitness" and competitiveness as a species? Since the sucrose from photosynthesis is the source of rubber, it might therefore be thought of as a form of nutrient storage. However, rubber formation is irreversible, and the tree has no means to break down rubber into components that it can metabolize. This is somewhat akin to putting money in a bank that welcomes deposits but disallows withdrawals. Another suggestion for the use of rubber in nature is protection from insect pests. While various biting and sucking insects are encountered in rubber planting (Rao 1965), they are not usually serious pests, arguably an attestation to the efficacy of the latex defense. There is also the suggestion that the most abundant soluble protein in latex, hevein, which has anti-fungal properties, might offer protection against pathogens (Van Parijs et al. 1991). Other potential anti-fungals in latex include hevamine and beta glucanase. Latex does seem ubiquitous in practically all parts of the rubber tree, at least on the macroscale. But whereas it would be difficult for a biting or boring insect to avoid puncturing laticifers; the invasion by fungal hyphae takes place on a far more microscopic scale. There is ample tissue outside of the laticifers that fungal pathogens can and do invade. Calculating from the data of Gomez et al. (1972), the space occupied by laticifers in the bark is only about 2%. Unsurprisingly, fungal infections of the bark such as "black stripe" (by *Phytophthora palmivora*) or "pink disease" (by *Corticium salmonicolor*) are not prevented.

1.2 Domestication of the Rubber Tree

As the twentieth century rolled in, demand for rubber in the industrialized world increased rapidly. Brazil was a major beneficiary, producing 95% of the world's supply of rubber at the time, all of it procured from wild jungle trees of the Amazon. Just over a quarter of a century later, however, production from the same region plummeted to barely 2.3% of world demand (Jackson 2008), not because supply from the jungle failed, but owing to competition from new and cheaper supplies from cultivated trees. The pivotal moment for this transition was when germplasm of the rubber tree was transferred from the Amazon jungle to plantations in the Orient. The year was 1876. With the sesquicentenary of this milestone approaching, it is timely to look back on the events that led to the advent of plantation rubber.

1.2.1 From the Forest to the Plantation

The story of Henry Wickham appropriating 70,000 rubber seeds from the Santarem region of Brazil and transporting them to the Royal Botanic Gardens at Kew where 2,700 germinated is well-known among rubber planters and researchers. The accolades accorded to him in changing the course of the rubber industry are well earned. Yet, without diminishing Wickham's contribution to rubber planting, another name less heard of, but deserving of note, would be that of Robert Cross. In 1876, the Kew Gardens presided over not one, but two expeditions to the Amazon. The first was the celebrated expedition of Wickham that brought home a consignment of seeds in June that year, while the second was the less historically heralded mission of Cross who returned in November with over a thousand *Hevea* specimens (Burkill 1935; Jones and Allen 1992; Kew Guild 2017). Whereas Wickham found renown as the cavalier "explorer," "adventurer," and "bio-pirate" in various colorful accounts of his exploits, Robert Cross was the staid and seasoned collector from Kew who had earlier made repeated forages into South America in search of botanical specimens such as *Cinchona* for anti-malarial quinine and *Castilla*, another rubber-bearing plant. Cross's collection matters. Besides adding to the numerical count of planting materials secured, the Wickham and Cross materials were collected from different locations of the Amazon and might, therefore, have been genetically dissimilar (Jones and Allen 1992).

The 1876 collections are important not only because they were historic firsts, but also because the trees descended from the original plants have never been bettered in rubber productivity by collections from later expeditions that covered far more extensive areas of the Amazon. For example, the 1981 expedition into the Brazilian states of Acre, Rondonia, and Matto Grosso undertaken under the auspices of the International Rubber Research and Development Board brought together scientists from China, Cote d'Ivoire, Indonesia, Malaysia, Nigeria, and Thailand and the host country, Brazil. Interesting genotypes—some with good potential as timber cultivars, for example—were secured. However, no cultivar bred from these later collections has matched the existing plantation clones in terms of rubber yield. Efforts at incorporating new germplasm into rubber planting are nevertheless still very much an ongoing effort (de Souza et al. 2015). A large gene resource is always advantageous for the improvement of any crop plant. Many breeders consider this to be especially critical in the case of the cultivated rubber tree.

There is an oft-repeated assertion (e.g., de Souza et al. 2015) that all commercially planted rubber trees are derived from just 22 seedlings sent from Ceylon in 1877 to the Singapore Botanic Gardens, nine of which were subsequently transferred north to Peninsular Malaya. (Were these Wickham materials, or Cross materials, or a mixture of the two?) If the crop gene pool were truly this limited, the amount of variation that rubber tree breeders has encountered from such a narrow gene base is remarkable, even allowing for the fact that the rubber tree is an out-breeder (Wycherley 1971; Simmonds 1986). Indeed, this claim has been questioned (Priyadarshan 2017). Accepting the "22 seedlings" axiom without reservation would be to ignore the 1,900 Wickham seedlings (Jones and Allen 1992) supplemented with the subsequent 100 Cross seedlings (Kew Guild 2017) sent to the Henarathgoda Botanic Gardens in Ceylon, which went on to become the de facto center of early rubber research. Burkill (1935) maintains, nevertheless, that none of Cross's materials reached the Orient. As the original rubber plants grew older, propagation by cuttings became difficult and their multiplication would therefore have required the availability of rubber seeds. That would have been a significant bottleneck in obtaining sufficient trees for commercial planting. Rubber trees start flowering when they are five or six years old. In time, plants of the subsequent generations would themselves set seed, resulting in an exponential increase in seed availability. Yet, in terms of the numbers of seeds available for collection, the starting population of seed-bearing plants is obviously a relevant consideration. The first nine seeds were obtained from trees growing in Ceylon in 1881. Seven years later, in 1888, 20,000 seeds were collected from the first-generation trees that had grown much larger by then, and from trees of the second generation that had newly come into flowering (Burkill 1935). In comparison, the 22 seedlings originally sent to Singapore and Malaya would have been a very small starting base indeed if this number was not augmented subsequently. Whether it is trees originating from the Wickham seeds or the Cross plants that make up present-day commercial plantings, there is no doubt that they all have their beginnings in the collections of 1876.

1.2.2 Expansion of Rubber Planting

The fate of the *Hevea* germplasm brought to the Old World was strongly influenced by circumstances on the ground in Ceylon and in Malaya that colored planters' views of the new crop before them. At the time rubber arrived, planters in both countries were looking for a replacement for coffee that they had found problematic because of disease. While the Malayans placed their bet on rubber, the Ceylonese were on the cusp of expanding tea planting for which they saw good prospects—both winning decisions, as history would show, for the respective countries. It must be remembered that although there was already a brisk demand for rubber for use in hoses, rainwear, braces, and so forth, the "real" rubber boom pushed by the automotive industry had not arrived yet, nor could it be foreseen at the time. Re-invention of the pneumatic tire by Dunlop was still a decade away, while Ford's mass-produced motor cars would not emerge until the beginning of the twentieth century. Tea, on the other hand, had already a secure market in Britain and in many parts of Europe. Moreover, a bad experience with *ceará* rubber (*Manihot glaziovii* rubber) that was introduced around the same time in Ceylon did not endear *Hevea* rubber to skeptical investors (Wenzlhuemer 2008). Whereas plantation rubber did take off in Ceylon at a steady pace with some 450,000 acres planted by 1920 (Herath 1984), its expansion in Malaya was nothing short of phenomenal, with 2,180,000 acres of rubber on the ground by the same year (Barlow 1978). At that point, Malaya was already producing virtually half of the global supply of natural rubber, lending at least some measure of credence to the notion that the world's entire rubber planting industry has its foundation in a very small number of plants originating from Malaya and Singapore (whether or not that number is actually the much touted 22). Malaya (later Malaysia) would hold on to its position as the world's premier natural rubber producer for the next half-century. Meanwhile, Ceylon remained a relatively minor player in world rubber production despite its head start in having received the lion's share of the Wickham–Cross materials from the New World. As for Brazil, its share of natural rubber production never recovered from being supplanted by the new rubber growing territories in the East. With jungle rubber no longer harvested on any large scale, natural rubber production in Brazil today accounts for only about 1% of world output (Lieberei 2007).

It is undeniably true that the British reaped handsome rewards for their effort in transferring rubber from Brazil to its colonies in the Orient. What should be noted, nonetheless, is that Britain did not prosper at the expense of the country from which rubber originated because Brazil never really stood a chance in commercial rubber planting. The reason for this was—as it remains today—the South American Leaf Blight (SALB) caused by *Microcyclus ulei*, a devastating fungal pathogen confined to the Americas (Lieberei 2007). So, ironically, the country of the rubber tree's origin struggles to support its commercial planting. It is not for want of trying, either. To wean himself from dependency on Asian rubber for his automobile production line, Henry Ford set up in the Amazon a whole township, Fordlandia, dedicated to rubber production in the 1920s. After pumping millions into the project, Ford threw up his hands in despair in the face of SALB, among various reasons. Even today, with modern fungicide regimes subsumed in advanced agricultural systems, rubber plantations in Brazil can be sustained only in certain "disease escape" regions (Benevenuto et al. 2017). Was it purely a stroke of luck that SALB was not transferred to the Orient together with the materials originally brought back from the Amazon? It might have been more than happenstance that spared Asian planters the misery of SALB. Spores of the fungus, in the form of conidia and ascospores, are generally short-lived (Chee 1976), and would probably not have survived the lengthy sea journey had any been carried on Wickham's seeds. It is unclear whether the materials procured by Cross had been plants (Burkill 1935; Jones and Allen 1992; Kew Guild 2017) or seeds (Thomas 2001). Had they been the former, they would have constituted a greater risk, but did not result in harm in the end.

Watching from the sidelines, the success of rubber planting outside of their country may understandably lead some Brazilians to lament an opportunity lost. Yet, they can take pride in Brazil having gifted the world with one of nature's most remarkable and precious crop plants.

1.3 Extracting Latex and Rubber from a Unique Laticifer Network

Hevea brasiliensis is the sole global source of natural rubber today because its productivity is unmatched by that of any other rubber-producing crop plant. Two plant species that yield *cis*-1,4 polyisoprene, guayule (*Parthenium argentatum*) and the Russian dandelion (*Taraxacum kok-saghyz*) have been considered as alternative sources of natural rubber. However, the yield of rubber from guayule (300–1,000 kg/ha/yr), and that from the Russian dandelion (110–200 kg/ha/yr) are far from the modern yields achievable by the rubber tree (around 3,000 kg/ha/yr) (van Beilen and Poirier 2007). Moreover, rubber in guayule occurs in the cells of the bark and woody tissue, and that in the Russian dandelion is found mainly in laticifers in the roots. Lacking a medium of free-flowing latex to convey it outside of the plant, rubber from these two species cannot be tapped. Instead, rubber is accumulated and stored in the vegetative tissue which has to be harvested periodically for processing, adding substantially to production costs.

Exuded latex which flows out of the tapped rubber tree, on the other hand, facilitates non-destructive latex harvest. Tapping allows for continual harvesting, with the operation commonly carried out on the same tree every two or three days. The method of excision tapping of the rubber tree remains basically unchanged from the system devised by Henry Ridley, Director of the Singapore Botanic Gardens, in 1889 (Wycherley 1958). Rather than making repeated incisions (slashes) into the bark to extract the latex as in the exploitation of jungle rubber trees, Ridley's method of excision involved making an initial sloping incision on the bark along which the exuded latex would flow into a collection cup. In subsequent tappings, a sliver of bark 1.5 mm thick would be planed off (excised from) the same upward-facing surface of the existing cut to re-open the laticifers and allow latex flow. In this manner, the position of the tapping cut gradually moves down the trunk of the tree as the bark is consumed. This system enables maximum conservation of the productive bark.

In the rubber tree trunk, articulated anastomosing laticifers are laid down in concentric cylindrical networks ('mantles') in the phloem region of the bark around the tree trunk. In a cross-section of the trunk, these cylinders of laticifers appear as concentric rings, with the youngest laticifers positioned innermost, next to the cambium from which they are differentiated. Because the laticifers within each ring are inter-connected in a network, the exuded latex would come not only from the bark in the immediate vicinity of the tapping cut, but also from the laticifers some distance away. However, laticifers in different rings are not connected. Therefore, to maximize yield, it is essential to tap the bark as deeply as possible without damaging the cambium so as to sever as many laticifer rings as possible. Experienced tappers cut the bark to within 1 mm of the cambium based on feel alone. Even then, about 40% of the laticifer rings (the tightly spaced innermost rings nearest the cambium) would have been missed by the tapping knife (Gomez et al. 1972). A skilled tapper can obtain from a high-yielding tree more than 100 ml of latex with each tapping, a third of this being rubber. A tree can be tapped 120–180 times a year. Bark renewed from the cambium would be available for exploitation again after about 10 years.

1.3.1 Regulation of Latex Flow from the Tapped Tree

To obtain high latex yield from the tree by tapping, it is not sufficient that the latex is available as a free-flowing fluid ensconced in its laticifer reservoir. High turgor pressure has to be maintained in the laticifer system in order to expel the latex when the laticifers are severed by the tapping knife. The turgor pressure that builds up in the *Hevea* laticifer system is truly prodigious, being of an order of 10 atmospheres (atm) (Buttery and Boatman 1966). By comparison, the pressure in an inflated passenger car tire is about 2 atm and the systolic blood pressure (120 mm mercury) of the human body is 0.16 atm. Turgor pressure is influenced by the ambient relative humidity, high humidity being conducive to high turgor. The highest yields are obtained when the tree is tapped at night, between 2000 and 0700 h, and turgor pressure subsequently drops as evapotranspiration increases in the daylight hours. Consequently, the yield can decline as much as 30% by 1300 h (Paardekooper and Soomark 1969). The fact that rubber content is slightly higher at mid-day cannot compensate for the loss in the latex volume collected. Night tapping used to be a common practice in the past, but working at such unsociable hours is not so much the rule in the present day.

While the initial high turgor pressure in the laticifers falls in the course of latex exudation after tapping, the final cessation of flow is not due to loss of pressure. In fact, the turgor pressure of around 6 atm at the cessation of flow is more than sufficient to expel latex from the tapping cut. Indeed, if the tree is re-tapped immediately after flow stops, there would be an immediate resurgence of flow, attesting to the residual turgor. The reason latex flow eventually ceases is that plugs form at the ends of the cut laticifers to seal the vessels and prevent further latex loss. The fact that latex can be extracted from the rubber tree by removing a 1.5 mm thick shaving from the surface of the tapping cut implies that the plugs form very close to the cut ends of the laticifers. An analogy to this would be the clotting of blood to stem further loss in a wound. The process of laticifer plugging is complex and various proposals to elucidate its mechanism have generally centered around the destabilization of latex as it emerges from the tapping cut to form plugs. These explanations have implicated the lutoids (which are vacuole-like organelles in the latex), "bark sap" that is released from bark tissue wounded by the tapping procedure, the small rubber particles (SRPs) which are especially susceptible to destabilization, and the latex serum (the C-serum) in which the rubber particles and other organelles are suspended. (See reviews by d'Auzac (1989) and Berthelot et al. (2016).)

Taking latex flow as a whole from the time of tapping to cessation of flow, the abovementioned two main variables, viz., the turgor pressure of the laticifer system and laticifer plugging, account for 99% of the change in flow rate over the duration of flow (Yeang 2005). The decrease in flow rate immediately after tapping is due mainly to the loss of turgor. On the other hand, laticifer plugging occurs throughout the flow, but becomes more pronounced toward its end when most of the laticifers are already plugged. Among the laticifers that are still yielding, the plugging rate rises steeply just before flow cessation. It is deduced that about half of the laticifers on the tapping cut are plugged after one-quarter of the total flow duration has elapsed. Just over 70% of the laticifers on the tapping cut are plugged by the mid-point of the flow duration (Yeang 2005).

1.3.2 The Rubber Component of Latex

Hevea latex is characterized as a colloidal dispersion of various cellular organelles in an aqueous serum. Unlike plant sap which contains water, minerals, sugars, and hormones, *Hevea* latex is living cytoplasm. Besides all of the abovementioned constituents, latex also contains various organelles, nucleic acids, proteins (including enzymes), and other organic molecules. The constituent of commercial interest is, of course, rubber which makes up one-third of the latex by volume or weight. Natural rubber, which has the unique ability to be repeatedly stretched without compromising its strength, is a hydrocarbon comprising long chains of *cis*-1,4 isoprene units $(C_5H_8)_n$.

Rubber in latex takes the form of specialized organelles, the rubber particles. These are generally spherical and range in size from 0.02 to 3 μm (Southorn and Yip 1968). Size variation is, of course, to be expected in any biological sampling. The rubber particle size distribution in the latex shows most particles to be around 0.1 μm in diameter, with an extended "tail" of considerably larger particles that appear on cursory inspection to be outliers. In fact, however, there are two distinct classes of rubber particles, viz., the large rubber particles (LRPs) and the small rubber particles (SRPs), with each population of particles having its own normal distribution (Yeang et al. 1995). SRPs have a mean diameter of about 0.07 μm, whereas LRPs have a mean diameter of 0.5–0.6 μm. While there does not appear to be many SRPs larger than 0.4 μm in diameter, the smallest LRPs would be comparable in size to SRPs, although the latter would, of course, vastly outnumber the former. Another indication that SRPs and LRPs are separate classes of rubber particles is the fact that the main proteins on their respective surfaces while related, are dissimilar. The rubber elongation factor (REF) on the LRP surface is a truncated version of the small rubber particle protein (SRPP) (Oh et al. 1999). These observations imply that (a) LRPs and SRPs are different entities; LRPs are not SRPs that have simply grown to an exceptionally large size; and (b) LRPs may have started off small, initially growing from particles the same size as SRPs, but the latter are capped off early in the size they attain.

If all rubber particles in the latex were grouped together as a single population, then the "typical" rubber particles would be represented by the SRPs which constitute 94% of all particles. But because SRPs are so tiny, they make up no more than 7% of the rubber crop. LRPs, on the other hand, have a far greater presence in the harvested rubber. Despite constituting only 6% by number, they comprise 93% of the rubber crop. In fact, the largest 1% of rubber particles alone account for 50% of the total rubber yield (Yeang et al. 1995). As a commodity, therefore, natural rubber is a product essentially of LRPs, with the far more numerous SRPs making only a very minor contribution. Without the LRPs, the natural rubber industry as we know it today would not exist. Because the median-sized SRPs are so numerous, it might have been tempting to speculate the prospects whereby just 1% of these tiny rubber particles "growing" to 0.5 or 0.6 μm would increase rubber content in the latex by 10%. However, such a proposition is of course not viable because the size variation in the total population of rubber particles in latex is, in fact, not a single Gaussian (normal) distribution.

What is it in LRPs, but not in SRPs, that allow the former to grow to a large size? The rubber compositions in SRPs and LRPs are not too different. Both types of rubber particles contain polyisoprene chains of molecular weight ranging from 10^4 to 10^7, although SRPs tend to have rubber with higher average molecular weight. The difference could be associated with their monolayer rubber particle membranes. Besides the main rubber particle membrane proteins being dissimilar as mentioned above, there may also be differences in their location and orientation with respect to the membrane lipids (Berthelot et al. 2014).

1.3.3 Latex and Rubber Regeneration

The success of the rubber tree as a crop plant is not attributed solely to its capacity to exude copious amounts of latex. For rubber production to be sustainable, the tree must have the ability to regenerate and replace large amounts of rubber (and non-rubber latex constituents) that include enzymes and other proteins which participate in rubber biosynthesis. Such an attribute of the rubber tree should not be taken for granted since repeated loss of large amounts of latex from the tree is an anthropogenic intervention that does not occur in nature. There would have been no adaptive pressure to evolve mechanisms for substantial latex regeneration. For example, large amounts of latex can be obtained from tapping *Castilla* (Panama rubber tree), but extraction as such, it is said, cannot be undertaken more than once or twice a year (Imle 1978). In the case of *Hevea*, it is relevant to note that most of the nuclei of the laticifers are parietal in the laticifer, and are not lost in the latex outflow when the tree is tapped (Dickenson 1965). As such, the all-important DNAs that encode the genes related to the biosynthesis of rubber are available for its regeneration. Similarly retained in the laticifers are much of the endoplasmic reticulum that serve as scaffolds for protein and lipid synthesis, and the mitochondria that are the powerhouses for cellular metabolism (Sipat 1985).

Isopentenyl diphosphate (IDP) is the precursor of many isoprenoids, including rubber in *Hevea* latex. The cytosolic mevalonate (MVA) pathway is the conventionally accepted pathway for IDP generation in rubber biosynthesis (Kekwick 1989). More recently, the plastidic 2-C-methyl-D-erythritol 4-phosphate (MEP) pathway has also been viewed as an alternative source of IDP, with the Frey-Wyssling particles in the latex being the plastid of interest (Chow et al. 2012). Downstream from IDP, "washed rubber particles" have been used as the platform on which to demonstrate de novo synthesis of rubber (Archer and Audley 1987; Yusof et al. 1998). Washed rubber particles are among the smallest SRPs. They are so tiny that they do not separate out centripetally in the rubber cream on high-speed latex centrifugation, but remain suspended in the latex serum from which they are isolated by gel filtration. Such SRPs—but not the larger rubber particles—are highly reactive in rubber biosynthesis. Accordingly, much of what is known about the final steps of rubber biosynthesis is based on the SRP model, with the expectation that it applies also to the LRPs, the rubber particles that account mainly for the rubber crop.

1.4 Improving Productivity of the Rubber Tree

Much has been said about the phenomenal improvement in yield output of the rubber tree, from 500 kg/ha/yr in unselected seedlings to 3,000 kg/ha/yr in modern clones, a sixfold increase (Kadir 1994). While the achievement is impressive, it should nevertheless be clarified that this success is not due entirely to genetic improvement through breeding and selection. As with all tree species, *Hevea* breeding is a lengthy undertaking. Trees are slow growing and fixing genes by repeated back-crossing *Hevea* would be extremely time-consuming since the rubber tree has a juvenile period of about five years before it begins to flower. Fixing desired genes through sexual reproduction alone would therefore be an unwieldy process. Instead, once a desired rubber tree genotype has been selected, it is mass propagated through clonal multiplication. Hence, bud-grafting adopted as an agronomic practice in rubber planting from the early twentieth century (Tan et al. 1996), plays an important role in generating uniformly high-quality planting materials for the industry.

Another plantation practice that has had a huge impact in rubber planting is latex yield stimulation using the ethylene-releasing agent ethephon. The rubber content in the latex of stimulated trees drops somewhat, but this is more than compensated for by the significantly greater volume collected. How does ethylene treatment increase the latex yield? As stated earlier, the rate of latex exudation from rubber trees is determined by (a) the turgor pressure that expels the latex (this being more important in the initial flow) and (b) laticifer plugging that slows down the rate of flow, stopping it eventually. Since turgor pressure and the initial flow rate after tapping tend to be decreased in stimulated trees (Abraham et al. 1968; Pakianathan 1977), they do not explain the ethylene effect. Instead, it is the lengthened flow duration due to delayed laticifer plugging that accounts for the yield increase (Abraham and Tayler 1967). It can be seen in a comparison of unstimulated and stimulated trees (Yeang 2005), that the propensity to plugging is very similar immediately after tapping (moderately elevated) and just before termination of flow (greatly elevated). Plugging rate is lowest during mid-flow and it is the extended duration of low plugging in this period that accounts for the lengthened flow and thence increased yield in ethephon-stimulated trees.

How does ethylene affect laticifer plugging? The mechanism of laticifer plugging is complex, but most researchers agree that an interaction between the lutoid contents and rubber particles is requisite in plug formation. While it is hence reasonable to infer that ethylene would be acting on the lutoids or the rubber particles, other possible explanations should not be dismissed. Ethephon also promotes the exudation of gums and resins from various plant species (Nair 2001), such as *Shorea javanica* (Messer 1990); here, it is noteworthy that such exudates contain neither lutoids nor rubber particles. It is therefore conceivable that water balance and transportation may play a role—perhaps a major role—in the ethylene action that enhances and sustains the outflow of exudates, be they gums, resins or latex. In this regard, salient changes have been reported in laticifer aquaporin expression following ethephon treatment (Tungngoen et al. 2009; Zou et al. 2015), although it is unclear if they are the cause or effect of extended latex flow.

Whereas various novel agronomic practices are introduced from time to time to the industry to increase crop productivity, breeding of the rubber tree remains an R&D staple. Even though secondary characteristics such as disease resistance or wind resistance are valued, rubber productivity is the most important attribute of the rubber tree cultivar. In breeding for high rubber yield, the plant breeder looks for vigor as the first criterion. Occasionally, some deficiency in vigor may be compensated for by other desirable characteristics such as the density of laticifers in the bark. Such was the case for RRIM 600, one of the most successful and most widely planted rubber clones worldwide. Since latex originates in laticifers, high latex output can only be achieved if abundant laticifers are available for tapping. Seedling trees in the 1920s had a mean of about 12 laticifer rings and a maximum approaching 25 rings in the bark. In modern clones, the mean number is around 25, with elite clones bearing as many as 40 laticifer rings in the productive bark (Gomez et al. 1972).

Both the volume of exuded latex as well as its rubber content contribute to the yield of rubber in each tapping. However, the variation in latex rubber content from a tree in regular tapping falls within a relatively narrow range from, say, 27–40%, i.e., an approximately 13% point difference, and usually less. On the other hand, the volume of latex collected from a high-yielding tree versus a low yielder can easily be multiple-fold. Hence, it is mainly the volume of latex collected that determines whether a tree is a high yielder or otherwise. The breeder selecting for high latex volume essentially selects for long flow of the latex after tapping because it is the flow duration that makes the difference in the volume of latex collected.

While the plant breeder selects for multiple positive traits of the rubber tree, negative characteristics, such as susceptibility to various diseases, have also to be de-selected in arriving at a commercially viable cultivar. This is particularly important where the inherent deficiencies cannot be easily remedied by agronomic intervention in the field. Among the adverse traits, the breeder looks out for is poor resistance to wind. There is no remedy when a tree snaps or uproots due either to poor wood strength or an overly voluminous crown.

Another undesirable condition of the rubber tree that is difficult to correct is tapping panel dryness where the tapping cut stops yielding. A tree that experiences only transient dryness due to over-exploitation may recover by attenuating the tapping intensity and halting ethephon yield stimulation. However, the more severe cases of tapping panel dryness commonly referred to as "brown bast" are where dryness is accompanied by a brown discoloration and the appearance of cracks and nodules on the bark. This form of dryness cannot be cured because latex has irreversibly coagulated within the dead or dying laticifers. Accordingly, the affected bark does not recover even after tapping rest, although some yield could be extracted by deep tapping to reach the young renewing bark growing beneath the dry tissue. No pathogenic microorganism has been unequivocally shown to be involved in brown bast and its etiology remains uncertain. It is generally recognized as a physiological disorder arising from over-exploitation. The exact nature of this physiological origin is yet to be ascertained, although numerous conjectures have been advanced since the early years of the industry (Yeang 2007). Among the more recent non-pathogen hypotheses proposed for the cause of brown bast are reactive oxygen species damaging membranes of the lutoids or the rubber particles (Chrestin et al. 1986; Li et al. 2010; Zhang et al. 2017), and programed cell death (Li et al. 2010; Yuan et al. 2014).

Although the brown bast syndrome has been a concern to planters since the beginning of the rubber industry, its severity has nevertheless been ameliorated in modern rubber planting. Stringent selection against susceptibility to dryness is not the only reason for this. A radical change occurred when bud-grafting became a common plantation practice because budded trees are more resistant to brown bast as compared with seedling trees. In a comparison between budded trees and the genetically identical seedling mother trees from which they had been disbudded, the incidence of brown bast in the first five years of tapping was, respectively, 2.6% and 6.5% despite the seedlings having been tapped at a milder intensity (third daily) than the buddings (alternate daily) (Paranjothy and Yeang 1977). Another change in plantation practice has also contributed to relatively lower incidences of brown bast. Until around the 1980s, alternate day tapping used to be commonly adopted because it optimized land productivity. With the gradual increase in tapping labor costs, however, the industry in many producing countries such as Malaysia switched to third daily tapping. Since tapping panel dryness is very sensitive to over-exploitation, this has helped to bring the incidence of the disorder down to manageable and tolerable levels, as long as highly susceptible clones are avoided.

Since the 1990s, another selection criterion has been added to rubber tree breeding. As rubberwood came into demand for furniture manufacture, breeders began targeting cultivars that gave high yields of both latex and timber, i.e., the "latex-timber" clones. Both rubber and timber have photosynthate as the starting resource, and the partition of assimilates means that removal of rubber from the tree comes at the expense of tree growth and wood formation. One strategy for the planter is to set aside trees that will remain untapped to maximize growth for timber harvest later. Even so, should the price of rubber rise unusually high, these trees could be a useful reserve to be brought temporarily into tapping. Whether producing rubber or timber, the rubber tree plays a positive role in fixing carbon dioxide in the atmosphere that helps alleviate global warming.

Many important traits of the rubber tree including rubber yield and timber yield are polygenic in their control, as are many secondary crop characteristics such as disease resistance. For instance, rubber cultivation in the Americas against a backdrop of SALB would benefit from the acquisition of horizontal resistance against different races of the pathogen (Simmonds 1990). For the selection of desirable traits, rubber tree breeders would be relying not so much on identifying single genes that control a particular character, but combinations of genes that offer a more holistic and sustained effect in the cultivar. In this regard, elucidation of the rubber tree genome facilitates the development of microarrays that allow a large number of genes to be rapidly screened to select for polygenic traits. Such recent advances in genomics introduce new vistas for crop improvement of the rubber tree.

References

Abraham PD, Tayler RS (1967) Stimulation of latex flow in *Hevea brasiliensis*. Expt Agric 3(1):1–12

Abraham PD, Wycherley PR, Pakianathan SW (1968) Stimulation of latex flow *in Hevea brasiliensis* by 4-amino-3,5,6-trichloropicolinic acid and 2-chloroethanephosphonic acid. J Rubb Res Inst Malaya 20(5):291–305

Agrawal AA, Konno K (2009) Latex: a model for understanding mechanisms, ecology, and evolution of plant defense against herbivory. Annu Rev Ecol Evol Systematics 40:311–331

Archer BL, Audley BG (1987) New aspects of rubber biosynthesis. Bot J Linn Soc 94:181–196

d'Auzac J (1989) Factors involved in the stopping of flow after tapping, (d'Auzac J. In: Jacob JL, Chrestin H (eds) Physiology of rubber tree latex. CRC Press, Florida, pp 257–285

Barlow C (1978) The natural rubber industry: its development, technology and economy in Malaysia. Oxford University Press, p 26

van Beilen JB, Poirier Y (2007) Guayule and Russian dandelion as alternative sources of natural rubber. Critical Rev Biotech 27:217–231

Berthelot K, Lecomte S, Estevez Y, Zhendre V, Henry S et al (2014) Rubber particle proteins, HbREF and HbSRPP, show different interactions with model membranes. Biochem Biophys Acta 1838:287–299

Berthelot K, Peruch F, Lecomte S (2016) Highlights on *Hevea brasiliensis* (pro)hevein proteins. Biochimie 127:258–270

Benevenuto JAZ, Passos JRS, Furtado EL (2017) *Microcyclus ulei* races in Brazil. Summa Phytopathol 43(4). http://dx.doi.org/10.1590/0100-5405/172339

Burkill IH (1935) Dictionary of the economic products of the Malay Peninsula. Crown Agents of the Colonies, on behalf of Governments of the Straits Settlements and Federated Malay States, London

Buttery BR, Boatman SG (1966) Manometric measurement of turgor pressures in laticiferous phloem tissues. J Exp Bot 17(51):283–296

Chee KH (1976) Factors affecting discharge, germination and viability of spores of *Microcyclus ulei*. Transac British Mycol Soc 66(3):499–504

Chow K-S, Mat-Isa M-N, Bahari A, Ghazali A-K, Alias H et al (2012) Metabolic routes affecting rubber biosynthesis in *Hevea brasiliensis* latex. J Exp Bot 63 (5):1863–1871

Chrestin H, Jacob JL, d'Auzac J (1986) Biochemical basis for cessation of latex flow and occurrence of physiological bark dryness. In: Proceedings of the International Rubber Conference 1985, Kuala Lumpur 3:20–42

Dickenson PB (1965) The ultrastructure of the latex vessel of *Hevea brasiliensis*. In: Mullins L (ed) Proceedings of the natural rubber producers research association jubilee conference, Cambridge 1964. Maclaren & Sons Ltd., London, pp 52–66

Gomez JB, Narayanan R, Chen KT (1972) Some structural factors affecting the productivity of *Hevea brasiliensis*. I. Structural determination of the laticiferous tissue. J Rubb Res Inst Malaya 23(3):193–203

Goodman A, Schilder H, Aldrich W (1974) The thermomechanical properties of gutta-percha. II. The history and molecular chemistry of gutta-percha. Oral Surgery, Oral Med, Oral Pathol, Oral Radiol 37(6):954–961

Herath HMG (1984) The rubber industry in Sri Lanka. Asian Surv 24(8):817–827

Imle EP (1978) *Hevea* rubber: past and future. Econ Bot 32(3):264–277

Jackson J (2008) The thief at the end of the world: rubber, power, and the seeds of empire. Penguin

Jones KP, Allen PW (1992) Historical development of the world rubber industry. In: Sethuraj MR, Mathew NM (eds) Natural rubber: biology, cultivation and technology. Elsevier Science Publishers, Amsterdam

Kadir AASA (1994) Advances in natural rubber production. Rubber Chem Technol 67(3):537–548

Kekwick RGO (1989) The formation of isoprenoids in *Hevea* latex. In: d'Auzac J, Jacob JL, Chrestin L (eds) Physiology of rubber tree latex. CRC Press, Florida, pp 145–164

Kew Guild (2017) Robert MacKenzie Cross (1836–1911). J Kew Guild. https://kewguild.org.uk/2017/01/27/robert-mackenzie-cross-1836-1911/

Li D, Deng Z, Chen C, Xia Z, Wu M, He, P, Chen S (2010) Identification and characterization of genes associated with tapping panel dryness from *Hevea brasiliensis* latex using suppression subtractive hybridization. BMC Plant Biol 10:140. http://www.biomedcentral.com/1471-2229/10/140

Lieberei R (2007) South American Leaf Blight of the rubber tree (*Hevea* spp.): new steps in plant domestication using physiological features and molecular markers. Ann Bot 100:1125–1142

Messer AC (1990) Traditional and chemical techniques for stimulation of *Shorea javanica* (Dipterocarpaceae) resin exudation in Sumatra. Econ Bot 44(4):463–469

Nair MNB (2001) Sustainable utilization of gums and resins by improved tapping techniques in some species. In: Proceedings: harvesting of non-wood forest products. Ministry of Forestry, Turkey. http://www.fao.org/3/Y4496E/Y4496E29.htm. Accessed 8 July 2019

Oh SK, Kang H, Shin DH, Yang J, Chow K-S et al (1999) Isolation, characterization, and functional analysis of a novel cDNA clone encoding a small rubber particle protein from *Hevea brasiliensis*. J Biol Chem 274 (24):17132–17138

Paardekooper EC, Soomark S (1969) Diurnal variation in latex yield and dry rubber content, and relation to saturation deficit of air. J Rubb Res Inst Malaya 21:241–347

Pakianathan SW (1977) Some factors affecting yield response to stimulation with 2-chloroethylphosphonic acid. J. Rubb Res Inst 25(1):50–60

Paranjothy K, Yeang HY (1977) A consideration of the nature and control of brown bast. In: Proceedings of the RRIM Planters' Conference, Kuala Lumpur, pp 74–90

Priyadarshan PM (2017) Refinements to *Hevea* rubber breeding. Tree Genet Genomes 13:1–17

Rao BS (1965) Pests of *Hevea* plantations in Malaya. Rubber Research Institute, Kuala Lumpur

Simmonds NW (1986) Theoretical aspects of synthetic/polycross populations of rubber seedlings. J Nat Rubb Res 1(1):1–15

Simmonds NW (1990) Breeding horizontal resistance to South American Leaf Blight of rubber. J Nat Rubb Res 5(2):102–113

Sipat A (1985) 3-Hydroxy-3-methylglutaryl-CoA reductase in the latex of *Hevea brasiliensis*. Methods Enzym 110:40–51

Southorn WA, Yip E (1968) Latex flow studies. III. Electrostatic considerations in the colloidal stability of fresh *Hevea* latex. J Rubb Res Inst Malaya 20(4): 201–215

de Souza LM, Le Guen V, Cerqueira-Silva CBM, Silva CC, Mantello CC, Conson ARO et al (2015) Genetic diversity strategy for the management and use of rubber genetic resources: more than 1,000 wild and cultivated accessions in a 100-genotype core collection. PLoS ONE 10(7):e0134607. https://doi.org/10.1371/journal.pone.0134607

Tan H, Khoo SK, Ong SH (1996) Selection of advanced poly cross progenies in *Hevea* improvement. J Nat Rubb Res 11(3):215–225

Tanaka Y, Nunogaki K, Kageyu A, Mori M, Sato Y (1988) Structure and biosynthesis mechanism of trans-polyisoprene from chicle. J Nat Rubb Res 3(3):177–183

Thomas KK (2001) Role of Clements Robert Markham in the introduction of *Hevea* rubber into the British India. Planter 77(902):287–292

Tungngoen K, Kongsawadworakul P, Viboonjun U, Katsuhara M, Brunel N, Sakr S, Narangajavana J, Chrestin H (2009) Involvement of *HbPIP2;1* and *HbTIP1;1* aquaporins in ethylene stimulation of latex yield through regulation of water exchanges between inner liber and latex cells in *Hevea brasiliensis*. Plant Physiol 151:843–856

Van Parijs J, Broekaert WF, Goldstein IJ, Peumans WJ (1991) Hevein: an antifungal protein from rubber-tree (*Hevea brasiliensis*) latex. Planta 183:258–264

Wenzlhuemer R (2008) From coffee to tea cultivation in Ceylon, 1880–1900. An economic and social history. Brill, Leiden

Wycherley PR (1958) The Singapore Botanic Gardens and rubber in Malaya. Gardens Bull Singapore 17:175–186

Wycherley PR (1971) *Hevea* seed, Part 1. Planter 47:291–298

Yeang HY (2005) The kinetics of latex flow from the rubber tree in relation to latex vessel plugging and turgor pressure. J Rubb Res 8(3):160–181

Yeang HY (2007) The brown bast syndrome of *Hevea brasiliensis*. Malaysian rubber board monograph 21. Malaysian Rubber Board, Kuala Lumpur

Yeang HY, Yip E, Hamzah S (1995) Characterisation of Zone 1 and Zone 2 rubber particles in *Hevea brasiliensis* latex. J Nat Rubb Res 10(2):108–123

Yuan K, Wang Z, Zhou X, Zou Z, Yang L (2014) The identification of differentially expressed latex proteins in healthy and tapping panel dryness (TPD) *Hevea brasiliensis* trees by iTRAQ and 2D LC-MS/MS. Acta Agric Univ Jiangxiensis 36(3):650–655 (Chinese text with English summary)

Yusof F, Audley BG, Ismail F, Walker JM (1998) A rapid assay for the incorporation of isopentenyl phosphate in rubber biosynthesis. J Rubb Res 1(1):48–56

Zhang Y, Leclercq J, Montoro P (2017) Reactive oxygen species in *Hevea brasiliensis* latex and relevance to tapping panel dryness. Tree Physiol 37:261–269

Zou Z, Gong J, An F, Xie G, Wang J, Mo Y, Yang L (2015) Genome-wide identification of rubber tree (*Hevea brasiliensis* Muell. Arg.) Aquaporin genes and their response to ethephon stimulation in the laticifer, a rubber producing tissue. BMC Genomics 16:1001

2 The Reyan 7-33-97 Rubber Tree Genome: Insight into Its Structure, Composition and Application

Han Cheng, Chaorong Tang, and Huasun Huang

Abstract

The *Hevea brasiliensis* rubber tree is an important industrial crop that is widely planted in the tropical regions. China has a rubber planting history of more than 100 years and has successfully established three planting districts: Yunnan, Guangdong, and Hainan. More than 100 clones raised by Chinese breeders are widely cultivated. However, it is increasingly more difficult to raise yield through the conventional breeding approach due to the narrow genetic pool of Wickham germplasm and the low quality of genome assemblies of the rubber tree. With rapid advances in sequencing technology, it is now more convenient to sequence the genome of non-model plant species. The rubber tree genome is huge in size and high in heterozygosity, thus making genome assembly more challenging. To date, four versions of genomes have been published for the rubber tree, thus providing deep insights into genome structure and composition, species adaptation, rubber biosynthesis, and many agronomic traits. In this chapter, we describe the Reyan 7-33-97 genome and the gene families identified in it.

2.1 Introduction

2.1.1 *Hevea* as a Commercial Source of Natural Rubber

Natural rubber is an important industrial product derived from plants. More than 2,000 plant species in the world can produce natural rubber (Priyadarshan and Goncalves 2003); however, very few of them can be successfully commercialized. *Hevea brasiliensis*, commonly known as the rubber tree, produces more than 90% of natural rubber in the world (Priyadarshan and Goncalves 2003). When Columbus discovered the New World in the fifteenth century, he found the native Americans playing with an elastic ball that was made from the hardened sap of a kind of tree (Priyadarshan 2017a, b). The sample was brought back to Europe and was called "rubber" by a British chemist, Joseph Priestley, in 1770 as the material could be used to erase pencil marks. In 1839, Charles Goodyear invented the vulcanization process which makes rubber less sticky and have superior mechanical properties. In 1888, John Boyd Dunlop invented the first inflatable tire (Saha and Priyadarshan 2012). These inventions greatly accelerated the development of the rubber

H. Cheng (✉) · H. Huang
Key Laboratory of Rubber Biology, Ministry of Agriculture, Rubber Research Institute, Chinese Academy of Tropical Agricultural Sciences, Haikou, Hainan, People's Republic of China
e-mail: forcheng@gmail.com

C. Tang
School of Agricultural Science, Hainan University, Haikou, Hainan, People's Republic of China

M. Matsui and K.-S. Chow (eds.), *The Rubber Tree Genome*, Compendium of Plant Genomes,
https://doi.org/10.1007/978-3-030-42258-5_2

industry. However, rubber trees were only found in Latin America at that time and these could not supply sufficient rubber. In 1876, Henry Alexander Wickham collected over 70,000 rubber tree seeds from the central Amazon Basin and sent them to the Kew Gardens, London. About 2,700 rubber seedlings were germinated from these seeds. The bulk of seedlings were dispatched to Henarathgoda, from which 22 plants were transported to Singapore in 1877. Nine were sent to Kuala Kangsar, Malaya (now Malaysia) and the remaining 13 were retained at the Singapore Botanic Gardens (Priyadarshan 2017a). The rubber trees at Kuala Kangsar flowered in 1884 and the offsprings were planted near the Perak River (He and Huang 1987).

The *Hevea* genus includes 10 species, i.e., *H. brasiliensis*, *H. benthamiana*, *H. camargoana*, *H. camporum*, *H. guianensis*, *H. microphylla*, *H. nitida*, *H. pauciflora*, *H. rigidifolia* and *H. spruceana* (Priyadarshan 2017a). All of these species originated from the Amazon Basin in South America. Only *H. brasiliensis*, also known as the para rubber tree, is widely grown in commercial plantations. The modern rubber clones are all mostly descended from the Wickham collection in 1876, and are therefore denoted as the Wickham germplasm (Saha and Priyadarshan 2012). Though there were other wild rubber germplasm collections in 1981 and 1995 (Withanage et al. 2015; Adifaiz et al. 2018), the genetic background of rubber tree clones today is still narrow due to the long cycle of the breeding program.

2.1.2 *Hevea* Cultivation and Breeding History in China

Rubber trees were first introduced to China by Zeng Wangyuan in 1902 (Wang 2015), but these tropical trees could not survive at Inde, a county located at the north of Guangdong Province. Two years later, Dao Anren successfully planted rubber trees at Yingjiang, Yunnan Province (Fig. 2.1) (He and Huang 1987). After that, several rubber cultivations were established in Hainan Province, including Qiong'an Garden, Qiaoxin Garden, Caihui Garden and Tianren Garden. By 1949, there were about 2,800 ha of rubber totaling 1.06 million trees in China, of which 0.64 million were tapped, yielding 200 tons of dry rubber per year (He and Huang 1987).

At this stage, there was no breeding institution in China, and the rubber gardens only planted seedling trees. During the 1950s, the rubber tree clones, PB 86 (Malaysian pedigree) and PR 107 (Indonesian pedigree), were introduced into China and became widely accepted by plantations (He and Huang 1987). In 1954, the South China Institute of Tropical Forest Science (now Chinese Academy of Tropical Agricultural Sciences, CATAS) was founded to focus specially on the study of rubber breeding and planting technology (Fig. 2.2). Since then, about eighteen institutions have participated in rubber breeding and 78 clones have been bred and widely planted in China (Table 2.1) (Huang 2005).

As the south of China geographically spans tropical and cold temperate zones, there are limited regions suitable for rubber cultivation. As a result of development efforts since the 1950s, three rubber planting districts have been established: Yunnan, Guangdong, and Hainan (He and Huang 1987). The overall rubber planted area in China reached 1.13 million ha in 2017. These districts have totally different climate conditions. In Hainan district, the major environmental challenge is typhoon, followed by cold temperatures while Guangdong district has to contend with cold as the most major issue, followed by typhoon. In Yunnan, low temperature is the only major environmental restriction for rubber planting. As rubber trees originate from the tropical Amazon Region, cold waves in winter pose a major environmental challenge to rubber planted areas in China. Temperatures above freezing can destroy rubber trees within weeks. In Guangdong Province, 80% of rubber tree seedlings were severely injured in the winter of 1954/1955 (He and Huang 1987), and about 61.6% of mature trees in tapping were severely damaged in 1975/1976 (Wang et al. 2008). The most recent cold damages happened in 2007/2008, during which all of the tapped trees in the east of Yunnan Province were injured, and more than 70% of trees in Hainan were damaged by low temperature (Kang et al. 2009). Therefore,

Fig. 2.1 The first rubber tree planted in China (Yingjiang, Yunnan 1904). This photo was taken by Yanshi Hu, Rubber Research Institute, CATAS in 2017

Fig. 2.2 Original site of the South China Institute of Tropical Forest Science (Guangzhou 1954)

cold resistance is one of the most important traits for the rubber breeding program in China. To date, a large number of cold-resistant clones such as CATAS 93-114, Yunyan 77-4, Yunyan 77-2, and Zhanshi 366 have been bred and are widely recommended in the Guangdong and Yunnan districts (Cheng et al. 2018a).

Besides low temperature, typhoon is another climatic threat to rubber cultivation in Hainan and Guangdong. Every year, tropical cyclones form in the Northwestern Pacific Basin and may proceed in the direction of a straight track (westward) or a parabolic recurving path. The former path affects southern China and causes great damage to rubber trees in the Hainan and Guangdong districts, especially in the regions near the east or southeast coasts. In Hainan, the most favorable land for rubber cultivation is located in the typhoon-affected regions. Since 1950s, wind resistance has been an essential agronomic trait for rubber clone improvement in China. Several clones, including Haiken 1, Wenchang 217, Wenchang 11, Reyan 7-33-97, Haiken 2, Wenchang 33-24, and Xuyan 141-2 have been demonstrated to be wind-resistant during several decades of planting in the typhoon regions (Huang 2005).

2.1.3 Reyan 7-33-97, the Most Popular *Hevea* Clone Bred in China

Reyan 7-33-97 is the most popular rubber tree clone in China and was bred by CATAS (Fig. 2.3). By 2017, this clone was planted in an acreage of more than 0.16 million ha in China; this representing more than 30% of the replanting

Table 2.1 Institutions in China that have participated in rubber breeding since 1949

	Institute	Duration	Location	Clones/Topics
1	Rubber Research Institute, Chinese Academy of Tropical Agricultural Sciences	From 1954	Hainan	Reyan 7-33-97, Reyan 7-20-59, Reyan 8-79, Reyan 88-13, Reyan 7-18-55, Reyan 8-333, Reyan 2-14-39, Reyan 4, Reyan 6-881, Reyan 78-3-5, Guangxi 6-68[a]
2	South Asia Tropical Crop Research Institute, Chinese Academy of Tropical Agricultural Sciences	1954–1987	Guangdong	CATAS 93-114, Youxian 9-3, Zhanshi 366
3	Institute of Tropical Bioscience and Biotechnology, Chinese Academy of Tropical Agricultural Sciences	From 1984	Hainan	Tissue culture
4	Baoting Institute of Tropical Crops	1954–1996	Hainan	Haiken 2, Baoting 911, Baoting 155, Baoting 235, Baoting 3410, Baoting 032-33-10, Baoting 34-13, Baoting 933, Baoting 79-017, Haiken 6
5	Rubber Research Institute of Hainan State Farms (HSF)	1955–2008	Hainan	Haiken 1, Wenchang 217, Wenchang 11, Wenchang 65-8-502, Wenchang 8-32-9, Hai 44, Wenyan 172
6	Daling Farm of HSF	1955–2000	Hainan	Daling 64-36-101, Daling 68-35, Tianren 31-45, Daling 17-155, Daling 79-8, Daling 64-21-65, Daling 76-1
7	Nanfeng Farm of HSF	1955–2000	Hainan	Nanfeng 37, Hekou 3-11, Nanfeng 70, Nanfeng 3
8	Dafeng Farm of HSF	1954–1966	Hainan	Dafeng 95, Dafeng 99, Dafeng 78-14, Dafeng 78-50, Dafeng 78-138, Dafeng 78-184, Dafeng 318
9	Xuwen Institute of Rubber Research	1956–1995	Guangdong	Hongxin 1, Nanqiang 1-97, Nanhua 1, Xuyan 141-2
10	Huazhou Institute of Rubber Research	1956–1994	Guangxi	Hua 59-2, Hua 1, Hua 2
11	Yunnan Institute of Tropical Crops	From 1959	Yunnan	Yunyan 77-2, Yunyan 77-4, Yunyan 277-5, Yunyan 73-46, Yunyan 68-273, Yunyan 1, Yunyan 73-477, Yunyan 72-324, Yunyan 74-625, Yunyan 75-11, Yunyan 3
12	Honghe Institute of Tropical Agriculture	1988–1996	Yunnan	
13	Dehong Institute of Tropical Agriculture	1988–1996	Yunnan	Deken 22
14	Fujian Institute of Tropical Crops	1961–1991	Fujian	Minlin 71-22
15	Guangxi Institute of Rubber Research	1974–1992	Guangxi	Guiyan 74-1, Guiyan 73-165
16	Hainan Station, Institute of Genetics, Chinese Academy of Sciences (CAS)	1960–1963, 1977–1979	Hainan	Tissue culture
17	South Chinese University of Tropical Agriculture[b]	1958	Hainan	Tissue culture Polyploidy induction
18	Institute of Tropical Forest, Chinese Academy of Forestry Sciences		Hainan	Wood property

Note [a]collaborate with Xianfeng farm of Longzhou; [b]now Hainan University

acreage in the last 10 years. The word "Reyan" in Chinese means "bred by CATAS." Due to institutional name changes over time, this clone is also known as CATAS 7-33-97 or SCATC 7-33-97. "SCATC" is the acronym for the South China Academy of Tropical Crops (changed from the South China Institute of Tropical Forest Science in 1965) which was subsequently renamed as the Chinese Academy of Tropical Agricultural Sciences (CATAS) in 1994. The digit "7" refers to the parental breeding cross of RRIM 600 × PR 107 in 1965, whereas "33" and "97" indicate the planting line and row, respectively, of the F1 progenies.

Reyan 7-33-97 had gone through small scale trials from 1966–1974. After that, large-scale trials were set up at the CATAS Farm in 1975 and 1976. RRIM 600 was used as a control for these trials. At the CATAS Farm, the average annual yield of Reyan 7-33-97 reached 1491 kg/ha for the

Fig. 2.3 A plantation of Reyan 7-33-97 trees

first five tapping years (87.7% higher than the RRIM 600 control), and 1959 kg/ha for the first ten tapping years (68.5% higher than the RRIM 600 control) (Huang 2005). On-farm trial and adaptability tests were also set up at Donghong Farm (1979), Licai Farm (1979), and Chongpoling Farm (1980) (Huang 2005).

Reyan 7-33-97 also has strong resistance to wind damage and is fairly resistant to low temperature. Reyan 7-33-97 has a windthrow rate of about 2.4%, which is significantly lower than that of RRIM 600 (6.02%). This clone has also been proven to be a quick starter with good yield, high dry rubber content, fast growth, low incidence of tapping panel dryness, and other good agronomic traits. For these reasons, Reyan 7-33-97 has been a Large-Scale Recommended rubber clone since 1995 (Huang 2005).

2.2 An Overview of the Reyan 7-33-97 Genome

2.2.1 Assessment of Genome Size of the Reyan 7-33-97 Clone

With the current advances in next-generation sequencing (NGS) technology, it is possible to sequence genomes of non-model plant species (Cheng et al. 2018b). In China, the genome sequencing project was first proposed in March 2010 by the Rubber Research Institute, CATAS. The roadmap of this sequencing project was planned: generation of a draft genome; construction of a high-density genetic map; and finally, a high-resolution genome assembly.

The first task was to assess the genome size of rubber tree, which was characterized by several methods previously. The first report was by Low and Bonner (1985). They calculated the composition of fast and slow annealing DNA, which showed that 48% of the genome was putative single copy (slow annealing DNA) and 32% consisted of highly repetitive or palindromic DNA (fast annealing DNA). The rubber genome size was then calculated as 600 Mb (Low and Bonner 1985). Bennett and Leitch used Feulgen micro-densitometry to produce a 2.15 Gb genome size by calculating the DNA C-value of *Hevea* (Bennett and Leitch 1997). Seguin et al. estimated the genome size of several *Hevea* species including the commercial rubber tree using flow cytometry, and suggested a genome size of 1.9 Gb (Seguin et al. 2003), which is threefold that calculated by Low and Bonner (1985). The distinctly different estimates of the rubber tree genome size were obviously not sufficient to support a genome sequencing project. To estimate the Reyan 7-33-97 genome size, 138 Gb processed shotgun sequences were obtained (Tang et al. 2016). The *Hevea* genome size was then re-estimated by the K-mer method, which is widely used for genome size estimation in NGS sequencing projects (Zhang et al. 2012; Xue et al. 2014). The Reyan 7-33-97 genome size was calculated by the KmerFreq program (Li et al. 2010), in which the number of K-mers for error-corrected reads was counted with the default 17-mer statistics. The genome size was then obtained by dividing the total number of 17-mers by the peak value of the 17-mer frequency distribution (Fig. 2.4). Finally, the genome size of Reyan 7-33-97 was estimated to be 1.46 Gb (Tang et al. 2016).

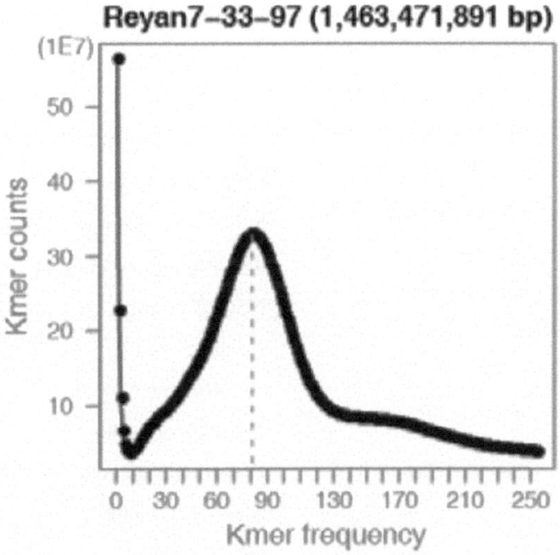

Fig. 2.4 Distribution of 17-mer frequency in raw sequence data of the Reyan 7-33-97. The x-axis and y-axis indicate the 17-mer frequency and number, respectively. The dashed line points to the peak value of 17-mer frequency. This figure is adapted from Tang et al. (2016)

2.2.2 Strategies and Tools for Genome Sequencing and Assembly

To sequence the Reyan 7-3-97 genome, two types of genomic libraries were constructed: short fragment paired-end library (<600 bp) and long fragment mate-paired library (800 bp ~10 kb) (Tang et al. 2016). The paired-end sequencing library was constructed with Bio Scientific NEXTflex™ Rapid DNA-Seq kit, while the mate-paired library was according to the SOLiD long terminal sequencing method. The libraries were then paired-end sequenced with Illumina GAII and Hiseq 2000 platforms. The generated 138 Gb (94 × genome coverage) paired-end 69-mer reads were assembled with the SOAPdenovo program to produce primary contigs. Then the long fragment mate-paired end sequences were used to construct scaffolds against the primary contigs with the SSPACE software (Boetzer et al. 2011). Finally, the gaps in the assembled scaffolds were filled using the GapFiller program (Nadalin et al. 2012).

To improve the assembly quality, a bacterial artificial chromosome (BAC) library was constructed (Haiminen et al. 2011). The BAC library was then sequenced and the data were utilized to assist scaffold construction. A total of 54,912 BAC clones with a mean insert size of 135 kb was stored in 143 384-well plates (or 143 pools). The BAC DNAs from each pool were sequenced with Illumina Hiseq 2000 by barcoding indexing. After removing contaminations from *E. coli*

Table 2.2 Statistics for the *Hevea* Reyan 7-33-97 genome and gene annotation (Tang et al. 2016)

Estimate of genome size	1.46 Gb
Number of scaffolds	7,453
Total length of scaffolds	1.37 Gb
N50 of scaffolds	1.28 Mb
Longest scaffold	6.41 Mb
Number of contigs	84,285
Total length of contigs	1.29 Gb
N50 of contigs	30.6 kb
Longest contigs	312.7 kb
GC content	34.84%
Number of genes	43,792
Percentage of gene length in genome	12.47%
Mean gene length	3,913 bp
Gene density	31.9 Mb^{-1}
Transcripts number	46,631
Mean transcript length	1,483 bp
Mean coding sequence length	1,123 bp
Mean exon length	308 bp
Exons GC content	41.54%
Mean intron length	677 bp
Intron GC content	32.61%
Masked repeat sequence length	977.5 Mb
Repeats percentage of total scaffolds	71.18%
Repeats percentage of total contigs	75.69%

genome sequences and the pCC1BAC BAC vector, 124 pools of data were finally validated, assembled and assigned onto scaffolds using SOAPdenovo and SSPACE software, respectively. About 3.2 Gb raw data were obtained from each pool to ensure that the pre-processed data reached at least 60× coverage. The assembly size of all the 124 pools was 5.95 Gb (about 4× the genome size).

A series of pseudo libraries (10, 20, 30, 40, 50, 70, 100, 150, and 200 kb in insert sizes) was then randomly generated from all of the 124 pool assemblies. The SSPACE software was used to link shotgun scaffolds with these pseudo reads, and then all gaps were replaced with BAC scaffold sequences where ever appropriate. The remaining gaps were closed by GapFiller again and were checked manually. Assembly statistics for the final refined genome such as genome size, number of scaffolds, and N50 values are shown in Table 2.2.

2.2.3 Genome Annotation

Three approaches were used by Tang et al. (2016) to annotate the Reyan 7-33-97 genome: ab initio prediction, and predictions based on 84,241 unique assembled transcripts (from RNA-seq transcriptomes generated in the same study) and protein homologues. Protein-coding genes were predicted ab initio or based on alignment with conserved protein homologues and assembled transcripts. For ab initio prediction, four HMM-based predictor programs were used, namely, AUGUSTUS (Stanke et al. 2004), GlimmerHMM (Majoros et al. 2004), SNAP (Korf 2004), and FGENESH++ (Solovyev et al.

2006). The first three programs were trained with PASA-built training sets and FGENESH++ was run with pre-trained parameters specialized for *Hevea*. SPALN (Iwata and Gotoh 2012) was used for protein homologue search with the parameter "*–Q4 –O0 –M10 –H180*" against proteins in Malpighiales species from the NCBI non-redundant (nr) and Uniprot databases. For prediction based on transcript assemblies, the transcripts from NGS transcriptomes were used to construct gene models using the PASA program for training the predictors, as well as extracting the most possible coding sequences (CDS) with the PASA in-built Transdecoder program (Haas et al. 2003). All results from the three types of prediction were integrated by EVM software (Haas et al. 2008). Next, potential mistakes caused by frame-shifting problems were rectified according to the following criteria (1) The order of priority was SPALN (Conserved), PASA (Expressed), and EVM (Predicted). (2) If a gene overlapped on the same strand, the longer one was kept. (3) Gene overlaps on a different strand orientation were allowed. (4) Gene nesting in another gene's intron was allowed. Finally, all gene models were updated and curated by the PASA software to confirm the UTR regions and alternative splicing forms. Highly repetitive genes, such as "Retro-transposon," were manually removed from the candidates. In total, we annotated 43,792 coding genes with 46,631 isoforms, composing about 12.47% of the assembled genome sequences (Table 2.2). The average lengths of genes (with introns) and transcripts (without introns) were 3,913 bp and 1,483 bp, respectively, and the average exon number per gene was 4.8. The average GC content of exons and introns was 41.54% and 32.61%, respectively. The length distribution of CDSs, exons, and introns was similar among *Hevea*, *Ricinus communis*, *Manihot esculenta*, *Populus trichocarpa*, and *Linum usitatissimum* (Fig. 2.5). A total of 39,919 of the protein-coding genes were found in the NCBI non-redundant protein database. The survey of gene families clustered 16,315 gene families, among which 1,077 are *Hevea*-specific and 10,675 can be found in *M. esculenta*, *R. communis*, *P. trichocarpa*, and *L. usitatissimum*.

The *Hevea* non-coding RNAs were annotated with the INFERNAL software (Nawrocki and Eddy 2013) package (version: 1.1) to search against the Rfam (version 11.0) database. Both "cmsearch" and "cmscan" programs in INFERNAL were used to identify the non-coding RNAs using an e-value cutoff of 0.01. Non-coding RNAs verified by both programs were merged to a concordant version. MicroRNAs (miRNAs) predicted by INFERNAL were analyzed for their secondary stem-loop structures using the RNAfold program (Gruber et al. 2008), and those with a minimum free energy >−25.0 kcal/mol were discarded. To predict microRNAs, the mature sequences from the miRbase database were used to search against the *Hevea* genome. The results showed 1,039 hits having one mismatch at most. The flanking 300 bp sequences of these hits were picked up, and subjected to Miranda algorithm prediction (Enright et al. 2004) which produced 279 miRNA candidates. Secondary structures were then calculated in these candidates with the RNAfold software, and finally, 260 miRNAs candidates were predicted in the Reyan 7-33-97 genome. In addition, a homologue search for non-coding RNAs predicted 167 rRNAs (ribosomal RNAs), 10 SRPRNAs (signal recognition particle RNAs), 3,445 snoRNAs (small nucleolar RNAs), 4 tmRNAs (transfer-messenger RNAs), 697 tRNAs (transfer RNAs), 219 snRNAs (small nuclear RNAs), and 217 RNAs of other types.

Repeat sequences were annotated de novo using RepeatMasker (http://www.repeatmasker.org) by RMBlast search against the Repbase database and custom repeat library building with RepeatModeler (http://www.repeatmasker.org/RepeatModeler.html). The GMATo software (Wang et al. 2013) was employed to scan the whole genome for SSR annotation. The annotation results showed that more than 71% of the Reyan 7-33-97 genome are repeat sequences, among which Gypsy and Copia are the most abundant, occupying about 50% of the total genome (0.7 Gb).

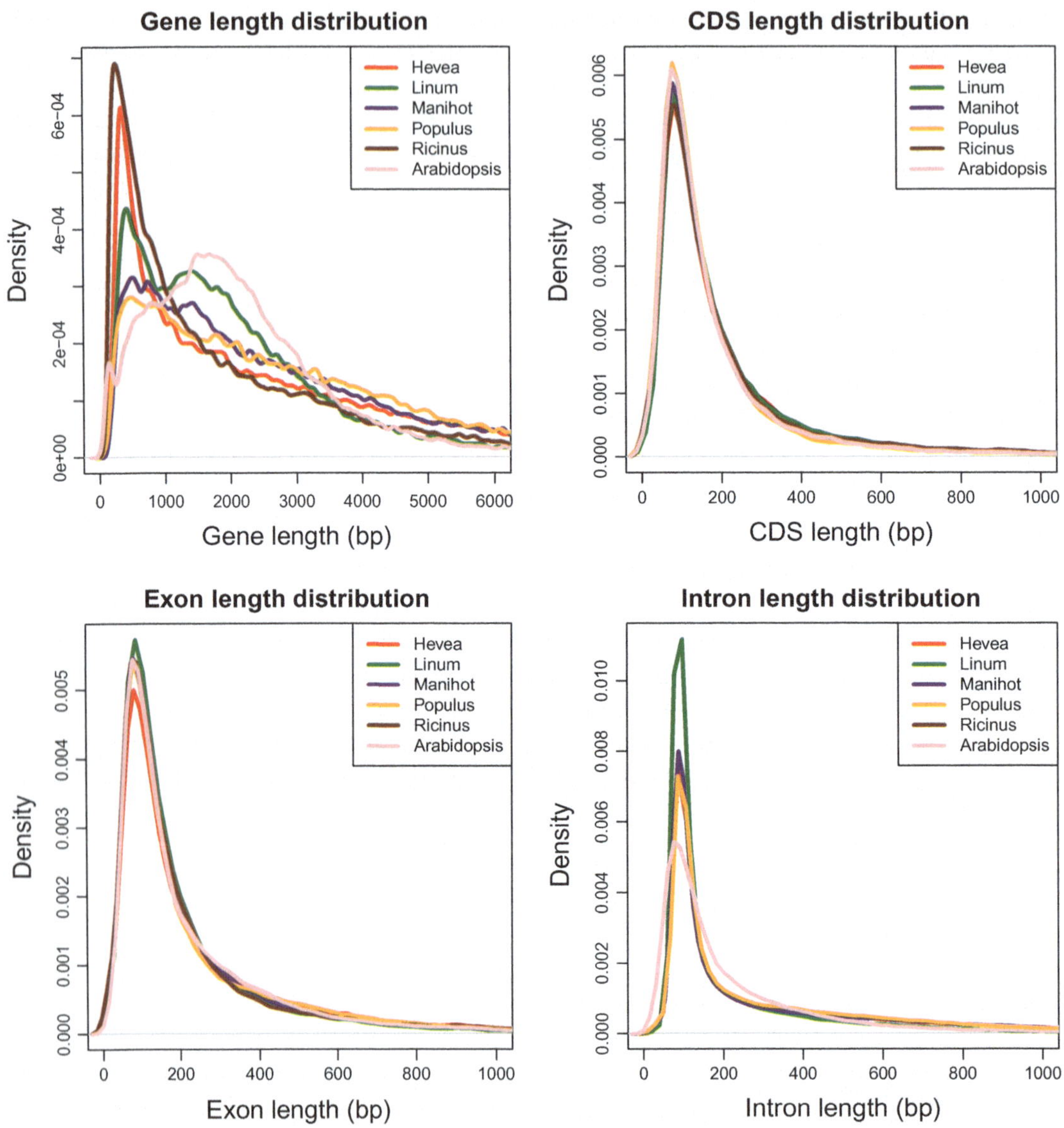

Fig. 2.5 Cross-species comparisons in length distribution of genes, CDSs, exons and introns. The six species compared were *Hevea brasiliensis*, *Linum usitatissimum*, *Manihot esculenta*, *Populus trichocarpa*, *Ricinus communis*, and *Arabidopsis thaliana*. This figure is adapted from Tang et al. (2016)

The long terminal repeat retrotransposons (LTR-RTs) at whole genome level were identified using the LTR_FINDER software (Xu and Wang 2007) for *Hevea*, *R. communis*, *M. esculenta*, *L. usitatissimum*, and *P. trichocarpa*. The genome data of the four non-*Hevea* species were obtained from the Phytozome database (version 10.1). To classify the types of LTR-RTs, the repeat regions for the four genomes were masked using RepeatMasker as the method for *Hevea*, and the LTR-RTs were identified by Ltr-finder based on the RepeatMasker results. LTR insertion time was estimated by computing the nucleotide substitution rate of the two LTRs in an intact LTR-RT since they were assumed to be identical at the retroelement insertion time. LTRs in each pair were aligned to calculate the substitution rate using the baseml algorithm of the

PAML package with the TN93 model (Yang 2007). The LTRs' insertion time was estimated by the formula T = R/2 × u (T, insertion time; R, substitution rate; u, mutation rate). The mutation rate was 1.5e−8 substitutions per site per year, twofold higher than the synonymous substitution rate of the coding region (Ma and Bennetzen 2004).

2.2.4 Gene Family Classification and Phylogenetic Analysis

Gene families were also classified and phylogenetically analyzed (Tang et al. 2016). Proteins of 13 species examined *(H. brasiliensis, P. trichocarpa, R. communis, L. usitatissimum, M. esculenta, Arabidopsis thaliana, Oryza sativa, Citrus sinensis, Medicago truncatula, Glycine max, Fragaria vesca, Vitis vinifera, Carica papaya)* were compared against each other in an all-to-all approach using the BLASTP program. The OrthoMCL program (Li et al. 2003) was utilized to cluster the alignment results into family groups. A total of 32,237 family groups were generated, of which 5,193 were conserved in all 13 species.

Seventy-two single-copy families were selected from these conserved groups to construct a phylogenetic tree (Fig. 2.6). For each family, the protein orthologs were aligned by MAFFT (Katoh and Standley 2013), and transformed to Nexus format. The JTT, as the best fitting model determined by Prottest3 (Darriba et al. 2011), was used for the super alignments, and a phylogenetic tree was constructed with PhyML software for 1,000 bootstrap replicates. The tree files from PhyML and the binary profile from OrthoMCL were exploited to determine the gene family gain and loss for the 13 species using the DOLLOP program from the PHYLIP package. The divergence time of each node in the phylogenetic tree was estimated based on JC69 model in the MCMCTree program from the PAML package. The "usedata" parameter was set to 1 for calculating the likelihood function in a normal way. For the "clock" parameter, the correlated rates were used following a log-normal distribution. In total, MCMCTree ran for 85,000 generations, with a "burn-in" 5,000 iterations to a stable state. Three reported divergence times were used as calibration. The first divergence time was between monocots and eudicots, which took place about 130–240 million years ago (mya), and formed the root. The second was for *A. thaliana* and *C. papaya* about 72 mya (Ming et al. 2008), and the last was for *M. truncatula* and *G. max* about 50–60 mya (Lavin et al. 2005).

2.3 Genome-Guided Discovery of Gene Families

Before the genomics era, rubber tree genes were usually identified through homologous cloning strategies. This is a tedious and onerous task and takes months, sometimes even years to obtain the full-length cDNA of a gene with 5' and 3' rapid amplification of cDNA ends (RACE). In most cases, scientists could only clone genes which show high expression and were not able to classify genes into families. A high-quality genome enables more accurate identification of genes and members of gene families based on sequence alignment and genome annotations. Since the emergence of rubber genomes, several gene families have been identified on a genome-wide scale, such as for aquaporin (Zou et al. 2015), metacaspase (Liu et al. 2016), MYB transcriptional factor (Wang et al. 2017), calcium-dependent protein kinase (CDPK) and CDPKs dependent kinase (Xiao et al. 2017), cysteine protease (Zou et al. 2017a), SnRK2 (Guo et al. 2017a), papain-like cysteine protease (Zou et al. 2017b), abscisic acid receptor (Guo et al. 2017b), and SWEET genes (Sui et al. 2017). These studies classified genes into families and characterized their expression and described their possible functions in their respective bioprocesses. In the rest of this section, we describe examples of gene families related to rubber biosynthesis and ethylene

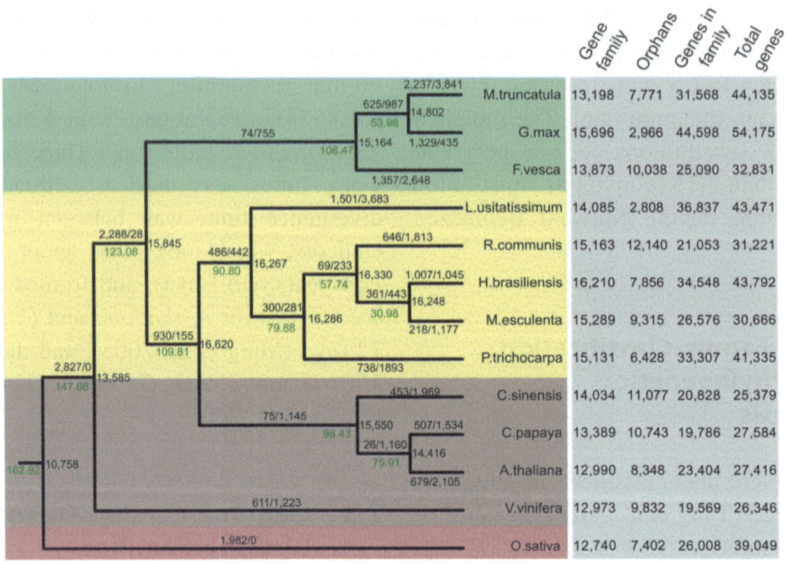

Fig. 2.6 Phylogenetic tree of 13 species based on orthologues of single-gene families. The number in each node indicates the number of gene families. The number at the root (10,758) represents the number of gene families in the common ancestor. The values above each branch denote gene family gain/loss number at each round of genome duplication after diversifying from the common ancestor. The green numbers below each branch denote speculated divergent time of each node. The clades are marked by four different block colors in the tree: the last one (pink) is a monocot species, *Oryza sativa*, used as an outgroup; the first three are Rosids clades in dicots, including Fabidae (green), COM (yellow) and Malvidae (brown). Bootstrap values for each node are above 50%. This figure is adapted from Tang et al. (2016)

biosynthesis and signaling, two physiological processes which are most important for the rubber tree as an economic crop (Tang et al. 2016).

Sucrose is the raw material for rubber biosynthesis in *Hevea* laticifers which is first metabolized into pyruvate and then into acetyl-CoA. The process from acetyl-CoA to rubber (*cis*-1,4-polyisoprene) can be divided into three major steps, i.e., isopentenyl pyrophosphate (IPP) synthesis, initiator synthesis, and rubber elongation, while all the genes responsible are traditionally termed as rubber biosynthesis genes (Tang et al. 2010). IPP, a common precursor for numerous isoprenoids, can be generated via the cytosolic mevalonic acid (MVA) or the plastidic methylerythritol 4-phosphate (MEP) pathway in plants. The involvement of the MVA pathway in rubber biosynthesis was established in the 1960s, and was for decades, conventionally regarded as the only pathway providing IPP for rubber biosynthesis in *Hevea* (Archer and Audley 1987). More recently, the identification of 1-deoxy-D-xylulose 5-phosphate synthase genes (DXS) from the latex transcriptome indicated the existence of the MEP pathway in laticifers (Ko et al. 2003; Chow et al. 2007). A recent study suggested that IPP can also be supplied through the MEP pathway in latex for the synthesis of rubber (Chow et al. 2012). To better understand the possible routes of rubber biosynthesis in *Hevea* via either the MVA or MEP pathway, we identified all rubber biosynthesis genes in the Reyan 7-33-97 genome assembly. In total, 20 rubber biosynthesis gene families were investigated, including six in the MVA pathway (ACAT, HMGS, HMGR, MVK, MVD, PMD), seven in the MEP pathway (DXS, DXR, CMS, CMK, MCS, HDS, HDR), four in initiator synthesis (IPPI, GPS, FPS, GGPS), and three in elongation of the polyisoprene chain (CPT, REF, SRPP). All predicted proteins were then subjected to InterPro analysis, and the candidate

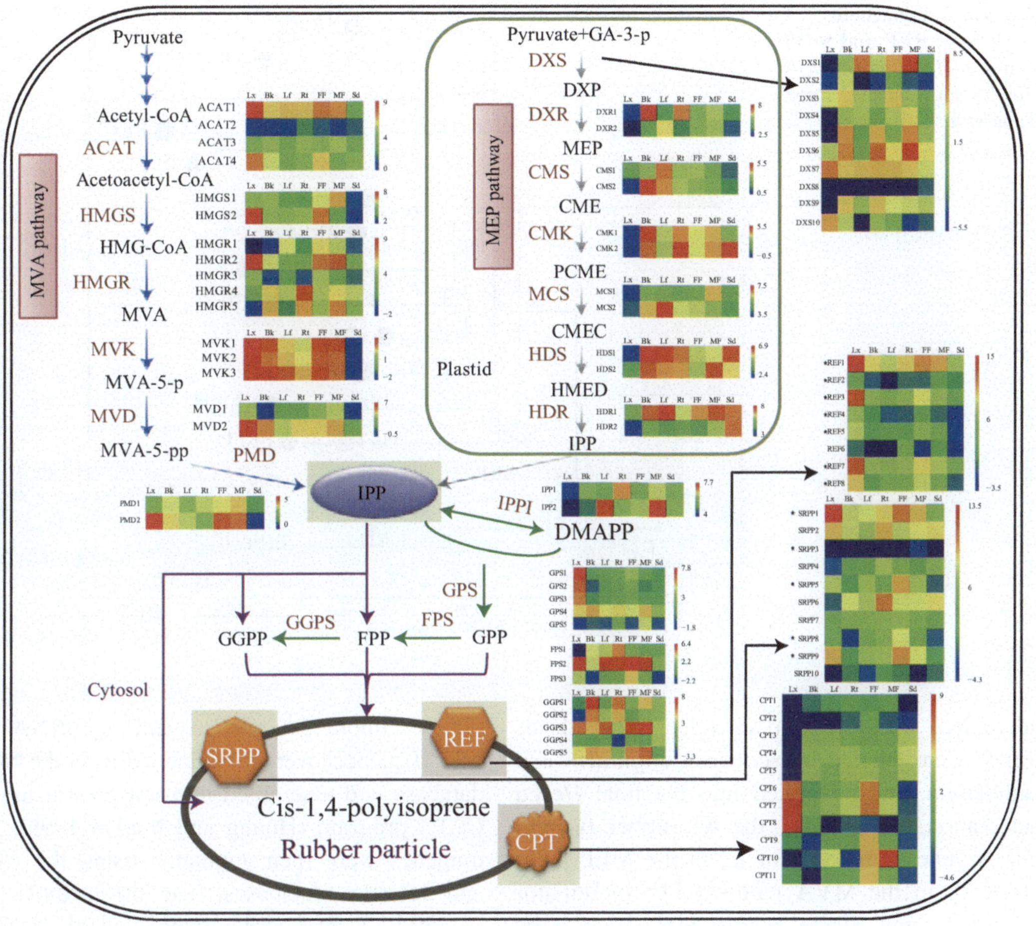

Fig. 2.7 Rubber biosynthesis pathway genes and gene expression in *Hevea brasiliensis*. Two pathways contribute to the synthesis of IPP, the precursor for rubber (*cis*-1,4-polyisoprene) biosynthesis: the MVA pathway in the cytosol (top left) and the MEP pathway in the plastid (top right). The expression of each pathway gene (in RPKM, reads per kilobase per million reads mapped) is demonstrated in the heatmaps. Tissue abbreviations: Lx, latex; Bk, bark; Lf, leaf; FF, female flower; MF, male flower; Sd, seed. Gene abbreviations: ACAT, acetyl-CoA acetyltransferase; HMGS, 3-hydroxy-3-methylglutaryl-CoA synthase; HMGR, 3-hydroxyl-3-methylglutaryl-CoA reductase; MVK, mevalonate kinase; MVD, phosphomevalonate kinase; PMD, diphosphomevalonate decarboxylase; DXS, 1-deoxy-D-xylulose 5-phosphate synthase; DXR,1-deoxy-D-xylulose 5-phosphate reductoisomerase; CMS, 2-C-methyl-D-erythritol 4-phosphate cytidylyltransferase; CMK, 4-(cytidine 5'-diphospho)-2-C-methyl-D-erythritol kinase; MCS, 2-C-methyl-D-erythritol 2,4-cyclodiphosphate synthase; HDS, 4-hydroxy-3-methylbut-2-enyl diphosphate synthase; HDR, 4-hydroxy-3-methylbut-2-enyl diphosphate reductase; IPPI, isopentenyl diphosphate isomerase; GPS, geranyl pyrophosphate synthase; FPS, farnesyl pyrophosphate synthase; GGPS, geranylgeranyl pyrophosphate synthase; CPT, *cis*-prenyltransferase ; REF, rubber elongation factor; SRPP, small rubber particle protein. This figure is adapted from Tang et al. (2016)

proteins containing certain domains of the respective gene families were checked manually and verified through PSI-blast (Altschul et al. 1997). In this way, 83 rubber biosynthesis genes were obtained from 20 gene families in the Reyan 7-33-97 genome. The SPALN program (version 2.1.0) (Iwata and Gotoh 2012) was subsequently used to map the 83 deduced rubber biosynthesis proteins to the genome assembly that had been masked for these 83 rubber biosynthesis genes. As a result, one additional candidate rubber biosynthesis gene was

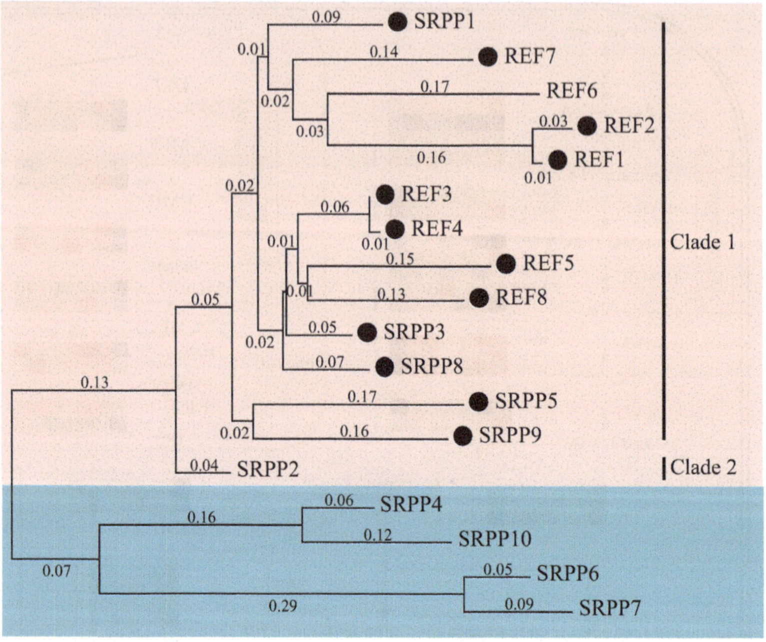

Fig. 2.8 Phylogenetic analysis of REF and SRPP genes in *Hevea brasiliensis*. This figure is adapted from Tang et al. (2016)

identified. All the candidate rubber biosynthesis genes were curated by sequence alignment and annotation, and integrated into the final *Hevea* annotation file. Finally, the 84 rubber biosynthesis genes consisting of 22 in the MEP pathway, 18 in the MVA pathway, 15 in initiator synthesis, and 29 in rubber elongation were identified in the Reyan 7-33-97 genome (Fig. 2.7).

The distribution of the REF/SRPP genes were investigated in 20 other plants: four rubber-producing plants (*Lactuca sativa*, *Parthenium argentatum*, *Taraxacum brevicorniculatum*, and *Taraxacum officinale*), five latex-producing plants (*M. esculenta*, *R. communis*, *C. papaya*, *Jatropha curcas*, and *Cannabis sativa*) and 11 non-latex-producing plants (*A. thaliana*, *P. trichocarpa*, *V. vinifera*, *L. usitatissimum*, *O. sativa*, *Gossypium hirsutum*, *Gossypium arboreum*, *Gossypium raimondii*, *Capsicum annuum*, *Zea mays*, and *Triticum aestivum*). Of these plants, 17 have published genomes, whose protein sequences were retrieved from GFF annotation files or directly downloaded from submitted proteome files, and then re-annotated using NCBI and InterPro database. Of the three plants without genome sequences, mRNA and EST sequences were downloaded from the NCBI database and assembled into unigenes using the CAP3 program (Huang and Madan 1999). The unigenes were then annotated using the NCBI and InterPro databases. The 50 mRNAs and 11,759 ESTs of *P. argentatum* available in NCBI were assembled into 8,164 unigenes, from which six REF/SRPP genes were identified. The 409 mRNAs and 41,301 ESTs of *T. officinale* were assembled into 17,181 unigenes, from which three REF/SRPP genes were detected. A total of 33 proteins of *T. brevicorniculatum* were downloaded from NCBI, and six of them were identified as REF/SRPP genes. The number of REF/SRPP genes ranged from one to ten in the 21 examined plants except *Hevea*, which had 18 REF/SRPP genes (Fig. 2.8, Table 2.3). Further phylogenetic analysis suggested that REF/SRPP proteins were homologous proteins derived from one common ancestor gene, and belonged to a large plant stress-related protein family with conserved motifs and high sequence similarity (Fig. 2.9).

Latex production is greatly increased by ethephon application and 509 differentially

Table 2.3 Gene number of the REF/SRPP family in different plant species (Tang et al. 2016)

	Plant species	Genome size	Data origin	REF/SRPP number	Remarks
Rubber-producing	*Hevea brasiliensis*	1.5 Gb	Tang et al. (2016)	18	–
	Lactuca sativa	2.59 Gb (Reyes-Chin-Wo et al. 2017)	NCBI	8	–
	Parthenium argentatum	NA	NCBI (mRNA, EST)	6	A total of 50 mRNA sequences and 11,759 ESTs were assembled into 8,164 unigenes for gene identification
	Taraxacum brevicorniculatum	NA	NCBI (Proteins) and reference (Laibach et al. 2015)	6	A total of 33 protein sequences were available in NCBI
	Taraxacum officinale	1.25 Gb (Lin et al. 2018)	NCBI (mRNA, EST)	3	A total of 409 mRNA sequences and 41,301 ESTs were assembled into 17,181 unigenes
Latex-producing	*Manihot esculenta*	742 Mb (Bredeson et al. 2016)	Phytozome	6	–
	Ricinus communis	350.6 Mb (Chan et al. 2010)	Phytozome	3	–
	Jatropha curcas	410 Mb (Carvalho et al. 2008)	NCBI	3	–
	Carica papaya	370 Mb (Ming et al. 2008)	Phytozome	3	–
	Cannabis sativa	534 Mb (van Bakel et al. 2011)	Phytozome	2	–
Non-rubber or latex-producing	*Arabidopsis thaliana*	125 Mb (Lamesch et al. 2012)	Phytozome	4	–
	Populus trichocarpa	480 Mb (Ye and Zhong 2015)	NCBI	6	–
	Gossypium hirsutum	2.25-2.43 Gb (Li et al. 2015a)	CGP	10	http://cgp.genomics.org.cn/
	Gossypium arboreum	1.7 Gb (Li et al. 2014)	CGP	8	http://cgp.genomics.org.cn/
	Gossypium raimondii	775 Mb (Wang et al. 2012b)	CGP	7	http://cgp.genomics.org.cn/
	Capsicum annuum	3.48 Gb (Ahn et al. 2016)	Phytozome	3	–
	Vitis vinifera	490 Mb (Jaillon et al. 2007)	Phytozome	2	–
	Linum usitatissimum	373 Mb (Wang et al. 2012a)	Phytozome	4	–
	Oryza sativa	466 Mb (Ouyang et al. 2007)	Phytozome	2	–
	Zea mays	2.3 Gb (Hirsch et al. 2016)	Phytozome	1	–
	Triticum aestivum	17 Gb (International Wheat Genome Sequencing Consortium (IWGSC) 2014)	IWGSC	4	The International Wheat Genome Sequencing Consortium

NA, not available

Fig. 2.9 Phylogenetic analysis of the REF/SRPP gene family in 13 plant species. A total of 63 REF/SRPP proteins from 13 plants were aligned using MAFFT v7.205, and the tree was constructed using MEGA5.1. Solid circles (●) indicate the *H. brasiliensis* REF/SRPP genes located in Scaffold1222. Empty circles (○) indicate the *H. brasiliensis* REF/SRPP scattered in other scaffolds. Solid triangles (▲) indicate the REF/SRPP genes from other rubber-producing plants. Note that most of the rubber REF/SRPP proteins including all those located in Scaffold1222 are clustered into an independent clade (light purple), and most of the REF/SRPP proteins from other rubber-producing plants into another independent clade (light blue). This figure is adapted from Tang et al. (2016)

expressed genes (DEGs) were identified in latex that respond to ethylene treatment (Fig. 2.10). These DEGs are either related to ethylene biosynthesis or signaling. About one-quarter of the 342 annotated DEGs (fold change >2) are transcription factors or protein kinases, both of which are essential in ethylene signaling and response in higher plants (Klee 2004; Yang et al. 2015; Dubois et al. 2018).

The genes responsible for ethylene biosynthesis in *Hevea* were identified and their expression patterns were examined. The three enzymes, i.e., S-adenosyl-L-methionine synthase (SAMS), 1-aminocyclopropane-1-carboxylic acid (ACC) synthase (ACS), and ACC oxidase (ACO), which act sequentially to synthesize ethylene from methionine (Yang and Hoffman 1984; Sauter et al. 2013), have 8, 14, and 16 genes, respectively, in the Reyan 7-33-97 genome (Table 2.4).

Next, the expression of the primary ethylene signaling components, namely, ETR (ETHYLENE RESISTANCE 1), CTR1, EIN2 (ETHYLENE INSENSITIVE 2), and EIN3/EIL1 (EIN3-LIKE 1) (Lacey and Binder 2014; Shakeel et al. 2015; Li et al. 2015b; Merchante et al. 2015), were investigated in response to ethylene treatment (Table 2.5). Four out of the eight receptor genes (ETR) showed enhanced expression in response to ethylene treatment, indicating that an

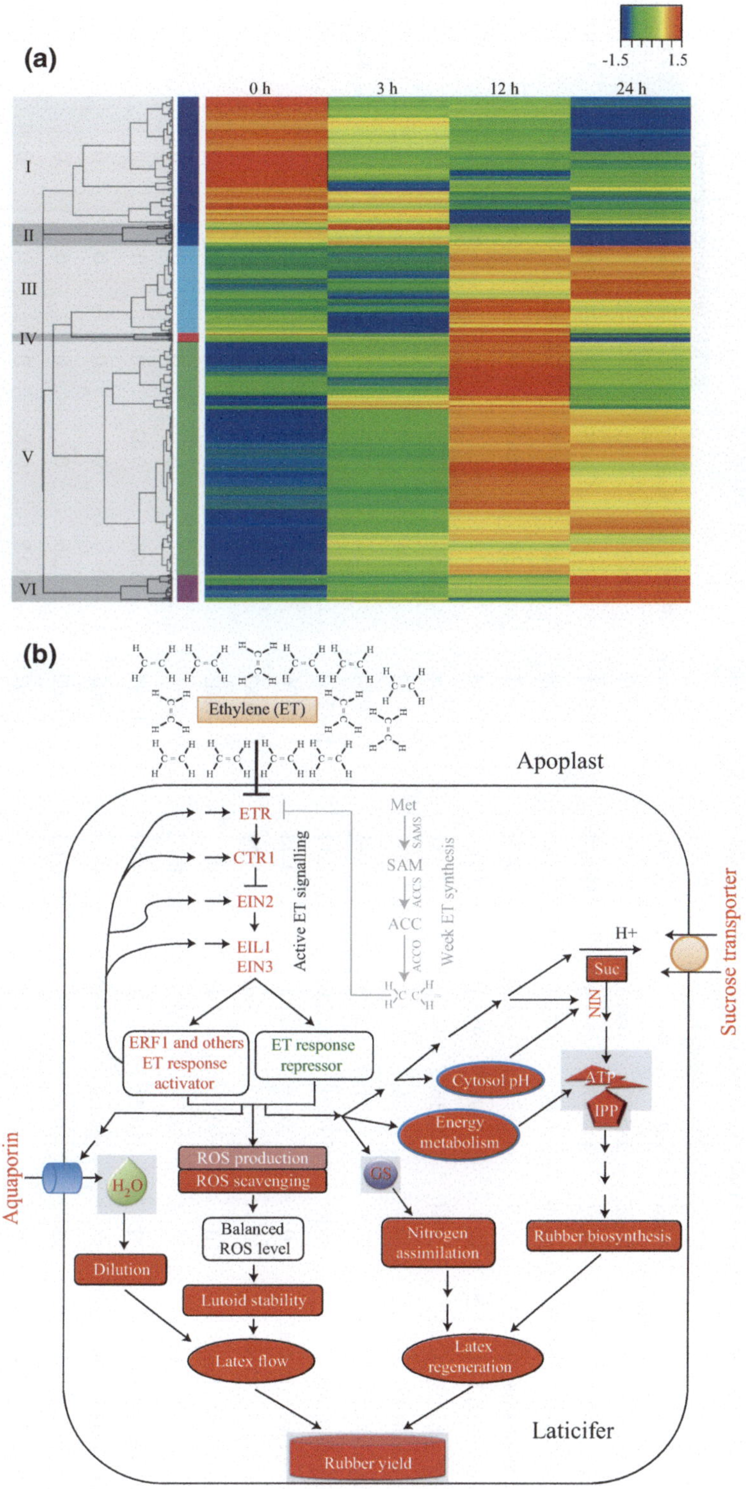

Fig. 2.10 Ethylene stimulation of rubber production in *Hevea*. **a** Expression dynamics of differentially expressed genes (DEGs) within 24 h of ethylene treatment. The expression levels (RPKM) of the 509 DEGs are normalized based on the Z-score method (scale bar). The DEGs are clustered into six clades; two (clades I and II) of them are down-regulated genes whereas the remainder (clades III to VI) are up-regulated. **b** A model summarizing the mechanisms of ethylene stimulation of rubber production. The red and green text used for genes and enzymes denotes that their expression/activity is activated or inhibited by ethylene, respectively. The red background of molecules or events indicates that their quantity or strength is boosted by ethylene. Suc, sucrose; Met, methionine; ROS, reactive oxygen species; NIN, neutral invertase; GS, glutamine synthetase. This figure is adapted from Tang et al. (2016)

Table 2.4 List of the ethylene-synthesis pathway genes and their expression patterns in *H. brasiliensis* (Tang et al. 2016)

Gene name/ID	RPKM (different tissues)							RPKM (ethylene treatment)			
	Latex	Leaf	Bark	Female flower	Male flower	Root	Seed	0 h	3 h	12 h	24 h
SAMS1|scaffold0056_2339331	10.05	285.82	660.81	405.74	220.97	326.36	177.41	11.12	7.9	16.67	7.32
SAMS2|scaffold0005_28735	25.91	112.47	507.35	510.92	375.15	535.15	474.24	25.29	28.55	35.19	31.82
SAMS3|scaffold0341_898891	0.09	1.54	23.2	5.62	7.66	20.84	9.75	0.07	0.09	0	0.09
SAMS4|scaffold0658_418257	5.72	4.75	5.81	4.79	4.77	3.33	58.65	8.38	5.59	6.94	12.33
SAMS5|scaffold0214_1008145	35.3	59.98	956.15	280.6	215.48	227.22	62.56	61.97	49.6	81.39	48.79
SAMS6|scaffold0703_585160	3.44	20.16	197.27	133.51	242.48	166.99	622.96	13.89	16.08	34.82	11.92
SAMS7|scaffold0206_1589232	0	0.26	5.45	2.5	1.32	2.95	3.67	0.14	0.41	0.29	0.27
SAMS8|scaffold0251_1321761	17.58	17.41	265.98	68.45	614.84	53.56	9.26	21.95	29.93	155.26	88.44
Total of SAMS genes	98.09	502.39	2622.02	1412.13	1682.67	1336.4	1418.5	142.81	138.15	330.56	200.98
ACS1|scaffold0821_194762	0	0	17.5	1.18	18.76	9.04	0	0.06	0	0	0
ACS2|scaffold0552_13262	0	1.22	3.06	1.29	4.87	21	0.03	0	0	0	0
ACS3|scaffold0841_18083	0	0	3.15	0	0.02	0	0	0	0	0	0
ACS4|scaffold0283_1243386	0.08	0	0.2	0	0	0.1	0	0	0	0	0
ACS5|scaffold0128_105114	0	0.72	1.33	1.83	12.97	13.88	0.03	0	0	0	0
ACS6|scaffold0430_864152	0	0	3.15	0	0.02	0	0	0	0	0	0
ACS7|scaffold0514_160107	0	0	0	0.05	0.02	1.36	0	0	0	0	0
ACS8|scaffold0302_549632	0	0	12.82	0	0.02	16.15	0.06	0	0	0.04	0
ACS9|scaffold0733_347033	0	0	0.12	0.09	0	0.45	0.03	0.03	0	0	0
ACS10|scaffold0426_169227	13.5	40.18	27.46	9.34	13.66	18.36	49.65	16.5	13.49	10.16	16.36
ACS11|scaffold0580_716666	10.88	31.81	14.09	8.43	52.64	17.58	14.12	9.23	9.8	7.8	12.21
ACS12|scaffold0641_691601	0	0	0	0	0.09	0.25	0	0	0	0	0
ACS13|scaffold0357_415289	0	0	0.59	0	0.9	4.87	0	0.15	0.07	0	0

(continued)

Table 2.4 (continued)

Gene name/ID	RPKM (different tissues)							RPKM (ethylene treatment)			
	Latex	Leaf	Bark	Female flower	Male flower	Root	Seed	0 h	3 h	12 h	24 h
ACS14|scaffold0103_1952423	0.08	3.72	5.47	5.4	1.2	10.19	0	0	0	0	0
Total of ACS genes	24.54	77.65	88.94	27.61	105.17	113.23	63.92	25.97	23.36	18	28.57
ACO1|scaffold0113_1714149	0	8.08	1.57	12.44	6	4.25	0	0.05	0	0	0.12
ACO2|scaffold1057_27974	0	2.79	32.16	81.16	154.77	58.24	164.85	0.14	0	0.24	0.06
ACO3|scaffold2408_724	0	1125.6	2.22	148.21	972.77	77.52	0.65	0.09	0	0	0
ACO4|scaffold1057_109102	0	0.08	1.32	0.62	0.15	0.41	0	0	0	0	0
ACO5|scaffold0371_1076732	0	4.44	7.43	1.83	2.07	4.3	252.65	0	0	0.06	0.12
ACO6|scaffold0611_59152	0	0	0.04	0.74	0.67	4.94	0.53	0	0	0	0
ACO7|scaffold1057_68935	0	5.38	154.84	70.19	30.09	27.93	3.97	0.05	0.12	0.69	0
ACO8|scaffold1486_68672	0	0.89	74.89	30.34	67.32	3	0.64	0	0	0	0
ACO9|scaffold0574_604518	0	4.88	29.78	1.91	0.57	34.67	0.04	0.08	0.15	0.05	0
ACO10|scaffold0894_322678	0.06	0.72	202.02	1.18	0.64	1886.08	0.49	0.94	1.18	0.89	0.65
ACO11|scaffold0075_154737	0	79.74	0.04	646.25	404.49	5.22	0.57	0	0	0	0
ACO12|scaffold0574_601972	0	4.88	29.78	1.91	0.57	34.67	0.04	0.08	0.15	0.05	0
ACO13|scaffold0574_608451	0	73.14	50.26	13.04	1.67	86.96	0.23	0.08	0	0	0.05
ACO14|scaffold1486_104737	47.77	4.13	33.99	5.02	12.61	37.89	4.12	85.42	112.38	150.81	82.96
ACO15|scaffold1486_107054	21	0.13	0.89	0.73	0.28	1.24	0	19.31	25.14	81.59	61.9
ACO16|scaffold0429_498737	0.85	42.43	96.7	16.22	69.74	82.86	124.77	0.6	0.95	0.7	0.35
Total of ACO genes	69.68	1357.31	717.93	1031.79	1724.41	2350.18	553.55	106.84	140.07	235.08	146.21

Table 2.5 List of genes encoding the primary ethylene signaling pathway components in *H. brasiliensis*. The RPKM expression values of genes were extracted from analysis of transcriptomes of ethylene-treated samples by Tang et al. (2016)

Gene name	Gene ID	Fold change	RPKM (ethylene treatment)			
			0 h	3 h	12 h	24 h
ETR	scaffold0838_409025	5.9	8.94	15.45	34.22	52.32
ETR	scaffold1031_40698	3.0	32.45	57.8	79.81	96.16
ETR	scaffold0249_259397	1.8	28.09	45.16	51.52	47.3
ETR	scaffold0239_551511	1.7	11.27	10.98	14.99	18.54
ETR	scaffold0163_736728	−1.6	27.72	21.68	18.32	17.62
ETR	scaffold0979_291486	NS	18.88	21.47	15.53	17.55
ETR	scaffold1412_21002	NS	6.84	6.65	7.72	6.12
ETR	scaffold2238_4418	NS	19.19	18.43	18.45	24.11
CTR1	scaffold0284_306828	NS	9.33	9.77	7.53	9.71
CTR1	scaffold0269_1316207	1.6	14.15	15.94	19.03	23.11
CTR1	scaffold0652_429526	NS	3.22	3.1	2.83	3.76
CTR1	scaffold0840_366136	1.7	10.25	11.94	17.42	14.28
CTR1	scaffold1663_22838	NS	1.31	1.04	1.01	1.09
EIN2	scaffold0050_2714677	3.1	123.94	100.4	200.78	308.97
EIN2	scaffold0093_375297	2.2	22.38	21.28	19.26	41.82
EIN3	scaffold0525_635537	NS	442.76	369.49	482.69	516.21
EIL1	scaffold0716_283509	3.3	17.45	22.12	57.54	35.45

NS, not significant

ethylene response had been triggered in laticifers. Also, the expression of EIN2 and EIL1 genes were increased by ethylene treatment.

The AP2/ERF family represents a large group of genes playing crucial roles in plant development and responses to biotic and abiotic stresses. Previously, 142 AP2/ERF family proteins were identified through transcriptome sequencing (Duan et al. 2013). AP2/ERF transcription factors, and especially the Ethylene-Response Factors, are essentially involved in the ethylene signal pathway, which regulates rubber biosynthesis in *Hevea brasiliensis*. In this study, AP2/ERF family genes were identified in the Reyan 7-33-97 genome by detecting at least one AP2/ERF domain with 60 to 70 amino acids (pfam domain ID PF00847). There are three subfamilies in the AP2/ERF super family (Nakano et al. 2006; Piyatrakul et al. 2014), namely, the ERF family with a single AP2/ERF domain, the AP2 family with two AP2/ERF domains, and the RAV family with a B3 DNA-binding domain (pfam domain ID PF02362) in addition to one AP2/ERF domain. A total of 225 *Hevea* AP2/ERF members, including 181 ERFs, 35 AP2s, 7 RAVs, and 2 Soloists (Table 2.6) were identified, representing the largest grouping compared to *Arabidopsis*, *Populus*, and *Oryza* (Table 2.7).

2.4 Prospects and Challenges of Genome Sequence Improvement

The rubber tree has a cultivation history of more than one hundred years (Huang 2005); however, there are only five generations of breeding populations until now in China (Fig. 2.11). Conventional breeding in rubber tree are relying on phenotypic selection. Due to the long growth duration, the *Hevea* breeding program takes more

Table 2.6 List and classification of the AP2/ERF superfamily genes in *H. brasiliensis* (Tang et al. 2016)

Family	Group	Sub-group	Number	Gene ID
ERF	I	a	2	scaffold0417_245256; scaffold0389_394558
		b	8	scaffold0782_27868; scaffold0481_907908; scaffold0030_939945; scaffold0649_393073; scaffold0431_33484; scaffold0262_270275; scaffold0332_605093; scaffold0745_352219
	II	a	5	scaffold2594_7264; scaffold2542_5010; scaffold2594_1826; scaffold0468_498601; scaffold0521_13119
		b	7	scaffold1195_120325; scaffold0862_53363; scaffold2101_31639; scaffold5106_3474; scaffold0770_341419; scaffold0026_2812041; scaffold1195_112880
		c	3	scaffold0004_929094; scaffold0004_911172; scaffold0087_294786
	III	a	6	scaffold0005_4638525; scaffold0447_369296; scaffold0026_3462933; scaffold7396_661; scaffold5331_2465; scaffold0163_1226775
		b	6	scaffold0923_259522; scaffold0082_640554; scaffold0840_249513; scaffold1388_89485; scaffold0997_163021; scaffold0534_750561
		c	8	scaffold0082_667097; scaffold0923_363448; scaffold0997_153922; scaffold0534_729612; scaffold0923_269698; scaffold0082_669813; scaffold1736_70072; scaffold1388_58203
		d	5	scaffold0009_198930; scaffold1322_86975; scaffold0139_882380; scaffold0356_425272; scaffold1388_74719
		e	6	scaffold0453_47997; scaffold0864_20762; scaffold0744_127631; scaffold0754_396540; scaffold0910_42997; scaffold3296_4981
	IV	a	5	scaffold0407_253102; scaffold1276_47774; scaffold0017_3456921; scaffold0639_343253; scaffold0647_30372
		b	7	scaffold0029_1331495; scaffold0153_270083; scaffold1536_83237; scaffold0114_161003; scaffold1915_51923; scaffold1192_200225; scaffold0388_1093805
	V	a	9	scaffold0331_797745; scaffold0140_964395; scaffold0067_1480217; scaffold0729_165312; scaffold0031_1897872; scaffold0246_979192; scaffold0012_1669935; scaffold1920_13422; scaffold0082_1206394
		b	11	scaffold0014_3680080; scaffold0407_153963; scaffold1113_21581; scaffold0705_80500; scaffold0340_894004; scaffold0188_1252373; scaffold0038_2579791; scaffold0782_488129; scaffold4546_2802; scaffold0829_385836; scaffold0005_4286493
	VI		6	scaffold1237_168382; scaffold1206_96322; scaffold1063_23830; scaffold0802_444591; scaffold0024_3294824; scaffold0105_1350087
	VI-L		5	scaffold0457_96835; scaffold1199_1585; scaffold0747_514098; scaffold3683_8912; scaffold0641_391038
	VII		5	scaffold0163_747179; scaffold0677_520937; scaffold0803_514434; scaffold0625_358254; scaffold0566_721951
	VIII	a	10	scaffold0568_99917; scaffold0620_531922; scaffold0093_1433865; scaffold0342_1158497; scaffold0356_824429; scaffold0552_587201; scaffold0434_836266; scaffold0319_822155; scaffold0620_551771; scaffold0568_69784
		b	12	scaffold0395_189295; scaffold0395_213745; scaffold0137_12965; scaffold1542_55957; scaffold0943_347758; scaffold0266_975464; scaffold0784_417116; scaffold2017_23463; scaffold0105_2241880; scaffold0307_56707; scaffold0285_1314809; scaffold0566_132325
	IX	a	9	scaffold0444_736744; scaffold2391_28193; scaffold0731_292941; scaffold0731_155122; scaffold2391_31439; scaffold0139_695834; scaffold0426_906054; scaffold0668_444732; scaffold0206_1492583
		b	12	scaffold0426_858732; scaffold0770_519202; scaffold0426_862663; scaffold0770_505198; scaffold0668_421116; scaffold0668_410096; scaffold0426_899695; scaffold0206_1507257; scaffold0451_505634; scaffold0827_411643; scaffold0557_662337; scaffold0426_897571
		c	20	scaffold0204_440674; scaffold1015_293137; scaffold0485_569337; scaffold0827_408531; scaffold0898_249087; scaffold0425_12443; scaffold1128_256669; scaffold0204_466288; scaffold0444_56334; scaffold0095_1882802; scaffold5750_395; scaffold0444_58791; scaffold4973_247; scaffold1128_258258; scaffold0557_665510; scaffold0425_2148; scaffold1564_9190; scaffold0239_518050; scaffold0095_1888719; scaffold0204_503033
	X	a	10	scaffold0636_609282; scaffold0266_1127243; scaffold0320_39785; scaffold0246_1294475; scaffold0456_283714; scaffold0169_1298056; scaffold0841_297289; scaffold0807_400454; scaffold0755_107260; scaffold0079_40633
		b	4	scaffold0359_512732; scaffold1415_38342; scaffold5091_914; scaffold1415_50475
	Xb-L		0	

(continued)

Table 2.6 (continued)

Family	Group	Sub-group	Number	Gene ID
AP2			35	scaffold0059_2398611; scaffold0345_1032004; scaffold0425_294393; scaffold0354_490404; scaffold0015_435374; scaffold0430_646696; scaffold0714_395225; scaffold0650_631600; scaffold1342_95223; scaffold0001_3961914; scaffold1040_103569; scaffold1124_4554; scaffold0013_532383; scaffold0851_245869; scaffold1421_86909; scaffold1068_8360; scaffold0017_1455021; scaffold1355_138740; scaffold0037_175787; scaffold0065_537890; scaffold2818_16308; scaffold0978_156777; scaffold1296_154280; scaffold1911_48728; scaffold0494_232395; scaffold0635_298267; scaffold0681_114797; scaffold0551_393258; scaffold0551_597280; scaffold0551_592198; scaffold0029_3291124; scaffold1267_104008; scaffold0551_387317; scaffold1884_28887; scaffold0086_1539772
RAV			7	scaffold0668_480899; scaffold0770_445357; scaffold0076_54889; scaffold1038_221486; scaffold1026_146817; scaffold0920_168200; scaffold1550_23116
Soloist			2	scaffold0525_854691; scaffold0498_38631
Total			225	

Table 2.7 Number of the AP2/ERF superfamily genes in *Hevea*, *Arabidopsis*, *Populus*, and *Oryza* based on genomic analyses (Tang et al. 2016)

Family	Group	*Hevea*	*Arabidopsis*	*Populus*	*Oryza*
ERF	I	10	10	5	9
	II	15	15	20	16
	III	31	23	35	27
	IV	12	9	6	6
	V	20	5	10	8
	VI	6	8	11	6
	VI-L	5	4	4	3
	VII	5	5	6	15
	VIII	22	15	17	15
	IX	41	17	42	18
	X	14	8	9	12
	Xb-L	0	3	4	10
AP2		35	18	26	29
RAV		7	6	6	5
Soloist		2	1	1	1
Total		225	147	202	180

than 30 years to produce new tree clones. In recent years, scientists developed marker-assisted selection (MAS) technique which helps breeders to select candidates in hybrid populations using molecular markers (Collard and Mackill 2008; Ceballos et al. 2015). This method facilitates the breeding program by accelerating the trait selection procedure. However, the MAS strategy is based on limited molecular markers, and is therefore only effective on the traits controlled by major genes (Ceballos et al. 2015). In the rubber tree, yield and abiotic resistance are quantitative traits and are controlled by numerous minor genes. The MAS strategy is therefore ineffective for early selection in *Hevea* breeding.

With the current progress in genome analysis, it is feasible to associate single nucleotide polymorphism (SNP) markers with quantitative traits on a genome-wide level. A genome-wide association mapping study identified 14,155 SNPs which are possibly associated with latex yield and girth traits in 170 Amazonian rubber germplasm

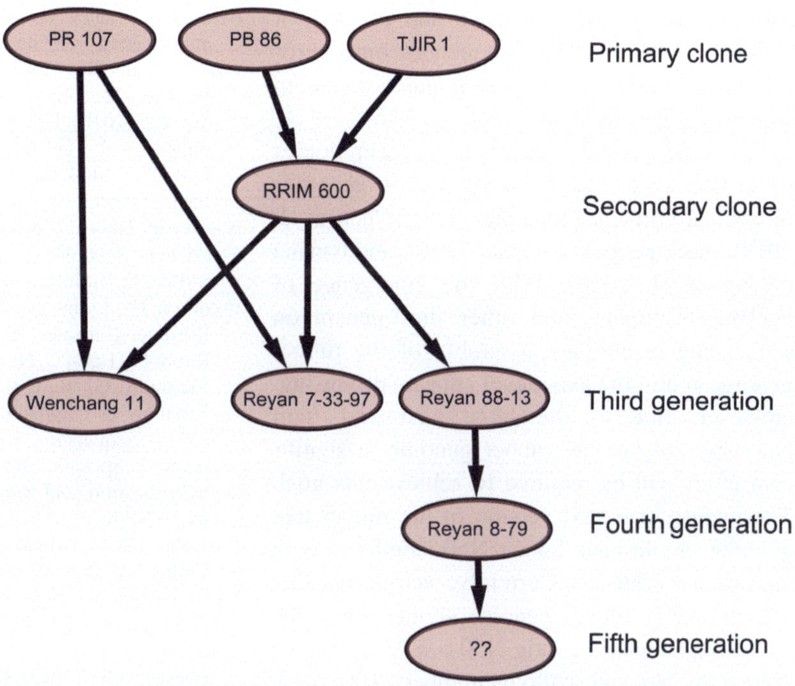

Fig. 2.11 Schematic diagram of rubber breeding generations of Reyan 7-33-97 and its siblings

accessions (Chanroj et al. 2017). With the discovery of much larger numbers of SNPs through genotyping-by-sequencing (GBS) and the development of new methods to genotype large numbers of germplasms efficiently, genome selection (GS) is becoming more feasible to assist genetic evaluation and improvement in a rubber tree breeding program (Goddard and Hayes 2007; Poland and Rutkoski 2016). However, genome quality is still a crucial prerequisite for efficient GS study in the rubber tree.

By 2017, rubber genome sequencing projects were established not only in China but also in rubber growing countries including Malaysia, Thailand, and India. Additionally, these projects included the participation of collaborators from organizations such as RIKEN, Japan and CIRAD, France. Draft genomes have now been published for tree clones Reyan 7-33-97 (Tang et al. 2016), RRIM 600 (Rahman et al. 2013; Lau et al. 2016), and BPM 24 (Pootakham et al. 2017), most of which were assembled through NGS technology. With the emergence of third-generation platforms such as the PacBio sequencing technique (Eid et al. 2009), it is feasible to obtain a higher quality genome assembly. The PacBio sequencing technique generates reads of about 10–15 kb in length, which is much longer than those by Illumina Hiseq 2000. PacBio long reads are beneficial for handling repeat sequences during genome assembly. Optical mapping is another recently developed sequencing technique that gathers long-range sequence information of genomes (Zhou et al. 2007). This technique is similar to the ordered restriction maps, and does not include any bias caused by cloning, amplification, hybridization, or sequencing manipulations. It is therefore suited for improving the quality of fragmented genome assemblies that can no longer be improved by classical methods. Another approach to enhance sequence assembly is Hi-C (van Berkum et al. 2010). This approach allows unbiased genome-wide identification of chromatin interactions by coupling proximity ligation and subsequent massively parallel sequencing. Hi-C helps to determine the order and orientation of each scaffold sequence on chromosomes, and thus upgrade a genome assembly to chromosome level. This technique can potentially save both time and effort as it could mount 95% of the sequences onto chromosomes and does not require population data to construct a genetic map.

Third-generation sequencing technologies have recently produced higher quality genomes in tropical crops, such as pineapple (Chen et al. 2019), black pepper (Hu et al. 2019), and banana (Wang et al. 2019). With the emergence of PacBio, Nanopore, and other third-generation sequencing techniques, assembly of the rubber genome at chromosome level appears promising. However, due to the large size and high heterozygosity of the rubber genome, a significant effort will be required to achieve this goal. The quality of a next version of the rubber tree genome should have longer N50 lengths of both contigs and scaffolds. Currently, we propose that a high-quality rubber genome should attain the following standard: contig N50 > 5 Mb, scaffold N50 $\approx$ 20 Mb, and scaffolds number <100. Also, more than 95% of the scaffolds should be mounted and ordered onto the 18 chromosomes of the rubber tree.

Acknowledgements This work was supported by the Central Public-interest Scientific Institution Basal Research Fund for Innovative Research Team Program of CATAS (NO. 17CXTD-28).

References

Adifaiz AF, Maiden NA, Aizat Shamin N, Roslinda S, Suleiman Z, Rafii MY (2018) Genetic diversity of the 1995 RRIM *Hevea* germplasm collection for utilisation in the rubber breeding programme. J Rubber Res 21:153–164

Ahn Y-K, Karna S, Jun T-H, Yang E-Y, Lee H-E, Kim J-H, Kim J-H (2016) Complete genome sequencing and analysis of *Capsicum annuum* varieties. Mol Breeding 36:140

Altschul SF, Madden TL, Schäffer AA, Zhang J, Zhang Z, Miller W, Lipman DJ (1997) Gapped BLAST and PSI-BLAST: a new generation of protein database search programs. Nucleic Acids Res 25:3389–3402

Archer BL, Audley BG (1987) New aspects of rubber biosynthesis. Bot J Linn Soc 94:181–196

van Bakel H, Stout JM, Cote AG, Tallon CM, Sharpe AG, Hughes TR, Page JE (2011) The draft genome and transcriptome of *Cannabis sativa*. Genome Biol 12: R102

Bennett M, Leitch I (1997) Nuclear DNA amounts in angiosperms—583 new estimates. Ann Bot 80:169–196

van Berkum NL, Lieberman-Aiden E, Williams L, Imakaev M, Gnirke A, Mirny LA, Dekker J, Lander ES (2010) Hi-C: a method to study the three-dimensional architecture of genomes. J Vis Exp. https://www.ncbi.nlm.nih.gov/pmc/articles/PMC3149993/

Boetzer M, Henkel CV, Jansen HJ, Butler D, Pirovano W (2011) Scaffolding pre-assembled contigs using SSPACE. Bioinformatics 27:578–579

Bredeson JV, Lyons JB, Prochnik SE, Wu GA, Ha CM, Edsinger-Gonzales E, Grimwood J, Schmutz J, Rabbi IY, Egesi C, Nauluvula P, Lebot V, Ndunguru J, Mkamilo G, Bart RS, Setter TL, Gleadow RM, Kulakow P, Ferguson ME, Rounsley S, Rokhsar DS (2016) Sequencing wild and cultivated cassava and related species reveals extensive interspecific hybridization and genetic diversity. Nat Biotechnol 34:562–570

Carvalho CR, Clarindo WR, Praça MM, Araújo FS, Carels N (2008) Genome size, base composition and karyotype of *Jatropha curcas* L., an important biofuel plant. Plant Sci 174:613–617

Ceballos H, Kawuki RS, Gracen VE, Yencho GC, Hershey CH (2015) Conventional breeding, marker-assisted selection, genomic selection and inbreeding in clonally propagated crops: a case study for cassava. Theor Appl Genet 128:1647–1667

Chan AP, Crabtree J, Zhao Q, Lorenzi H, Orvis J, Puiu D, Melake-Berhan A, Jones KM, Redman J, Chen G, Cahoon EB, Gedil M, Stanke M, Haas BJ, Wortman JR, Fraser-Liggett CM, Ravel J, Rabinowicz PD (2010) Draft genome sequence of the oilseed species *Ricinus communis*. Nat Biotechnol 28:951–956

Chanroj V, Rattanawong R, Phumichai T, Tangphatsornruang S, Ukoskit K (2017) Genome-wide association mapping of latex yield and girth in Amazonian accessions of *Hevea brasiliensis* grown in a suboptimal climate zone. Genomics 109:475–484

Chen L-Y, VanBuren R, Paris M, Zhou H, Zhang X, Wai CM, Yan H, Chen S, Alonge M, Ramakrishnan S, Liao Z, Liu J, Lin J, Yue J, Fatima M, Lin Z, Zhang J, Huang L, Wang H, Hwa T-Y, Kao S-M, Choi JY, Sharma A, Song J, Wang L, Yim WC, Cushman JC, Paull RE, Matsumoto T, Qin Y, Wu Q, Wang J, Yu Q, Wu J, Zhang S, Boches P, Tung C-W, Wang M-L, d'Eeckenbrugge GC, Sanewski GM, Purugganan MD, Schatz MC, Bennetzen JL, Lexer C, Ming R (2019) The bracteatus pineapple genome and domestication of clonally propagated crops. Nat Genet 51:1549–1558

Cheng H, Chen X, Fang J, An Z, Hu Y, Huang H (2018a) Comparative transcriptome analysis reveals an early gene expression profile that contributes to cold resistance in *Hevea brasiliensis* (the para rubber tree). Tree Physiol 38:1409–1423

Cheng H, Wang Y, Sun M (2018b) Comparison of gene expression profiles in nonmodel eukaryotic organisms with RNA-seq. In: Wang Y, Sun M (eds) Transcriptome data analysis. Springer New York, New York, NY, pp 3–16

Chow K-S, Mat-Isa Mohd-N, Bahari A, Ghazali A-K, Alias H, Mohd.-Zainuddin Z, Hoh C-C, Wan K-L (2012) Metabolic routes affecting rubber biosynthesis in *Hevea brasiliensis* latex. J Exp Bot 63:1863–1871

Chow K-S, Wan K-L, Isa MNM, Bahari A, Tan S-H, Harikrishna K, Yeang H-Y (2007) Insights into rubber biosynthesis from transcriptome analysis of *Hevea brasiliensis* latex. J Exp Bot 58:2429–2440

Collard BCY, Mackill DJ (2008) Marker-assisted selection: an approach for precision plant breeding in the twenty-first century. Philos Trans R Soc Lond B Biol Sci 363:557–572

Darriba D, Taboada GL, Doallo R, Posada D (2011) ProtTest 3: fast selection of best-fit models of protein evolution. Bioinformatics 27:1164–1165

Duan C, Argout X, Gébelin V, Summo M, Dufayard J-F, Leclercq J, Kuswanhadi Piyatrakul P, Pirrello J, Rio M, Champion A, Montoro P (2013) Identification of the *Hevea brasiliensis* AP2/ERF superfamily by RNA sequencing. BMC Genom 14:30

Dubois M, Van den Broeck L, Inzé D (2018) The pivotal role of ethylene in plant growth. Trends Plant Sci 23:311–323

Eid J, Fehr A, Gray J, Luong K, Lyle J, Otto G, Peluso P, Rank D, Baybayan P, Bettman B, Bibillo A, Bjornson K, Chaudhuri B, Christians F, Cicero R, Clark S, Dalal R, deWinter A, Dixon J, Foquet M, Gaertner A, Hardenbol P, Heiner C, Hester K, Holden D, Kearns G, Kong X, Kuse R, Lacroix Y, Lin S, Lundquist P, Ma C, Marks P, Maxham M, Murphy D, Park I, Pham T, Phillips M, Roy J, Sebra R, Shen G, Sorenson J, Tomaney A, Travers K, Trulson M, Vieceli J, Wegener J, Wu D, Yang A, Zaccarin D, Zhao P, Zhong F, Korlach J, Turner S (2009) Real-time DNA sequencing from single polymerase molecules. Science 323:133–138

Enright AJ, John B, Gaul U, Tuschl T, Sander C, Marks DS (2004) MicroRNA targets in Drosophila. Genome Biol 5:R1

Goddard ME, Hayes BJ (2007) Genomic selection. J Anim Breed Genet 124:323–330

Gruber AR, Lorenz R, Bernhart SH, Neuböck R, Hofacker IL (2008) The Vienna RNA websuite. Nucleic Acids Res 36:W70–W74

Guo D, Li H-L, Zhu J-H, Wang Y, An F, Xie G-S, Peng S-Q (2017a) Genome-wide identification, characterization, and expression analysis of SnRK2 family in *Hevea brasiliensis*. Tree Genet Genomes 13:86

Guo D, Zhou Y, Li H-L, Zhu J-H, Wang Y, Chen X-T, Peng S-Q (2017b) Identification and characterization of the abscisic acid (ABA) receptor gene family and its expression in response to hormones in the rubber tree. Sci Rep 7:45157

Haas BJ, Delcher AL, Mount SM, Wortman JR, Smith RK, Hannick LI, Maiti R, Ronning CM, Rusch DB, Town CD, Salzberg SL, White O (2003) Improving the *Arabidopsis* genome annotation using maximal transcript alignment assemblies. Nucleic Acids Res 31:5654–5666

Haas BJ, Salzberg SL, Zhu W, Pertea M, Allen JE, Orvis J, White O, Buell CR, Wortman JR (2008) Automated eukaryotic gene structure annotation using EVidenceModeler and the program to assemble spliced alignments. Genome Biol 9:R7

Haiminen N, Feltus FA, Parida L (2011) Assessing pooled BAC and whole genome shotgun strategies for assembly of complex genomes. BMC Genom 12:194

He K, Huang Z (1987) Rubber culture in the northern part of tropical area. Guang Dong Science and Technology Press

Hirsch CN, Hirsch CD, Brohammer AB, Bowman MJ, Soifer I, Barad O, Shem-Tov D, Baruch K, Lu F, Hernandez AG, Fields CJ, Wright CL, Koehler K, Springer NM, Buckler E, Buell CR, de Leon N, Kaeppler SM, Childs KL, Mikel MA (2016) Draft assembly of elite inbred line PH207 provides insights into genomic and transcriptome diversity in maize. Plant Cell 28:2700–2714

Hu L, Xu Z, Wang M, Fan R, Yuan D, Wu B, Wu H, Qin X, Yan L, Tan L, Sim S, Li W, Saski CA, Daniell H, Wendel JF, Lindsey K, Zhang X, Hao C, Jin S (2019) The chromosome-scale reference genome of black pepper provides insight into piperine biosynthesis. Nat Commun 10:1–11

Huang X, Madan A (1999) CAP3: a DNA sequence assembly program. Genome Res 9:868–877

Huang H (2005) Fifty years of rubber breeding in China. China Agricultural Press

International Wheat Genome Sequencing Consortium (IWGSC) (2014) A chromosome-based draft sequence of the hexaploid bread wheat (*Triticum aestivum*) genome. Science 345:1251788

Iwata H, Gotoh O (2012) Benchmarking spliced alignment programs including Spaln2, an extended version of Spaln that incorporates additional species-specific features. Nucleic Acids Res 40:e161

Jaillon O, Aury J-M, Noel B, Policriti A, Clepet C, Casagrande A, Choisne N, Aubourg S, Vitulo N, Jubin C, Vezzi A, Legeai F, Hugueney P, Dasilva C, Horner D, Mica E, Jublot D, Poulain J, Bruyère C, Billault A, Segurens B, Gouyvenoux M, Ugarte E, Cattonaro F, Anthouard V, Vico V, Del Fabbro C, Alaux M, Di Gaspero G, Dumas V, Felice N, Paillard S, Juman I, Moroldo M, Scalabrin S, Canaguier A, Le Clainche I, Malacrida G, Durand E, Pesole G, Laucou V, Chatelet P, Merdinoglu D, Delledonne M, Pezzotti M, Lecharny A, Scarpelli C, Artiguenave F, Pè ME, Valle G, Morgante M, Caboche M, Adam-Blondon A-F, Weissenbach J, Quétier F, Wincker P, French-Italian Public Consortium for Grapevine Genome Characterization (2007) The grapevine genome sequence suggests ancestral hexaploidization in major angiosperm phyla. Nature 449:463–467

Kang L, Xie G, Tao Z, Yang L, Cui Z (2009) Analysis on rubber tree cold injury in 2007/2008 winter in Hainan. Chin Agric Sci Bull 251–257

Katoh K, Standley DM (2013) MAFFT multiple sequence alignment software version 7: improvements in performance and usability. Mol Biol Evol 30:772–780

Klee HJ (2004) Ethylene signal transduction. Moving beyond *Arabidopsis*. Plant Physiol 135:660–667

Ko J-H, Chow K-S, Han K-H (2003) Transcriptome analysis reveals novel features of the molecular events occurring in the laticifers of *Hevea brasiliensis* (para rubber tree). Plant Mol Biol 53:479–492

Korf I (2004) Gene finding in novel genomes. BMC Bioinform 5:59

Lacey RF, Binder BM (2014) How plants sense ethylene gas–the ethylene receptors. J Inorg Biochem 133:58–62

Laibach N, Hillebrand A, Twyman RM, Prüfer D, Gronover CS (2015) Identification of a *Taraxacum brevicorniculatum* rubber elongation factor protein that is localized on rubber particles and promotes rubber biosynthesis. Plant J 82:609–620

Lamesch P, Berardini TZ, Li D, Swarbreck D, Wilks C, Sasidharan R, Muller R, Dreher K, Alexander DL, Garcia-Hernandez M, Karthikeyan AS, Lee CH, Nelson WD, Ploetz L, Singh S, Wensel A, Huala E (2012) The *Arabidopsis* information resource (TAIR): improved gene annotation and new tools. Nucleic Acids Res 40:D1202–1210

Lau N-S, Makita Y, Kawashima M, Taylor TD, Kondo S, Othman AS, Shu-Chien AC, Matsui M (2016) The rubber tree genome shows expansion of gene family associated with rubber biosynthesis. Sci Rep 6

Lavin M, Herendeen PS, Wojciechowski MF (2005) Evolutionary rates analysis of leguminosae implicates a rapid diversification of lineages during the tertiary. Syst Biol 54:575–594

Li F, Fan G, Lu C, Xiao G, Zou C, Kohel RJ, Ma Z, Shang H, Ma X, Wu J, Liang X, Huang G, Percy RG, Liu K, Yang W, Chen W, Du X, Shi C, Yuan Y, Ye W, Liu X, Zhang X, Liu W, Wei H, Wei S, Huang G, Zhang X, Zhu S, Zhang H, Sun F, Wang X, Liang J, Wang J, He Q, Huang L, Wang J, Cui J, Song G, Wang K, Xu X, Yu JZ, Zhu Y, Yu S (2015a) Genome sequence of cultivated upland cotton (*Gossypium hirsutum* TM-1) provides insights into genome evolution. Nat Biotechnol 33:524–530

Li F, Fan G, Wang K, Sun F, Yuan Y, Song G, Li Q, Ma Z, Lu C, Zou C, Chen W, Liang X, Shang H, Liu W, Shi C, Xiao G, Gou C, Ye W, Xu X, Zhang X, Wei H, Li Z, Zhang G, Wang J, Liu K, Kohel RJ, Percy RG, Yu JZ, Zhu Y-X, Wang J, Yu S (2014) Genome sequence of the cultivated cotton *Gossypium arboreum*. Nat Genet 46:567–572

Li W, Ma M, Feng Y, Li H, Wang Y, Ma Y, Li M, An F, Guo H (2015b) EIN2-directed translational regulation of ethylene signaling in *Arabidopsis*. Cell 163:670–683

Li L, Stoeckert CJ, Roos DS (2003) OrthoMCL: identification of ortholog groups for eukaryotic genomes. Genome Res 13:2178–2189

Li R, Zhu H, Ruan J, Qian W, Fang X, Shi Z, Li Y, Li S, Shan G, Kristiansen K, Li S, Yang H, Wang J, Wang J (2010) De novo assembly of human genomes with massively parallel short read sequencing. Genome Res 20:265–272

Lin T, Xu X, Ruan J, Liu S, Wu S, Shao X, Wang X, Gan L, Qin B, Yang Y, Cheng Z, Yang S, Zhang Z, Xiong G, Huang S, Yu H, Li J (2018) Genome analysis of *Taraxacum kok-saghyz* Rodin provides new insights into rubber biosynthesis. Natl Sci Rev 5:78–87

Liu H, Deng Z, Chen J, Wang S, Hao L, Li D (2016) Genome-wide identification and expression analysis of the metacaspase gene family in *Hevea brasiliensis*. Plant Physiol Biochem 105:90–101

Low FC, Bonner J (1985) Characterization of the nuclear genome of *Hevea brasiliensis*. In: Kuala Lumpur, Malaysia, pp 1–9

Ma J, Bennetzen JL (2004) Rapid recent growth and divergence of rice nuclear genomes. PNAS 101:12404–12410

Majoros WH, Pertea M, Salzberg SL (2004) TigrScan and GlimmerHMM: two open source ab initio eukaryotic gene-finders. Bioinformatics 20:2878–2879

Merchante C, Brumos J, Yun J, Hu Q, Spencer KR, Enríquez P, Binder BM, Heber S, Stepanova AN, Alonso JM (2015) Gene-specific translation regulation mediated by the hormone-signaling molecule EIN2. Cell 163:684–697

Ming R, Hou S, Feng Y, Yu Q, Dionne-Laporte A, Saw JH, Senin P, Wang W, Ly BV, Lewis KLT, Salzberg SL, Feng L, Jones MR, Skelton RL, Murray JE, Chen C, Qian W, Shen J, Du P, Eustice M, Tong E, Tang H, Lyons E, Paull RE, Michael TP, Wall K, Rice DW, Albert H, Wang M-L, Zhu YJ, Schatz M, Nagarajan N, Acob RA, Guan P, Blas A, Wai CM, Ackerman CM, Ren Y, Liu C, Wang J, Wang J, Na J-K, Shakirov EV, Haas B, Thimmapuram J, Nelson D, Wang X, Bowers JE, Gschwend AR, Delcher AL, Singh R, Suzuki JY, Tripathi S, Neupane K, Wei H, Irikura B, Paidi M, Jiang N, Zhang W, Presting G, Windsor A, Navajas-Pérez R, Torres MJ, Feltus FA, Porter B, Li Y, Burroughs AM, Luo M-C, Liu L, Christopher DA, Mount SM, Moore PH, Sugimura T, Jiang J, Schuler MA, Friedman V, Mitchell-Olds T, Shippen DE, dePamphilis CW, Palmer JD, Freeling M, Paterson AH, Gonsalves D, Wang L, Alam M (2008) The draft genome of the transgenic tropical fruit tree papaya (*Carica papaya Linnaeus*). Nature 452:991–996

Nadalin F, Vezzi F, Policriti A (2012) GapFiller: a de novo assembly approach to fill the gap within paired reads. BMC Bioinform 13:S8

Nakano T, Suzuki K, Fujimura T, Shinshi H (2006) Genome-wide analysis of the ERF gene family in *Arabidopsis* and rice. Plant Physiol 140:411–432

Nawrocki EP, Eddy SR (2013) Infernal 1.1: 100-fold faster RNA homology searches. Bioinformatics 29:2933–2935

Ouyang S, Zhu W, Hamilton J, Lin H, Campbell M, Childs K, Thibaud-Nissen F, Malek RL, Lee Y, Zheng L, Orvis J, Haas B, Wortman J, Buell CR

(2007) The TIGR rice genome annotation resource: improvements and new features. Nucleic Acids Res 35:D883–887

Piyatrakul P, Yang M, Putranto R-A, Pirrello J, Dessailly F, Hu S, Summo M, Theeravatanasuk K, Leclercq J, Kuswanhadi null, Montoro P (2014) Sequence and expression analyses of ethylene response factors highly expressed in latex cells from *Hevea brasiliensis*. PLoS ONE 9:e99367

Poland J, Rutkoski J (2016) Advances and challenges in genomic selection for disease resistance. Annu Rev Phytopathol 54:79–98

Pootakham W, Sonthirod C, Naktang C, Ruang-Areerate P, Yoocha T, Sangsrakru D, Theerawattanasuk K, Rattanawong R, Lekawipat N, Tangphatsornruang S (2017) De novo hybrid assembly of the rubber tree genome reveals evidence of paleotetraploidy in *Hevea* species. Sci Rep 7:41457

Priyadarshan PM (2017a) Refinements to *Hevea* rubber breeding. Tree Genet Genomes 13:20

Priyadarshan PM (2017b) Biology of *Hevea* rubber, 1st ed. Springer, New York, NY

Priyadarshan PM, Goncalves P de S (2003) *Hevea* gene pool for breeding. Genet Resour Crop Evol 50:101–114

Rahman AYA, Usharraj AO, Misra BB, Thottathil GP, Jayasekaran K, Feng Y, Hou S, Ong SY, Ng FL, Lee LS, Tan HS, Sakaff MKLM, Teh BS, Khoo BF, Badai SS, Aziz NA, Yuryev A, Knudsen B, Dionne-Laporte A, Mchunu NP, Yu Q, Langston BJ, Freitas TAK, Young AG, Chen R, Wang L, Najimudin N, Saito JA, Alam M (2013) Draft genome sequence of the rubber tree *Hevea brasiliensis*. BMC Genomics 14:75

Reyes-Chin-Wo S, Wang Z, Yang X, Kozik A, Arikit S, Song C, Xia L, Froenicke L, Lavelle DO, Truco M-J, Xia R, Zhu S, Xu C, Xu H, Xu X, Cox K, Korf I, Meyers BC, Michelmore RW (2017) Genome assembly with in vitro proximity ligation data and whole-genome triplication in lettuce. Nat Commun 8:14953

Saha T, Priyadarshan PM (2012) Genomics of *Hevea* rubber. In: Schnell RJ, Priyadarshan PM (eds) Genomics of tree crops. Springer New York, New York, NY, pp 261–298. http://link.springer.com/10.1007/978-1-4614-0920-5_9

Sauter M, Moffatt B, Saechao MC, Hell R, Wirtz M (2013) Methionine salvage and S-adenosylmethionine: essential links between sulfur, ethylene and polyamine biosynthesis. Biochem J 451:145–154

Seguin M, Flori A, Legnaté H, Clément-Demange A (2003) Rubber tree. Genetic diversity of tropical crops. CIRAD. France and Science Publ., Enfield, pp 277–306

Shakeel SN, Gao Z, Amir M, Chen Y-F, Rai MI, Haq NU, Schaller GE (2015) Ethylene regulates levels of ethylene receptor/CTR1 signaling complexes in *Arabidopsis thaliana*. J Biol Chem 290:12415–12424

Solovyev V, Kosarev P, Seledsov I, Vorobyev D (2006) Automatic annotation of eukaryotic genes, pseudogenes and promoters. Genome Biol 7:S10

Stanke M, Steinkamp R, Waack S, Morgenstern B (2004) AUGUSTUS: a web server for gene finding in eukaryotes. Nucleic Acids Res 32:W309–W312

Sui J-L, Xiao X-H, Qi J-Y, Fang Y-J, Tang C-R (2017) The SWEET gene family in *Hevea brasiliensis*—its evolution and expression compared with four other plant species. FEBS Open Bio 7:1943–1959

Tang C, Huang D, Yang J, Liu S, Sakr S, Li H, Zhou Y, Qin Y (2010) The sucrose transporter HbSUT3 plays an active role in sucrose loading to laticifer and rubber productivity in exploited trees of *Hevea brasiliensis* (para rubber tree). Plant, Cell Environ 33:1708–1720

Tang C, Yang M, Fang Y, Luo Y, Gao S, Xiao X, An Z, Zhou B, Zhang B, Tan X, Yeang H-Y, Qin Y, Yang J, Lin Q, Mei H, Montoro P, Long X, Qi J, Hua Y, He Z, Sun M, Li W, Zeng X, Cheng H, Liu Y, Yang J, Tian W, Zhuang N, Zeng R, Li D, He P, Li Z, Zou Z, Li S, Li C, Wang J, Wei D, Lai C-Q, Luo W, Yu J, Hu S, Huang H (2016) The rubber tree genome reveals new insights into rubber production and species adaptation. Nature Plants 2:16073

Wang J (2015) Xilian, The watch of the century old rubber garden. http://www.hainanzuojia.com/InfoShow_7391.html

Wang Z, Hobson N, Galindo L, Zhu S, Shi D, McDill J, Yang L, Hawkins S, Neutelings G, Datla R, Lambert G, Galbraith DW, Grassa CJ, Geraldes A, Cronk QC, Cullis C, Dash PK, Kumar PA, Cloutier S, Sharpe AG, Wong GK-S, Wang J, Deyholos MK (2012a) The genome of flax (*Linum usitatissimum*) assembled de novo from short shotgun sequence reads. Plant J 72:461–473

Wang K, Wang Z, Li F, Ye W, Wang J, Song G, Yue Z, Cong L, Shang H, Zhu S, Zou C, Li Q, Yuan Y, Lu C, Wei H, Gou C, Zheng Z, Yin Y, Zhang X, Liu K, Wang B, Song C, Shi N, Kohel RJ, Percy RG, Yu JZ, Zhu Y-X, Wang J, Yu S (2012b) The draft genome of a diploid cotton *Gossypium raimondii*. Nat Genet 44:1098–1103

Wang X, Lu P, Luo Z (2013) GMATo: a novel tool for the identification and analysis of microsatellites in large genomes. Bioinformation 9:541–544

Wang Z, Miao H, Liu J, Xu B, Yao X, Xu C, Zhao S, Fang X, Jia C, Wang J, Zhang J, Li J, Xu Y, Wang J, Ma W, Wu Z, Yu L, Yang Y, Liu C, Guo Y, Sun S, Baurens F-C, Martin G, Salmon F, Garsmeur O, Yahiaoui N, Hervouet C, Rouard M, Laboureau N, Habas R, Ricci S, Peng M, Guo A, Xie J, Li Y, Ding Z, Yan Y, Tie W, D'Hont A, Hu W, Jin Z (2019) *Musa balbisiana* genome reveals subgenome evolution and functional divergence. Nat Plants 5:810–821

Wang S, Qian Y, Lan M, Li Q (2008) Preliminary report on cold damage of rubber tree in the east of Yunnan during winter–spring of 2007 to 2008. Trop Agric Sci Technol 4–8

Wang Y, Zhan D-F, Li H-L, Guo D, Zhu J-H, Peng S-Q (2017) Transcriptome-wide identification and characterization of MYB transcription factor genes in the laticifer cells of *Hevea brasiliensis*. Front Plant Sci 8. https://www.frontiersin.org/articles/10.3389/fpls.2017.01974/full

Withanage SP, Attanayake DPSTG, Jayasekara NEM, Liyanage KK, Karunasekara KB, Sarath Kumara IDMJ (2015) Evaluation and utilization of the *Hevea* germplasm collected from 1981 IRRDB expedition to the Amazon; a review. J Rubber Res Inst Sri Lanka 95:24

Xiao X-H, Yang M, Sui J-L, Qi J-Y, Fang Y-J, Hu S-N, Tang C-R (2017) The calcium-dependent protein kinase (CDPK) and CDPK-related kinase gene families in *Hevea brasiliensis*-comparison with five other plant species in structure, evolution, and expression. FEBS Open Bio 7:4–24

Xu Z, Wang H (2007) LTR_FINDER: an efficient tool for the prediction of full-length LTR retrotransposons. Nucleic Acids Res 35:W265–W268

Xue J, Zhou X, Zhang C-X, Yu L-L, Fan H-W, Wang Z, Xu H-J, Xi Y, Zhu Z-R, Zhou W-W, Pan P-L, Li B-L, Colbourne JK, Noda H, Suetsugu Y, Kobayashi T, Zheng Y, Liu S, Zhang R, Liu Y, Luo Y-D, Fang D-M, Chen Y, Zhan D-L, Lv X-D, Cai Y, Wang Z-B, Huang H-J, Cheng R-L, Zhang X-C, Lou Y-H, Yu B, Zhuo J-C, Ye Y-X, Zhang W-Q, Shen Z-C, Yang H-M, Wang J, Wang J, Bao Y-Y, Cheng J-A (2014) Genomes of the rice pest brown planthopper and its endosymbionts reveal complex complementary contributions for host adaptation. Genome Biol 15:521

Yang Z (2007) PAML 4: phylogenetic analysis by maximum likelihood. Mol Biol Evol 24:1586–1591

Yang SF, Hoffman NE (1984) Ethylene biosynthesis and its regulation in higher plants. Ann Rev Plant Physiol 35:155–189

Yang C, Lu X, Ma B, Chen S-Y, Zhang J-S (2015) Ethylene signaling in rice and *Arabidopsis*: conserved and diverged aspects. Mol Plant 8:495–505

Ye Z-H, Zhong R (2015) Molecular control of wood formation in trees. J Exp Bot 66:4119–4131

Zhang G, Fang X, Guo X, Li L, Luo R, Xu F, Yang P, Zhang L, Wang X, Qi H, Xiong Z, Que H, Xie Y, Holland PWH, Paps J, Zhu Y, Wu F, Chen Y, Wang J, Peng C, Meng J, Yang L, Liu J, Wen B, Zhang N, Huang Z, Zhu Q, Feng Y, Mount A, Hedgecock D, Xu Z, Liu Y, Domazet-Lošo T, Du Y, Sun X, Zhang S, Liu B, Cheng P, Jiang X, Li J, Fan D, Wang W, Fu W, Wang T, Wang B, Zhang J, Peng Z, Li Y, Li N, Wang J, Chen M, He Y, Tan F, Song X, Zheng Q, Huang R, Yang H, Du X, Chen L, Yang M, Gaffney PM, Wang S, Luo L, She Z, Ming Y, Huang W, Zhang S, Huang B, Zhang Y, Qu T, Ni P, Miao G, Wang J, Wang Q, Steinberg CEW, Wang H, Li N, Qian L, Zhang G, Li Y, Yang H, Liu X, Wang J, Yin Y, Wang J (2012) The oyster genome reveals stress adaptation and complexity of shell formation. Nature 490:49–54

Zhou S, Herschleb J, Schwartz DC (2007) Chapter 9 A single molecule system for whole genome analysis. In: Perspectives in bioanalysis. Elsevier, pp 265–300. https://linkinghub.elsevier.com/retrieve/pii/S187100690602009X

Zou Z, Gong J, An F, Xie G, Wang J, Mo Y, Yang L (2015) Genome-wide identification of rubber tree (*Hevea brasiliensis* Muell. Arg.) aquaporin genes and their response to ethephon stimulation in the laticifer, a rubber-producing tissue. BMC Genomics 16:1001

Zou Z, Liu J, Yang L, Xie G (2017a) Survey of the rubber tree genome reveals a high number of cysteine protease-encoding genes homologous to *Arabidopsis* SAG12 Araujo WL (ed). PLoS ONE 12:e0171725

Zou Z, Xie G, Yang L (2017b) Papain-like cysteine protease encoding genes in rubber (*Hevea brasiliensis*): comparative genomics, phylogenetic, and transcriptional profiling analysis. Planta 246:999–1018

The RRIM 600 Rubber Tree Genome: Sequencing and Analysis Strategies of a Premier Pedigree Clone

3

Nyok-Sean Lau, Yuko Makita, Ahmad Sofiman Othman, and Minami Matsui

Abstract

The field of plant genomics research has benefited immensely from the development and widespread adoption of next-generation sequencing techniques. As DNA sequencing has become orders of magnitudes cheaper and more efficient, genome sequencing of any organism of choice is now within reach of individual research groups. These advances led to an escalation of available genomic resources in the plant sciences. Although advances in sequencing technologies have enabled the decoding of the genetic blueprint of many plant species, these improvements have generated many challenges and tasks ahead. The software and algorithms used to analyse genomics data need to be developed in line with the pace of evolving sequencing platforms. The effective deployment of sequencing technologies to genome studies is also fraught with challenges including high levels of heterozygosity and repeat sequences, which are common to many plants with large genome sizes. Here, we review approaches used in the sequencing, assembly and annotation of the *Hevea brasiliensis* RRIM 600 genome as well as challenges in sequencing plant genomes.

N.-S. Lau · A. S. Othman
Centre for Chemical Biology, Universiti Sains Malaysia, 11900 Bayan Lepas, Penang, Malaysia

N.-S. Lau · Y. Makita · M. Matsui (✉)
Synthetic Genomics Research Group, RIKEN Center for Sustainable Resource Science (CSRS), 1-7-22 Suehiro-cho, Tsurumi-ku, Yokohama, Kanagawa 230-0045, Japan
e-mail: minami@riken.jp

A. S. Othman
School of Biological Sciences, Universiti Sains Malaysia, 11800 Minden, Penang, Malaysia

3.1 Introduction

3.1.1 It All Started with '22 Seedlings'

Natural rubber has been a major contributor to Malaysia's economic development throughout much of the twentieth century. Although Malaysia is no longer the top world producer of this commodity, rubber still has a niche in the country's socio-economy. For decades, more than 90% of the rubber planted area has been operated by smallholders. By 2018, the national production of about 0.6 mil tonnes of natural rubber was from an estimated 1.08 mil hectares of planted area (Malaysian Rubber Board 2018). Nonetheless, as an industry encompassing both upstream and downstream activities, the national export value in the same year totalled RM40.36 billion, of which rubber and rubberwood-based products constituted nearly 84%. Thus, from a

M. Matsui and K.-S. Chow (eds.), *The Rubber Tree Genome*, Compendium of Plant Genomes,
https://doi.org/10.1007/978-3-030-42258-5_3

purely agricultural activity, the Malaysian rubber industry today has evolved to become a producer of both raw rubber materials and finished products.

H. brasiliensis (the para rubber tree) originated from the Amazonian rain forest of South America. It was introduced to Malaysia, known as Malaya at that time, by the British in the 1870s. Henry Wickham collected 70,000 seeds from Brazil and transported them to the Royal Botanic Gardens at Kew, England in 1876. Of the seeds germinated, a selection of nearly about 2,000 was sent to Ceylon (now Sri Lanka). By 1877, 22 seedlings were despatched to the Singapore Botanic Gardens, from which nine were sent to Kuala Kangsar in Peninsular Malaya. Thus, it is believed that rubber tree genotypes cultivated extensively in Southeast Asia are derived from the '22 seedlings' of Wickham's collection (Baulkwill 1989). Henry Ridley, as the Director of the Singapore Botanic Gardens, was instrumental in promoting the planting of rubber trees in Peninsular Malaya. The massive boom in the rubber industry began by the early twentieth century. Large-scale planting started when the first rubber estate was established in Melaka in 1903. Buoyed by the high rubber prices, rubber cultivation expanded across Malaya located mostly in the states of Perak, Selangor, Negeri Sembilan and Johor.

One of the earliest research on rubber was carried out by the Department of Agriculture, but as from 1925, this responsibility was handed over to the newly established Rubber Research Institute of Malaya (RRIM). Since 1928, breeding programmes have been carried out to develop rubber clones having high latex yield along with other favourable characters such as good growth vigour, low disease incidence, tolerance to tree dryness and wind resistance (Ong et al. 1994). RRIM 600 is one of the top three performing genotypes bred from the RRIM 600 clone series, together with RRIM 623 and RRIM 628 (Planters' Bulletin 1974). Additionally, RRIM 600 was the most common clone planted in Malaysia between the 1970s and 80s (Ong and Bahri 2014). Not only was RRIM 600 a premier pedigree clone at the height of rubber production in Malaysia, it was also recommended for wide scale planting in countries such as Brazil and Thailand (Junior et al. 2015; Pethin et al. 2015). Additionally, RRIM 600 was demonstrated to be a clone with wide adaptability based on evaluation of its performance in sub-optimal areas in India, Vietnam, China and Brazil (Priyadarshan et al. 2005). In view of the importance of RRIM 600 in the Malaysian rubber breeding history and its valuable phenotypic traits, the clone was selected for genome sequencing (Lau et al. 2016; Rahman et al. 2013).

3.1.2 Plant Genome Sequencing

Advances made in DNA sequencing technology have brought about the exponential decrease in cost per base and increase in the throughput of sequencing machines. These advances, coupled with the recognition by funding agencies of the importance of genome science underpinning almost all aspects of modern biology, have led to the rapid increase in sequenced genomes. The first plant genome to be sequenced was that of the model species, *Arabidopsis thaliana*, in 2000 (Arabidopsis Genome Initiative 2000). Since then, the number of sequenced plant genomes has steadily increased, reaching more than 100 by 2015 (Michael and VanBuren 2015). Furthermore, the plant research community portrayed a vision of sequencing more than 1,000 plant species over the subsequent years (Abdurakhmonov 2016). The RRIM 600 rubber tree draft genome sequence was published amidst the first wave of plant genomes (Rahman et al. 2013). As the rubber tree is the sole crop that has been successfully exploited for commercial production of natural rubber, there is substantial economic justification to have its genome sequenced. The release of the first draft genome sequence of the rubber tree enabled functional genomics studies that helped to define the functions of genes (An et al. 2015; Liu et al. 2015;

Zhu et al. 2015), and provided unprecedented access to sequence-based marker discovery (Pootakham et al. 2015; Shearman et al. 2015).

Genome sequencing is a challenging endeavour that requires significant time, resources and expertise. Before starting a genome sequencing project, how a genome sequence may address the biological questions at hand should be thoroughly considered. Most of the plants sequenced meet specific criteria for being selected for genome sequencing, e.g. the size of the research community studying a particular plant; the role of the plant as a model organism or as an economic crop; low ploidy level; low heterozygosity; availability of genetic resources such as genetic maps, large-scale expressed sequence tag (EST) collections, transcriptomes, bacterial artificial chromosome (BAC) libraries and other genomic tools (Michael and Jackson 2013). Motivations for large-scale sequencing include gene annotation, single-nucleotide polymorphism discovery and comparative genomics. Each of these purposes requires different strategies for the acquisition and management of templates for the sequencing process as well as for downstream bioinformatics analysis.

3.2 Genome Sequencing Approaches of the Rubber Tree Clone, RRIM 600

This section discusses the approaches for sequencing of the *H. brasiliensis* RRIM 600 genome that had been selected in view of technology advances and project expectations. The genome size of *H. brasiliensis* is approximately 2.15 Gb, which is approximately 16-fold larger than that of *A. thaliana*. In 2013, the first draft genome assembly of the *H. brasiliensis* clone RRIM 600 was published (Rahman et al. 2013). At that time, there were two major genome sequencing strategies: whole-genome shotgun (WGS) sequencing and the BAC-by-BAC sequencing approach. The WGS method entails random shearing of genomic DNA into small fragments, followed by massive parallel sequencing of the fragments and assembly of the sequences into contigs. The latter approach involves the cloning of genomic fragments into BACs followed by sequencing of large numbers of overlapping BACs. *A. thaliana*, rice and maize were among the earliest plant genomes to be sequenced, completed using Sanger-based sequencing along the minimal tiling path of BACs (Arabidopsis Genome Initiative 2000; International Rice Genome Sequencing Project 2005; Schnable et al. 2009). Plant genomics research is rapidly changing pace with new technologies, and sequencing initiatives quickly adopted WGS approach to sequence their favourite genome. In comparison with the BAC-based sequencing strategy, WGS is relatively more straightforward and cost-effective. Early attempts at Sanger-based WGS sequencing in plants had been applied in sequencing of grape, papaya and the poplar tree (Jaillon et al. 2007; Ming et al. 2008; Tuskan et al. 2006). Consequently, Rahman et al. (2013) opted for a WGS sequencing approach for the first draft of the *H. brasiliensis* genome.

Despite the successful use of Sanger sequencing in many plant WGS projects, Sanger technology suffers from limitations such as low throughput, high cost and long turnaround time. There has been a shift in the plant genome sequencing paradigm since the emergence of next-generation sequencing (NGS) technologies. The first approach to high-throughput second generation sequencing, pyrosequencing, was reported in 2015 (Margulies et al. 2005). The platform was developed by 454 Life Sciences and later commercialized by Roche. Pyrosequencing utilizes the emulsion PCR technique where DNA fragments are clonally amplified within water-in-oil microreactors. The Roche 454 FLX system can generate one million reads per run with an average read length of 300–500 bp and single-read accuracy of >99.5%. A combination of Roche 454 and Sanger sequencing has been utilized for the de novo sequencing of several crop genomes including grape, cucumber and apple (Huang et al. 2009a; Velasco et al. 2007, 2010). Another high-throughput sequencing system is SoLID™, which employs emulsion PCR template preparation and unique sequencing by ligation (McKernan et al. 2009). The SOLiD system produces up to 320 Gb of data per run with a maximum read length of 75 bp. The SOLiD platform is mainly used for re-sequencing applications where comparison to a reference enables the identification and removal of erroneous reads (Ashelford et al. 2011). The most widely used NGS sequencing system was developed by Solexa, which was later acquired by Illumina. In the Illumina sequencing platform, sequencing templates are immobilized on a flow cell and bridge PCR is used to amplify the templates into clonal clusters. This is followed by sequencing via a sequencing-by-synthesis method. The high-throughput HiSeq system from Illumina can generate terabyte-sized data per run, with read lengths up to 150 bp. Illumina sequencing has been widely used in plant genomics for applications such as gene expression profiling (Davidson et al. 2011; Filichkin et al. 2010; Zhang et al. 2010), de novo sequencing (Dassanayake et al. 2011; Huang et al. 2009a; Xu et al. 2011) and re-sequencing (Cao et al. 2011; Huang et al. 2009b; Lai et al. 2010). To date, the majority of the plant genomes sequenced by only NGS platforms were completed using a hybrid strategy, i.e. more than one sequencing platform was employed to generate reads for the WGS assembly. For example, Roche 454 and Illumina technologies were used to sequence the genome of *Medicago truncatula*, while a combination of Roche 454, Illumina and SOLiD sequencing was used to characterize the genome of strawberry (Shulaev et al. 2010; Tang et al. 2014).

A similar hybrid approach was applied to produce the first draft assembly for the *H. brasiliensis* RRIM 600 genome. A combination of 43 × coverage of sequence data from Roche 454, Illumina and SOLiD platforms were used to assemble 1.12 Gb of the estimated 2.15 Gb genome (Rahman et al. 2013) (Fig. 3.1). The de novo assembly of Roche 454, Illumina and SOLiD reads was performed using the Newbler software since the majority of the sequencing reads were based on the Roche 454 platform. The draft assembly consists of 608,017 scaffolds with an N50 size of 2.97 Kb (Table 3.1). Although the draft assembly captured only 52% of the estimated genome size, the integrity and completeness of the assembly were validated by the alignment of publicly available *H. brasiliensis* EST sequences and RNA-seq reads. Most plant genomes have high repeat content and contain regions of segmental or whole genome duplications due to polyploidization events. These create problems in reconstructing genome sequences. More than 70% of the *H. brasiliensis* genome assembly was identified as repetitive DNA. Short-read sequencing could not effectively resolve the repeat regions in the first *H. brasiliensis* draft genome, resulting in a fragmented assembly with a large number of scaffolds and low N50 value. Like other plant genomes assemblies that have been performed using NGS only, this draft assembly is not 'finished' and contains gaps in the assembly. Genome finishing involves processes such as inspection and experimental resolution of inconsistencies and thus, can be time-consuming, laborious and expensive (Hamilton and Robin Buell 2012). So far, *A. thaliana* and rice sequenced using traditional Sanger technology remain as the only 'finished' plant genomes (Arabidopsis Genome Initiative 2000; International Rice Genome Sequencing Project 2005).

Technological revolutions in DNA sequencing began with the introduction of third-generation sequencing platforms that produce long single-molecule reads which offer the opportunity to crack complex genomes. A superior way to resolve gaps in a genome assembly is to generate long reads than can span the gaps, connecting contigs into scaffolds. Single-molecule-real-time (SMRT) sequencing developed by Pacific Biosciences (PacBio) is one of the most popular platform in the market that produce long reads. PacBio's SMRT technology harnesses the natural process of DNA replication and enables the observation of DNA synthesis in real-time by the incorporation of zero-mode waveguides and phospholinked nucleotides. The PacBio Sequel system with current P6-C4 chemistry can generate sequences with average read lengths of more than 10 kb and up to 10 Gb data per SMRT cell. Several high-quality plant genome assemblies with improved continuity

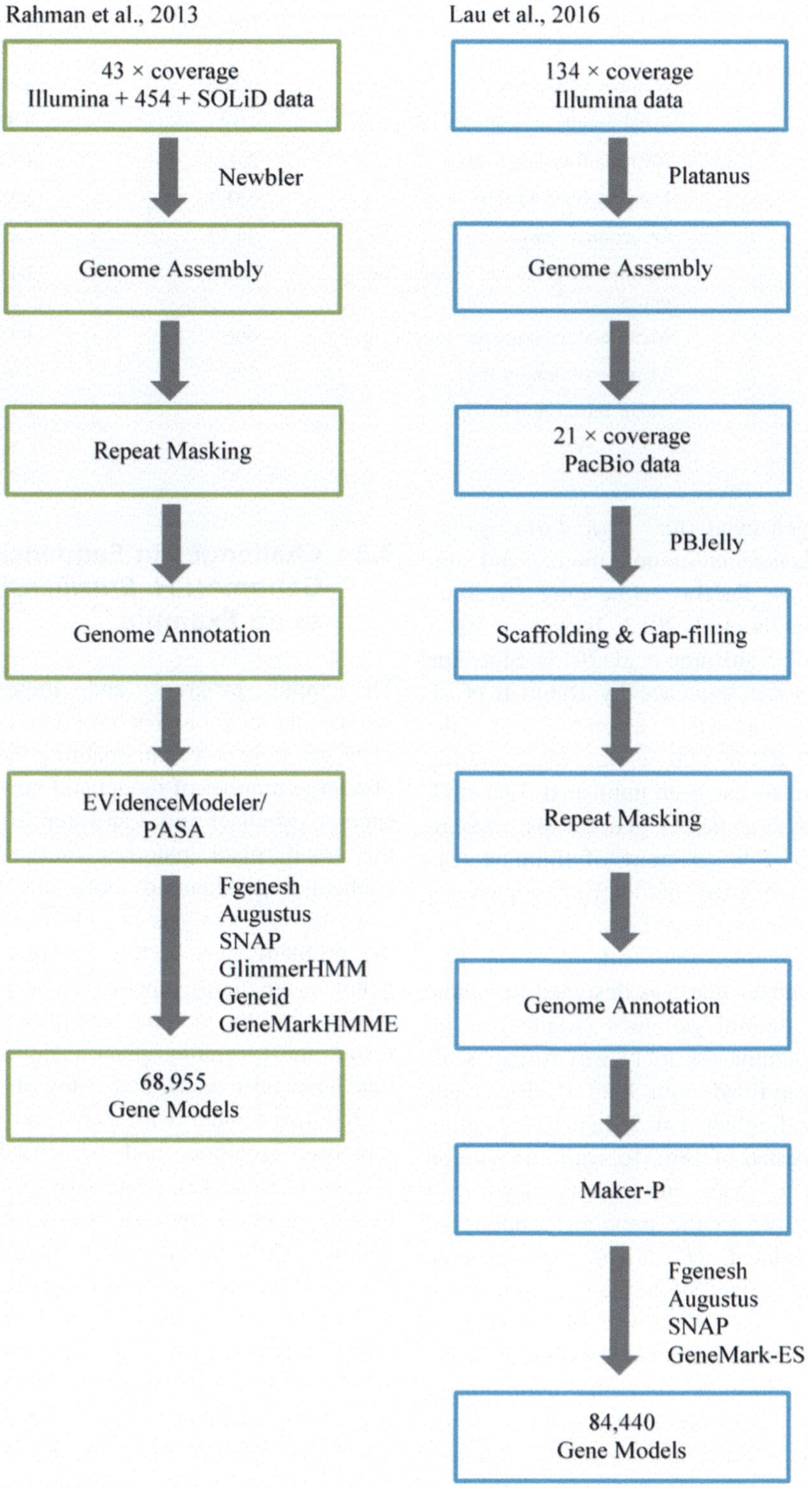

Fig. 3.1 Overview of the genome assembly and annotation workflow used in *H. brasiliensis* RRIM 600 genome sequencing projects

Table 3.1 *H. brasiliensis* RRIM 600 genome assembly and annotation statistics

Feature	Rahman et al. (2013)	Lau et al. (2016)
Number of scaffolds	608,017	189,316
Total length of scaffolds (Gb)	1.12	1.55
N50 scaffold length (kb)	2.97	67.24
Longest scaffold (kb)	531.47	871.19
GC content (%)	34.17	34.17
Number of predicted protein-coding genes	68,955	84,440
Mean coding sequence length (bp)	696	971
Mean exon length (bp)	238	196
Mean intron length (bp)	332	478

have been generated for *Utricularia gibba*, *Oropetium thomaeum*, quinoa, maize, and sunflower based on PacBio sequencing (Badouin et al. 2017; Jarvis et al. 2017; Jiao et al. 2017; Lan et al. 2017; VanBuren et al. 2015). Since the first draft genome sequence by Rahman et al. (2013), an improved assembly of the *H. brasiliensis* RRIM 600 genome incorporating PacBio long reads has been published (Lau et al. 2016). The *H. brasiliensis* genome was assembled using 134-fold coverage of Illumina data and 21-fold coverage of PacBio sequencing reads. The Illumina paired-end and mate-pair reads were assembled into scaffolds using Platanus, an assembler that was designed to handle heterozygous diploid genomes (Kajitani et al. 2014). The Illumina assembly was further scaffolded and gap-filled with long PacBio reads using PBJelly (English et al. 2012). The resulting assembly consisted of 189,316 scaffolds with an N50 size of 67.2 kb and a total length of 1.55 Gb. Relative to the previously published draft, the updated *H. brasiliensis* genome assembly presented a 23-fold increase in the N50 size and a three-fold reduction in the number of scaffolds. The improved genome assembly offers unprecedented insights into the evolutionary relationship of *H. brasiliensis* with other Euphorbiaceae species and allows the genome-wide characterization of genes responsible for valuable agronomic traits.

3.3 Challenges in Sequencing Plant Genomes: *H. Brasiliensis* as an Example

The rapid progress and improvement of sequencing technologies have greatly accelerated plant genomic research, enabling the sequencing of a large number of model and non-model plant species (Michael and VanBuren 2015). Despite this trend, plant genomes can present unique challenges to standard tools and pipelines for assembly and analysis of sequence data. One of the problems lies in the fact that many plant genomes are highly repetitive. This poses a major challenge to the genome assembly process. As a result, most existing plant genome assemblies that have been generated using short reads are highly fragmented with gaps corresponding to repetitive regions and structural variations (Claros et al. 2012). Repetitive DNA sequences occupy a major part of nearly all eukaryotic genomes. The repeat content in plant genomes are variable, ranging from 3% in the small genome of *U. gibba* to 85% in the large genome of maize (Ibarra-Laclette et al. 2013; Schnable et al. 2009). In the *H. brasiliensis* genome, repetitive sequences account for more than 70% of the assembled genome (Lau et al. 2016; Rahman et al. 2013). Transposable elements (TEs) account for the largest fraction of repeated

DNA sequences and they can be classified into two groups: class I and class II TEs. Class I transposons (including retrotransposons) can constitute a significant proportion of all TEs in plant genomes, while class II elements are typically less abundant. This trend is observed in *H. brasiliensis* in which retrotransposons occupy the majority of the genome (~57.3%) (Lau et al. 2016). Gypsy (~43.6%) and Copia (~11.7%) types constitute the major LTR retrotransposon components. In contrast, DNA transposons only constitute about ~1.2% of the *H. brasiliensis* genome. Some TE families have hundreds or thousands of copies in plant genomes, and exceed 10 kb in length. Owing to their abundance, and highly conserved and repetitive nature, TEs often complicate genome assembly.

The proliferation of repeats can also lead to increased genome size. There are several examples of genome expansion through the amplification of certain TE families. For instance, amplification of three TE families accounts for a two-fold increase in genome size in *Oryza australiensis* as compared to its nearest relatives (Piegu et al. 2006). Maize provided another example of genome obesity resulting from TE expansion (SanMiguel et al. 1996). The level of repetitiveness in *H. brasiliensis* (more than 70%) is much higher compared to that of castor bean (50.33%), cassava (25.7%) and Barbados nut (36.9%) within the Euphorbiaceae family (Chan et al. 2010; Sato et al. 2011; Wang et al. 2014). The expansion of the *H. brasiliensis* genome size seems to have resulted from the noticeable proliferation of TEs in the genome. The cost to sequence a genome increases as the genome size increases. Indeed, only a few plants with genome sizes larger than 10 Gb, e.g. wheat, *Ginkgo biloba*, spruce and pine have been sequenced to date (Birol et al. 2013; Guan et al. 2016; Neale et al. 2014; Nystedt et al. 2013; Zimin et al. 2017). While current sequencing technologies can generate reads to provide sufficient coverage for assembling large genomes in a cost-effective manner, the assembly of such a large amount of data could be technically challenging. The assembly of the loblolly pine genome, the largest genome ever sequenced, required high computational demand and could only be resolved by using novel assembly approaches (Neale et al. 2014).

Heterozygosity represents another substantial challenge for the de novo assembly of plant genomes. *H. brasiliensis* has a heterozygosity range of 0.05 to 0.96 and the clone RRIM 600 has an outcrossing rate of 62.1% (Gouvêa et al. 2010; Sunderasan et al. 1994). The assembly of heterozygous genomes requires the adoption or careful consideration of appropriate strategies and ideally, the resolution of individual haplotypes. An outcrossing species such as grape exhibits a high level of heterozygosity (up to 13% difference between heterozygous alleles) which impedes genome assembly (Jaillon et al. 2007). Problems introduced by high heterozygosity during assembly include the separation of heterozygous sequences from haplotype variants that ought to be merged into two contigs, thus creating erroneous segmental duplication (Kelley and Salzberg 2010). This kind of redundancy may result in duplicated gene content since some loci seem to have both possible alleles. To alleviate this problem, some plant genome sequencing consortiums have chosen to sequence the homozygous derivatives, even if they are not of agronomic importance. For example, a highly homozygous inbred line was utilized to generate the genome assembly for grape (Jaillon et al. 2007). Likewise, the potato reference genome assembly was derived from a doubled monoploid clone that was homozygous for a single set of 12 chromosomes (Xu et al. 2011).

3.4 Annotating the *H. Brasiliensis* Genome

Following genome assembly, annotation is performed to assign biologically relevant information to the genome sequence. Plant genome annotation could be challenging due to the fact that many plant genomes are large, complex and contain many repeat sequences and pseudogenes. Genome assembly quality has a major impact on gene annotation. Fragmented assemblies having small contig and scaffold sizes complicate annotation as a single gene may be split into multiple contigs, causing the over-estimation of gene numbers (Denton et al. 2014). Many genome assemblies and annotations have been updated over time as a result of continuous efforts to increase assembly contiguity and improve gene models. An updated genome assembly and annotation for the *H. brasiliensis* RRIM 600 clone (Lau et al. 2016) was published three years following the release of the first draft genome sequence (Lau et al. 2016; Rahman et al. 2013). In this section, annotation of the first draft and the subsequently improved genome assembly of RRIM 600 will be briefly reviewed.

TE sequences occupy a major fraction of the plant genomes and some of the TEs may be incorrectly annotated as genes. As an example, it was estimated that only 40,000 of the more than 55,000 annotated genes in the rice genome are actually genes while the other 10,000–15,000 are TEs (Bennetzen et al. 2004). Thus, repetitive sequences in the *H. brasiliensis* genome assemblies were masked prior to gene prediction. A repeat library of *H. brasiliensis* was established using both homology-based search and de novo predictions. For homology-based detection, RepeatMasker was used to screen the genome for known TEs in the RepBase database (Tarailo-Graovac and Chen 2009). De novo analysis of repeats within the assembly was performed using RepeatModeler. The repeats identified from both the homology-based and de novo predictions were combined to build a species-specific repeat library for *H. brasiliensis*.

The genome annotation process can be described as structural annotation (which involves the determination of the structures of genomic features) and functional annotation (which aims at assigning functions to these structures). Currently, a popular tool for structural annotation is Maker-P, an automated pipeline developed specifically to annotate novel plant genomes (Campbell et al. 2014). Other automated structural annotation pipelines include EVidenceModeler, which reports gene structures as weighted consensus of evidence from several sources (Haas et al. 2008). Both Maker-P and EVidenceModeler integrate ab initio gene predictions, and transcriptome and protein alignment evidence into the final gene models. Although automated genome prediction pipelines generally provide good results, qualitative validation and visual inspection of the annotations are sometimes necessary to detect systemic issues. While Evidence Modeler was used for annotation of the first draft of the *H. brasiliensis* genome sequence (Rahman et al. 2013); Maker-P was employed for annotation of the updated assembly (Lau et al. 2016). Plant protein sequences downloaded from PlantGDB were aligned to the first draft genome assembly to complement the predicted gene models. For annotation of the updated genome assembly, protein sequences from castor bean, cassava, Barbados nut and the Viridiplantae dataset from Swiss-Prot were used as homology-based evidence. RNA-seq reads generated from our study (Rahman et al. 2013) and from previous work (Chow et al. 2014) that were assembled de novo were provided as transcript evidence for EVidenceModeler and Maker-P analyses, respectively. RNA-seq sequences have great potential in resolving gene models as these data provide extrinsic evidence for the delimitation of exons, splice sites and alternatively spliced exons. Ab initio gene predictions were performed on the *H. brasiliensis* genome using the following software: Augustus, Snap, Fgenesh, GeneMark-ES, GlimmerHMM and Geneid (Blanco et al. 2007; Korf 2004; Lomsadze et al. 2005; Majoros et al. 2004; Salamov and Solovyev 2000; Stanke et al. 2006). Ab initio gene finders used in annotating the first draft genome sequence were mostly trained using PASA transcriptome alignment; whereas Augustus and Snap used in the subsequent updated genome annotation were trained using Maker-P in an iterative fashion following a previously described protocol (Cantarel et al. 2008).

The number of gene models annotated in plant genomes ranged from 20,000 to 94,000 with a median predicted gene count of 32,605 (Michael and Jackson 2013). A total of 68,955 gene models, with an average gene length of 1,332 bp, an average coding length of 696 bp and a highest detected number of 35 exons per gene were predicted from the first draft genome assembly of *H. brasiliensis*. An increase in the number of gene models was observed in the improved assembly whereby the updated genome sequence encoded 84,440 protein-coding genes, with an average transcript length of 3,069 bp, an average coding length of 971 bp and an average of 4.97 exons per transcript. Functional annotation of the predicted gene models was based on comparison with databases such as NCBI nr, TrEMBL, Swiss-Prot, KEGG, GO and InterPro. Despite the substantial body of knowledge accrued on plant gene functions over the years, there are still a large number of plant genes whose functions have not been elucidated. Of the predicted genes in the improved *H. brasiliensis* genome assembly (Lau et al. 2016), 81.1% could be assigned structured annotation. This figure is comparable with that reported for *A. thaliana*, which has been considered the best annotated plant (Lamesch et al. 2012).

3.5 Genome Database for the Rubber Tree

Upon completion of genome sequencing, one of the priorities is to share the genome data with the research community immediately after the genome is published. The importance of data sharing is well-recognized because it accelerates scientific progress by expanding the impact of valuable sequence data, and promoting collaboration. With the rapid increase in available genome sequences, the demand for dedicated databases to facilitate the management and scientific development of the enormous volume of sequence information is constantly growing. The functions of genome databases have also evolved from being warehouses that merely store sequence data to being providers of online analysis tools and a framework for comparative genomic analysis. These genome databases can be generally classified into three types: species- and taxa-specific databases, as well as comprehensive databases. Among the sequenced plant genomes, only a small number have species-specific databases to host their genome information while most of these genomes are deposited in NCBI without a customized database. The genome databases of the model plants *A. thaliana*, The Arabidopsis Initiative Resource (https://www.arabidopsis.org/), and rice, The Rice Genome Annotation Project (https://rice.plantbiology.msu.edu/), appear to be the most well-constructed single genome databases for plants (Huala et al. 2001; Ouyang et al. 2007). However, most plant species-specific genome databases offer only a limited range of analysis features, which therefore reduces their popularity and usability. Some examples of comprehensive plant genome databases are Gramene, PlantGDB, MIPS PlantsDB, EnsemblPlants, PLAZA, GreenPhylDB and Phytozome (Duvick et al. 2008; Goodstein et al. 2012; Kersey et al. 2012; Nussbaumer et al. 2013; Rouard et al. 2011; Van Bel et al. 2012; Youens-Clark et al. 2011). These comprehensive databases typically host a large collection of plant genomes and provide tools for comparative studies.

A rubber tree genome and transcriptome database was constructed based on the genome sequence of the *H. brasiliensis* RRIM 600 clone (Makita et al. 2018). This database contains genome information and includes gene functional annotations and multi-transcriptome data such as RNA-seq, full-length cDNAs, PacBio Isoform sequencing (Iso-seq), ESTs, and genome-wide transcription start sites (TSSs) derived from CAGE technology (Table 3.2). The database also provides analysis and visualization of co-expressed gene networks for prediction of functionally related gene groups from transcriptome data (Fig. 3.2). Three search functions are available for retrieving genes according to users' selected expression thresholds: keyword search, sequence homology search and gene expression search. The rubber tree genome and transcriptome database is useful for comprehensive understanding of the genome information and will assist both industrial and academic researchers in utilizing the information for improving important agronomic traits of the rubber tree.

Table 3.2 Annotation resources in the RIKEN Rubber Genome and Transcriptome Database (Nov 2019)

Category	Contents	Number of annotated genes	URL of the data source
Genome and proteins	Draft genome of RRIM 600	84,443	https://rubber.riken.jp/
Functional annotation	KEGG	61,453	https://www.genome.jp/
	NCBI Protein	64,585	https://www.ncbi.nlm.nih.gov/protein/
	Swiss-Prot	33,553	https://www.uniprot.org/uniprot/
	TrEMBLE	63,053	
	GO	35,247	https://www.geneontology.org/
Transcriptome	RNA-seq	42,614	https://rubber.riken.jp/
	CAGE	21,168	
	FL-cDNA	7,704	
	ESTs	23,790	https://www4a.biotec.or.th/rubber/ https://scarecrow.fmrp.usp.br/heveabr/
	Iso-seq	17,668	https://www4a.biotec.or.th/rubber/

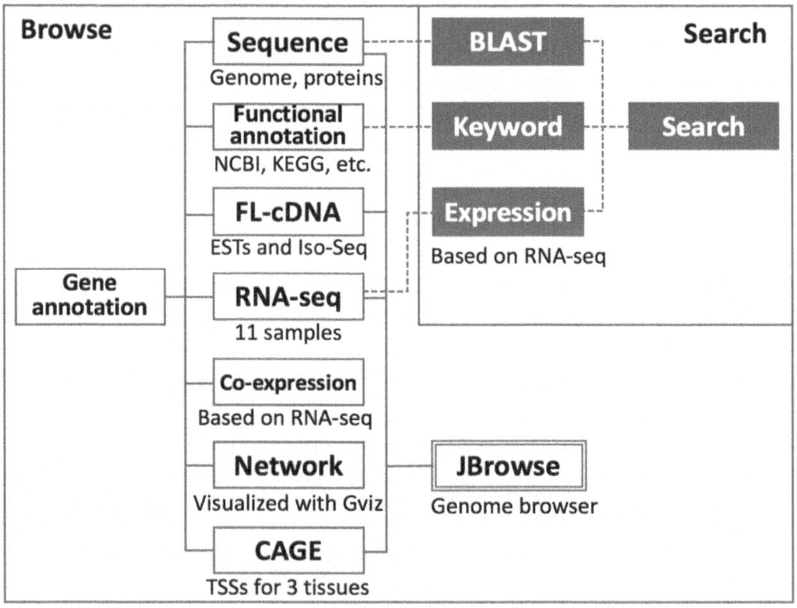

Fig. 3.2 Structure of the RIKEN Rubber Genome and Transcriptome Database

3.6 Outlook

The first wave of plant genome sequencing has long passed since the publication of the first plant genome sequence of *A. thaliana*. We are entering a new era in plant genomics research in which new sequencing technologies will revolutionize the way we approach future genome sequencing projects. Producing 'gold standard' reference genomes for important species will be the next target for the plant research community to address fundamental questions in plant biology. The generation of contiguous and complete genome assemblies has allowed for not only annotation of complete gene models, but also studies on the evolution of genome organization and structural variation caused by chromosome rearrangements. High-throughput technologies for physical mapping which have recently emerged such as BioNano, Chromosome Conformation Capture (Hi-C) and 10 × Genomics, have potentially big impact on plant genomics research. These technologies enable researchers to decipher genome organization at the chromosome level but at a fraction of the cost relative to reference genomes generated using traditional Sanger sequencing. Long-range scaffolding technologies have been used to produce chromosome-scale scaffolds for genomes of numerous plants, including maize, rice, sorghum and barley (Deschamps et al. 2018; Du et al. 2017; Jiao et al. 2017; Mascher et al. 2017). We foresee the application of these technologies in improving the contiguity of the *H. brasiliensis* genome, and in producing a chromosome-level assembly for *H. brasiliensis* in the near future.

New sequencing technologies are only part of the future of plant genomics research; the milestone of utilizing a high-quality genome sequence is the bridging of the knowledge gap between genotype and phenotype. The availability of genomic resources in many agricultural plants including specialty crop plants is a panacea for modern plant breeding. Modern breeding programs are powered by more efficient breeding approaches, in parallel with the reduction in breeding cycle duration. The generation of whole-genome sequence, transcriptome and single nucleotide polymorphism (SNP) marker information as well as the application of automated genotyping technologies have made it possible to screen multiple genotypes within a short time. Genome-wide association approaches are now providing opportunities for genomic selection and the use of expression quantitative trait loci (eQTLs) in breeding. Continuous efforts in improving the genetic and genomic resources for crop plants will make 'breeding by design' a realizable goal for efficient crop improvement. In the post-genome era of the rubber tree, applications of genomic technologies can offer opportunities for the improvement of agronomic traits to meet the ever increasing demand for natural rubber.

References

Abdurakhmonov IY (2016) Genomics era for plants and crop species—advances made and needed tasks ahead. In: Abdurakhmonov I (ed) Plant genomics. Rijeka, IntechOpen, pp 3–25

An F, Zou Z, Cai X, Wang J, Rookes J, Lin W, Cahill D, Kong L (2015) Regulation of HbPIP2;3, a latex-abundant water transporter, is associated with latex dilution and yield in the rubber tree (*Hevea brasiliensis* Muell. Arg.). PLoS One 10:e0125595

Arabidopsis Genome Initiative (2000) Analysis of the genome sequence of the flowering plant *Arabidopsis thaliana*. Nature 408:796

Ashelford K, Eriksson ME, Allen CM, D'Amore R, Johansson M, Gould P, Kay S, Millar AJ, Hall N, Hall A (2011) Full genome re-sequencing reveals a novel circadian clock mutation in *Arabidopsis*. Genome Biol 12:R28

Badouin H, Gouzy J, Grassa CJ, Murat F, Staton SE, Cottret L, Lelandais-Briere C, Owens GL, Carrere S, Mayjonade B et al (2017) The sunflower genome provides insights into oil metabolism, flowering and Asterid evolution. Nature 546:148–152

Baulkwill WJ (1989) The history of natural rubber production. In: Webster CC, Baulkwill WJ (eds) Rubber, Longman Scientific and Technical, Essex, England

Bennetzen JL, Coleman C, Liu R, Ma J, Ramakrishna W (2004) Consistent over-estimation of gene number in complex plant genomes. Curr Opin Plant Biol 7:732–736

Birol I, Raymond A, Jackman SD, Pleasance S, Coope R, Taylor GA, Yuen MMS, Keeling CI, Brand D, Vandervalk BP et al (2013) Assembling the 20 Gb white spruce (*Picea glauca*) genome from whole-genome shotgun sequencing data. Bioinformatics (Oxford, England) 29:1492–1497

Blanco E, Parra G, Guigo R (2007) Using geneid to identify genes. Curr Protoc Bioinform (Chapter 4, Unit 4 3)

Campbell MS, Law M, Holt C, Stein JC, Moghe GD, Hufnagel DE, Lei J, Achawanantakun R, Jiao D, Lawrence CJ et al (2014) MAKER-P: a tool kit for the rapid creation, management, and quality control of plant genome annotations. Plant Physiol 164:513–524

Cantarel BL, Korf I, Robb SM, Parra G, Ross E, Moore B, Holt C, Sanchez Alvarado A, Yandell M (2008) MAKER: an easy-to-use annotation pipeline designed for emerging model organism genomes. Genome Res 18:188–196

Cao J, Schneeberger K, Ossowski S, Gunther T, Bender S, Fitz J, Koenig D, Lanz C, Stegle O, Lippert C et al (2011) Whole-genome sequencing of multiple *Arabidopsis thaliana* populations. Nat Genet 43:956–963

Chan AP, Crabtree J, Zhao Q, Lorenzi H, Orvis J, Puiu D, Melake-Berhan A, Jones KM, Redman J, Chen G et al (2010) Draft genome sequence of the oilseed species *Ricinus communis*. Nat Biotechnol 28:951

Chow KS, Ghazali AK, Hoh CC, Mohd-Zainuddin Z (2014) RNA sequencing read depth requirement for optimal transcriptome coverage in *Hevea brasiliensis*. BMC Res Notes 7:69

Claros MG, Bautista R, Guerrero-Fernandez D, Benzerki H, Seoane P, Fernandez-Pozo N (2012) Why assembling plant genome sequences is so challenging. Biology (Basel) 1:439–459

Dassanayake M, Oh DH, Haas JS, Hernandez A, Hong H, Ali S, Yun DJ, Bressan RA, Zhu JK, Bohnert HJ et al (2011) The genome of the extremophile crucifer *Thellungiella parvula*. Nat Genet 43:913–918

Davidson RM, Hansey CN, Gowda M, Childs KL, Lin H, Vaillancourt B, Sekhon RS, de Leon N, Kaeppler SM, Jiang N et al (2011) Utility of RNA sequencing for analysis of maize reproductive transcriptomes. The Plant Genome 4:191–203

Denton JF, Lugo-Martinez J, Tucker AE, Schrider DR, Warren WC, Hahn MW (2014) Extensive error in the number of genes inferred from draft genome assemblies. PLoS Comput Biol 10:e1003998–e1003998

Deschamps S, Zhang Y, Llaca V, Ye L, Sanyal A, King M, May G, Lin H (2018) A chromosome-scale assembly of the sorghum genome using nanopore sequencing and optical mapping. Nat Commun 9:4844

Du H, Yu Y, Ma Y, Gao Q, Cao Y, Chen Z, Ma B, Qi M, Li Y, Zhao X et al (2017) Sequencing and de novo assembly of a near complete *indica* rice genome. Nat Commun 8:15324

Duvick J, Fu A, Muppirala U, Sabharwal M, Wilkerson MD, Lawrence CJ, Lushbough C, Brendel V (2008) PlantGDB: a resource for comparative plant genomics. Nucleic Acids Res 36:D959–965

English AC, Richards S, Han Y, Wang M, Vee V, Qu J, Qin X, Muzny DM, Reid JG, Worley KC et al (2012) Mind the gap: upgrading genomes with Pacific Biosciences RS long-read sequencing technology. PLoS ONE 7:e47768

Filichkin SA, Priest HD, Givan SA, Shen R, Bryant DW, Fox SE, Wong WK, Mockler TC (2010) Genome-wide mapping of alternative splicing in *Arabidopsis thaliana*. Genome Res 20:45–58

Goodstein DM, Shu S, Howson R, Neupane R, Hayes RD, Fazo J, Mitros T, Dirks W, Hellsten U, Putnam N et al (2012) Phytozome: a comparative platform for green plant genomics. Nucleic Acids Res 40:D1178–1186

Gouvêa LRL, Rubiano LB, Chioratto AF, Zucchi MI, de Souza Gonçalves P (2010) Genetic divergence of rubber tree estimated by multivariate techniques and microsatellite markers. Genet Mol Biol 33:308–318

Guan R, Zhao Y, Zhang H, Fan G, Liu X, Zhou W, Shi C, Wang J, Liu W, Liang X et al (2016) Draft genome of the living fossil *Ginkgo biloba*. Gigascience 5:49

Haas BJ, Salzberg SL, Zhu W, Pertea M, Allen JE, Orvis J, White O, Buell CR, Wortman JR (2008) Automated eukaryotic gene structure annotation using EVidenceModeler and the program to assemble spliced alignments. Genome Biol 9:R7–R7

Hamilton JP, Robin Buell C (2012) Advances in plant genome sequencing. Plant J 70:177–190

Huala E, Dickerman AW, Garcia-Hernandez M, Weems D, Reiser L, LaFond F, Hanley D, Kiphart D, Zhuang M, Huang W et al (2001) The Arabidopsis Information Resource (TAIR): a comprehensive database and web-based information retrieval, analysis, and visualization system for a model plant. Nucleic Acids Res 29:102–105

Huang S, Li R, Zhang Z, Li L, Gu X, Fan W, Lucas WJ, Wang X, Xie B, Ni P et al (2009a) The genome of the cucumber, *Cucumis sativus* L. Nat Genet 41:1275–1281

Huang X, Feng Q, Qian Q, Zhao Q, Wang L, Wang A, Guan J, Fan D, Weng Q, Huang T et al (2009b) High-throughput genotyping by whole-genome resequencing. Genome Res 19:1068–1076

Ibarra-Laclette E, Lyons E, Hernandez-Guzman G, Perez-Torres CA, Carretero-Paulet L, Chang TH, Lan T, Welch AJ, Juarez MJ, Simpson J et al (2013) Architecture and evolution of a minute plant genome. Nature 498:94–98

International Rice Genome Sequencing Project (2005) The map-based sequence of the rice genome. Nature 436:793

Jaillon O, Aury JM, Noel B, Policriti A, Clepet C, Casagrande A, Choisne N, Aubourg S, Vitulo N, Jubin C et al (2007) The grapevine genome sequence suggests ancestral hexaploidization in major angiosperm phyla. Nature 449:463–467

Jarvis DE, Ho YS, Lightfoot DJ, Schmöckel SM, Li B, Borm TJA, Ohyanagi H, Mineta K, Michell CT, Saber N et al (2017) The genome of *Chenopodium quinoa*. Nature 542:307

Jiao Y, Peluso P, Shi J, Liang T, Stitzer MC, Wang B, Campbell MS, Stein JC, Wei X, Chin CS et al (2017) Improved maize reference genome with single-molecule technologies. Nature 546:524–527

Junior HdJE, Ohto JM, da Silva LL, Palma HAL, Ballarin AW (2015) Potential of rubberwood (*Hevea brasiliensis*) for structural use after the period of latex extraction: a case study in Brazil. J Wood Sci 61:384–390

Kajitani R, Toshimoto K, Noguchi H, Toyoda A, Ogura Y, Okuno M, Yabana M, Harada M, Nagayasu E, Maruyama H et al (2014) Efficient de novo assembly of highly heterozygous genomes from whole-genome shotgun short reads. Genome Res 24:1384–1395

Kelley DR, Salzberg SL (2010) Detection and correction of false segmental duplications caused by genome mis-assembly. Genome Biol 11:R28–R28

Kersey PJ, Staines DM, Lawson D, Kulesha E, Derwent P, Humphrey JC, Hughes DS, Keenan S, Kerhornou A, Koscielny G et al (2012) Ensembl genomes: an integrative resource for genome-scale data from non-vertebrate species. Nucleic Acids Res 40:D91–97

Korf I (2004) Gene finding in novel genomes. BMC Bioinform 5:59

Lai J, Li R, Xu X, Jin W, Xu M, Zhao H, Xiang Z, Song W, Ying K, Zhang M et al (2010) Genome-wide patterns of genetic variation among elite maize inbred lines. Nat Genet 42:1027–1030

Lamesch P, Berardini TZ, Li D, Swarbreck D, Wilks C, Sasidharan R, Muller R, Dreher K, Alexander DL, Garcia-Hernandez M et al (2012) The Arabidopsis Information Resource (TAIR): improved gene annotation and new tools. Nucleic Acids Res 40:D1202–1210

Lan T, Renner T, Ibarra-Laclette E, Farr KM, Chang T-H, Cervantes-Pérez SA, Zheng C, Sankoff D, Tang H, Purbojati RW et al (2017) Long-read sequencing uncovers the adaptive topography of a carnivorous plant genome. Proc Natl Acad Sci 114:E4435

Lau N-S, Makita Y, Kawashima M, Taylor TD, Kondo S, Othman AS, Shu-Chien AC, Matsui M (2016) The rubber tree genome shows expansion of gene family associated with rubber biosynthesis. Sci Rep 6:28594

Liu S, Lan J, Zhou B, Qin Y, Zhou Y, Xiao X, Yang J, Gou J, Qi J, Huang Y et al (2015) HbNIN2, a cytosolic alkaline/neutral-invertase, is responsible for sucrose catabolism in rubber-producing laticifers of *Hevea brasiliensis* (para rubber tree). New Phytol 206:709–725

Lomsadze A, Ter-Hovhannisyan V, Chernoff YO, Borodovsky M (2005) Gene identification in novel eukaryotic genomes by self-training algorithm. Nucleic Acids Res 33:6494–6506

Majoros WH, Pertea M, Salzberg SL (2004) TigrScan and GlimmerHMM: two open source ab initio eukaryotic gene-finders. Bioinformatics 20:2878–2879

Makita Y, Kawashima M, Lau NS, Othman AS, Matsui M (2018) Construction of Para rubber tree genome and multi-transcriptome database accelerates rubber researches. BMC Genomics 19:922

Malaysian Rubber Board (2018) Natural rubber statistics 2018 (Kuala Lumpur: Malaysian Rubber Board)

Margulies M, Egholm M, Altman WE, Attiya S, Bader JS, Bemben LA, Berka J, Braverman MS, Chen YJ, Chen Z et al (2005) Genome sequencing in microfabricated high-density picolitre reactors. Nature 437:376–380

Mascher M, Gundlach H, Himmelbach A, Beier S, Twardziok SO, Wicker T, Radchuk V, Dockter C, Hedley PE, Russell J et al (2017) A chromosome conformation capture ordered sequence of the barley genome. Nature 544:427

McKernan KJ, Peckham HE, Costa GL, McLaughlin SF, Fu Y, Tsung EF, Clouser CR, Duncan C, Ichikawa JK, Lee CC et al (2009) Sequence and structural variation in a human genome uncovered by short-read, massively parallel ligation sequencing using two-base encoding. Genome Res 19:1527–1541

Michael TP, Jackson S (2013) The first 50 plant genomes. The Plant Genome 6

Michael TP, VanBuren R (2015) Progress, challenges and the future of crop genomes. Curr Opin Plant Biol 24:71–81

Ming R, Hou S, Feng Y, Yu Q, Dionne-Laporte A, Saw JH, Senin P, Wang W, Ly BV, Lewis KL et al (2008) The draft genome of the transgenic tropical fruit tree papaya (*Carica papaya* Linnaeus). Nature 452:991–996

Neale DB, Wegrzyn JL, Stevens KA, Zimin AV, Puiu D, Crepeau MW, Cardeno C, Koriabine M, Holtz-Morris AE, Liechty JD et al (2014) Decoding the massive genome of loblolly pine using haploid DNA and novel assembly strategies. Genome Biol 15:R59–R59

Nussbaumer T, Martis MM, Roessner SK, Pfeifer M, Bader KC, Sharma S, Gundlach H, Spannagl M (2013) MIPS PlantsDB: a database framework for comparative plant genome research. Nucleic Acids Res 41:D1144–1151

Nystedt B, Street NR, Wetterbom A, Zuccolo A, Lin Y-C, Scofield DG, Vezzi F, Delhomme N, Giacomello S, Alexeyenko A et al (2013) The Norway spruce genome sequence and conifer genome evolution. Nature 497:579

Ong CW, Shamsul Bahri AR (2014) The introduction of rubber planting recommendations by The Rubber Research Institute of Malaysia since 1925. J Biol Agric Healthc 4:49–53

Ong SH, Ramli O, Othman H, Masahuling B, Naimah I (1994) Rubber breeding, progress and strategies to meet future needs of the plantation industry. In: Paper presented at: proceedings of the international planters conference. Kuala Lumpur

Ouyang S, Zhu W, Hamilton J, Lin H, Campbell M, Childs K, Thibaud-Nissen F, Malek RL, Lee Y, Zheng L et al (2007) The TIGR rice genome annotation resource: improvements and new features. Nucleic Acids Res 35:D883–887

Pethin D, Nakkanong K, Nualsri C (2015) Performance and genetic assessment of rubber tree clones in Southern Thailand. Sci Agri 72:306–313

Piegu B, Guyot R, Picault N, Roulin A, Sanyal A, Kim H, Collura K, Brar DS, Jackson S, Wing RA et al (2006) Doubling genome size without polyploidization: dynamics of retrotransposition-driven genomic expansions in *Oryza australiensis*, a wild relative of rice. Genome Res 16:1262–1269

Planters' Bulletin (1974) RRIM 600 Series Clones: Final Report (Malaysian Rubebr Board).

Pootakham W, Ruang-Areerate P, Jomchai N, Sonthirod C, Sangsrakru D, Yoocha T, Theerawattanasuk K, Nirapathpongporn K, Romruensukharom P, Tragoonrung S et al (2015) Construction of a high-density integrated genetic linkage map of rubber tree (*Hevea brasiliensis*) using genotyping-by-sequencing (GBS). Front Plant Sci 6

Priyadarshan PM, Hoa TTT, Huasun H, de Gonçalves PS (2005) Yielding potential of rubber (*Hevea brasiliensis*) in sub-optimal environments. J Crop Improv 14:221–247

Rahman AYA, Usharraj AO, Misra BB, Thottathil GP, Jayasekaran K, Feng Y, Hou S, Ong SY, Ng FL, Lee LS et al (2013) Draft genome sequence of the rubber tree *Hevea brasiliensis*. BMC Genomics 14:75

Rouard M, Guignon V, Aluome C, Laporte MA, Droc G, Walde C, Zmasek CM, Perin C, Conte MG (2011) GreenPhylDB v2.0: comparative and functional genomics in plants. Nucleic Acids Res 39:D1095–1102

Salamov AA, Solovyev VV (2000) Ab initio gene finding in *Drosophila* genomic DNA. Genome Res 10:516–522

SanMiguel P, Tikhonov A, Jin YK, Motchoulskaia N, Zakharov D, Melake-Berhan A, Springer PS, Edwards KJ, Lee M, Avramova Z et al (1996) Nested retrotransposons in the intergenic regions of the maize genome. Science 274:765–768

Sato S, Hirakawa H, Isobe S, Fukai E, Watanabe A, Kato M, Kawashima K, Minami C, Muraki A, Nakazaki N et al (2011) Sequence analysis of the genome of an oil-bearing tree, *Jatropha curcas* L. DNA Res 18:65–76

Schnable PS, Ware D, Fulton RS, Stein JC, Wei F, Pasternak S, Liang C, Zhang J, Fulton L, Graves TA et al (2009) The B73 maize genome: complexity, diversity, and dynamics. Science 326:1112–1115

Shearman JR, Sangsrakru D, Jomchai N, Ruang-areerate P, Sonthirod C, Naktang C, Theerawattanasuk K, Tragoonrung S, Tangphatsornruang S (2015) SNP identification from RNA sequencing and linkage map construction of rubber tree for anchoring the draft genome. PLoS ONE 10:e0121961

Shulaev V, Sargent DJ, Crowhurst RN, Mockler TC, Folkerts O, Delcher AL, Jaiswal P, Mockaitis K, Liston A, Mane SP et al (2010) The genome of woodland strawberry (*Fragaria vesca*). Nat Genet 43:109

Stanke M, Schöffmann O, Morgenstern B, Waack S (2006) Gene prediction in eukaryotes with a generalized hidden Markov model that uses hints from external sources. BMC Bioinform 7:62–62

Sunderasan E, Wickneswari R, Abd. Aziz MZ, Yeang HY (1994) Incidence of self- and cross-pollination in two *Hevea brasiliensis* clones. J Nat Rubber Res 9

Tang H, Krishnakumar V, Bidwell S, Rosen B, Chan A, Zhou S, Gentzbittel L, Childs KL, Yandell M, Gundlach H et al (2014) An improved genome release (version Mt4.0) for the model legume *Medicago truncatula*. BMC Genomics 15:312

Tarailo-Graovac M, Chen N (2009) Using RepeatMasker to identify repetitive elements in genomic sequences. Curr Protoc Bioinform (Chapter 4, Unit 4 10)

Tuskan GA, Difazio S, Jansson S, Bohlmann J, Grigoriev I, Hellsten U, Putnam N, Ralph S, Rombauts S, Salamov A et al (2006) The genome of black cottonwood, *Populus trichocarpa* (Torr. & Gray). Science 313:1596–1604

Van Bel M, Proost S, Wischnitzki E, Movahedi S, Scheerlinck C, Van de Peer Y, Vandepoele K (2012) Dissecting plant genomes with the PLAZA comparative genomics platform. Plant Physiol 158:590–600

VanBuren R, Bryant D, Edger PP, Tang H, Burgess D, Challabathula D, Spittle K, Hall R, Gu J, Lyons E et al (2015) Single-molecule sequencing of the desiccation-tolerant grass *Oropetium thomaeum*. Nature 527:508

Velasco R, Zharkikh A, Troggio M, Cartwright DA, Cestaro A, Pruss D, Pindo M, Fitzgerald LM, Vezzulli S, Reid J et al (2007) A high quality draft consensus sequence of the genome of a heterozygous grapevine variety. PLoS ONE 2:e1326

Velasco R, Zharkikh A, Affourtit J, Dhingra A, Cestaro A, Kalyanaraman A, Fontana P, Bhatnagar SK, Troggio M, Pruss D et al (2010) The genome of the domesticated apple (*Malus* x *domestica* Borkh.). Nat Genet 42:833–839

Wang W, Feng B, Xiao J, Xia Z, Zhou X, Li P, Zhang W, Wang Y, Møller BL, Zhang P et al (2014) Cassava genome from a wild ancestor to cultivated varieties. Nat Commun 5:5110

Xu X, Pan S, Cheng S, Zhang B, Mu D, Ni P, Zhang G, Yang S, Li R, Wang J et al (2011) Genome sequence and analysis of the tuber crop potato. Nature 475:189–195

Youens-Clark K, Buckler E, Casstevens T, Chen C, Declerck G, Derwent P, Dharmawardhana P, Jaiswal P, Kersey P, Karthikeyan AS et al (2011) Gramene database in 2010: updates and extensions. Nucleic Acids Res 39:D1085–1094

Zhang G, Guo G, Hu X, Zhang Y, Li Q, Li R, Zhuang R, Lu Z, He Z, Fang X et al (2010) Deep RNA sequencing at single base-pair resolution reveals high complexity of the rice transcriptome. Genome Res 20:646–654

Zhu JH, Xu J, Chang WJ, Zhang ZL (2015) Isolation and molecular characterization of 1-aminocyclopropane-1-carboxylic acid synthase genes in *Hevea brasiliensis*. Int J Mol Sci 16:4136–4149

Zimin AV, Puiu D, Hall R, Kingan S, Clavijo BJ, Salzberg SL (2017) The first near-complete assembly of the hexaploid bread wheat genome, *Triticum aestivum*. Gigascience 6:1–7

The BPM 24 Rubber Tree Genome, Organellar Genomes and Synteny Within the Family Euphorbiaceae

4

Jeremy R. Shearman, Wirulda Pootakham, and Sithichoke Tangphatsornruang

Abstract

Rubber tree is an important crop species grown over an expansive area in southern Thailand to harvest natural latex. The most commonly grown clone here is RRIM 600, but other clones are also grown in some areas, including BPM 24 which is a cytoplasmic male sterile elite clone. The organellar genomes, from both mitochondria and chloroplast of BPM 24 have been analyzed. The chloroplast sequence of BPM 24 is nearly identical to that published for the clone RRIM 600 with the exception of 5 SNPs, which were either non-coding or silent. The mitochondrial genome is remarkably different from that of other rubber tree clones such as RRIM 600, and the cause of cytoplasmic sterility has been identified as a novel transcript containing a portion of *ATPase subunit 9* (*atp9*). The nuclear genome of BPM 24 was assembled using a combination of Illumina and PacBio sequence data and Chicago HiC scaffolding, which produced an assembly containing a relatively small number of fragmented contigs. A high-density SNP map was used to anchor almost one-third of the sequence into 18 linkage groups that captured two-thirds of the protein coding sequences. Comparing orthologous genes with other members of the Euphorbiaceae family identified cassava as the closest relative with a shared whole genome duplication event that took place following a split from a common ancestor with castor bean.

4.1 Introduction

Hevea brasiliensis, or rubber tree, is a deciduous perennial tree crop indigenous to the rain forests of the Amazonian basin in South America. Of the 10 *Hevea* species, *H. brasiliensis* is the only one which produces enough high-quality natural rubber on a commercial scale, this accounting for more than 98% of total global production (Priyadarshan and Clément-Demange 2004). Rubber tree is an outbreeding species with a long breeding cycle which may take around 30–40 years. Thus, its genome sequence is important for high-throughput marker discovery and subsequent identification of important QTLs, as well as the underlying genes that control important commercial traits. Only then can marker-assisted breeding be employed effectively, allowing large numbers of desirable traits to be introduced into a single clone using a gene pyramiding approach.

J. R. Shearman · W. Pootakham ·
S. Tangphatsornruang (✉)
National Omics Center, National Center for Genetic Engineering and Biotechnology (BIOTEC), National Science and Technology Development Agency (NSTDA), Science Park, Thailand, Phahonyothin Road, Khlong Nueng, Khlong Luang 12120, Pathumthani, Thailand
e-mail: sithichoke.tan@biotec.or.th

M. Matsui and K.-S. Chow (eds.), *The Rubber Tree Genome*, Compendium of Plant Genomes,
https://doi.org/10.1007/978-3-030-42258-5_4

For such a strategy, it is important to have markers that are either the causative mutation of the trait of interest, or those which are tightly linked with the causative genes.

Rubber farming in Thailand began in 1899 when Phraya Ratsadanupradit Mahison Phakdi, the commissioner of Monthon Phuket (consisting of the provinces Phuket, Thalang, Ranong, Phang Nga, Takua Pa, Krabi, Kelantan and Terengganu with Satun added later), visited Malaya (now Malaysia) and observed the rubber plantations and farming methods employed there. He did not obtain any of the young rubber trees from that visit, but later visited Indonesian rubber farms and brought some young rubber tree seedlings back to Thailand, which were planted in Kantang District, Trang Province. These initial plantations were used to show the profitability of rubber trees and to produce seeds for distribution to local small-scale farmers (Stifel 1973). Since around 1960, the most popular clone in Thailand has been RRIM 600. Additionally, some smallholder farms planted BPM 24, a clone recommended for small-scale planting. The clone BPM 24 has good latex production and response to wounding, and is best suited to cut tapping.

One of the rubber tree clones with a sequenced genome is BPM 24 (BPM = Balai Penelitian Perkebunan Medan) (Pootakham et al. 2017). The clone BPM 24 is a cytoplasmic male sterile descendant of a GT 1 (female) x AVROS 1734 (male) cross that was developed in the 1990s and is considered a notable clone (Priyadarshan and Clément-Demange 2004). The clone GT 1 (GT = Gondang Tapen, Indonesia) is a primary clone that can be traced back to the 1920s. GT 1 itself is male sterile, as is its offspring BPM 24, and the offspring of BPM 24. The male parent, AVROS 1734 (AVROS = Algemeene Vereeniging van Rubberplanters ter Oostkust van Sumatra), is a clone which originated from Indonesia that has high yield and good resistance to common rubber tree diseases. References to AVROS 1734 in the literature go as far back as the 1920s. Both parental strains tend to have higher yields than BPM 24 across a range of environments; yet BPM 24 is considered an elite clone and was recommended as a preferred clone for small-scale planting because it has good response to cut tapping and exhibits a high degree of resistance to two major fungal pathogens (*Phytophthora* and *Corynespora*) found throughout Southeast Asia.

4.2 Genome Assembly

Plants have three distinct genomes: nuclear, mitochondrial and plastid. Mitochondria and chloroplasts are thought to have originated from separate endosymbiotic events where the cell incorporated a proteobacterium or a cyanobacterium, respectively (Zimorski et al. 2014). Since these ancient events, the three genomes have undergone significant co-evolution involving gene transfer between them, primarily in the form of genes moving into the nuclear genome (Lloyd and Timmis 2011). Mitochondria and chloroplasts have mRNA transcript processing not seen in the nuclear genome, including polycistronic cleavage, mRNA editing and trans-splicing (Wicke et al. 2011).

4.2.1 Chloroplast Genome

The chloroplast genomes of land plants are quite conserved, they generally have a cyclic structure with two single copy regions separated by two copies of an inverted repeat, a large single copy region and a small single copy region (Wicke et al. 2011). The rubber tree chloroplast is no exception to this (Tangphatsornruang et al. 2011). The rubber tree RRIM 600 chloroplast genome was assembled de novo into a 161,191 bp circular DNA strand using 454 shotgun sequencing and Newbler (Tangphatsornruang et al. 2011, Fig. 4.1). The genome contains 112 unique genes including 30 transfer RNA (tRNA) genes, 4 ribosomal RNA (rRNA) genes and 78 predicted protein coding genes (Table 4.1). The large single copy region is 89,209 bp in size and the small single copy region is 18,362 bp in size, separated by the inverted repeats which are 26,810 bp each. Phylogenetic analysis of the chloroplast genes

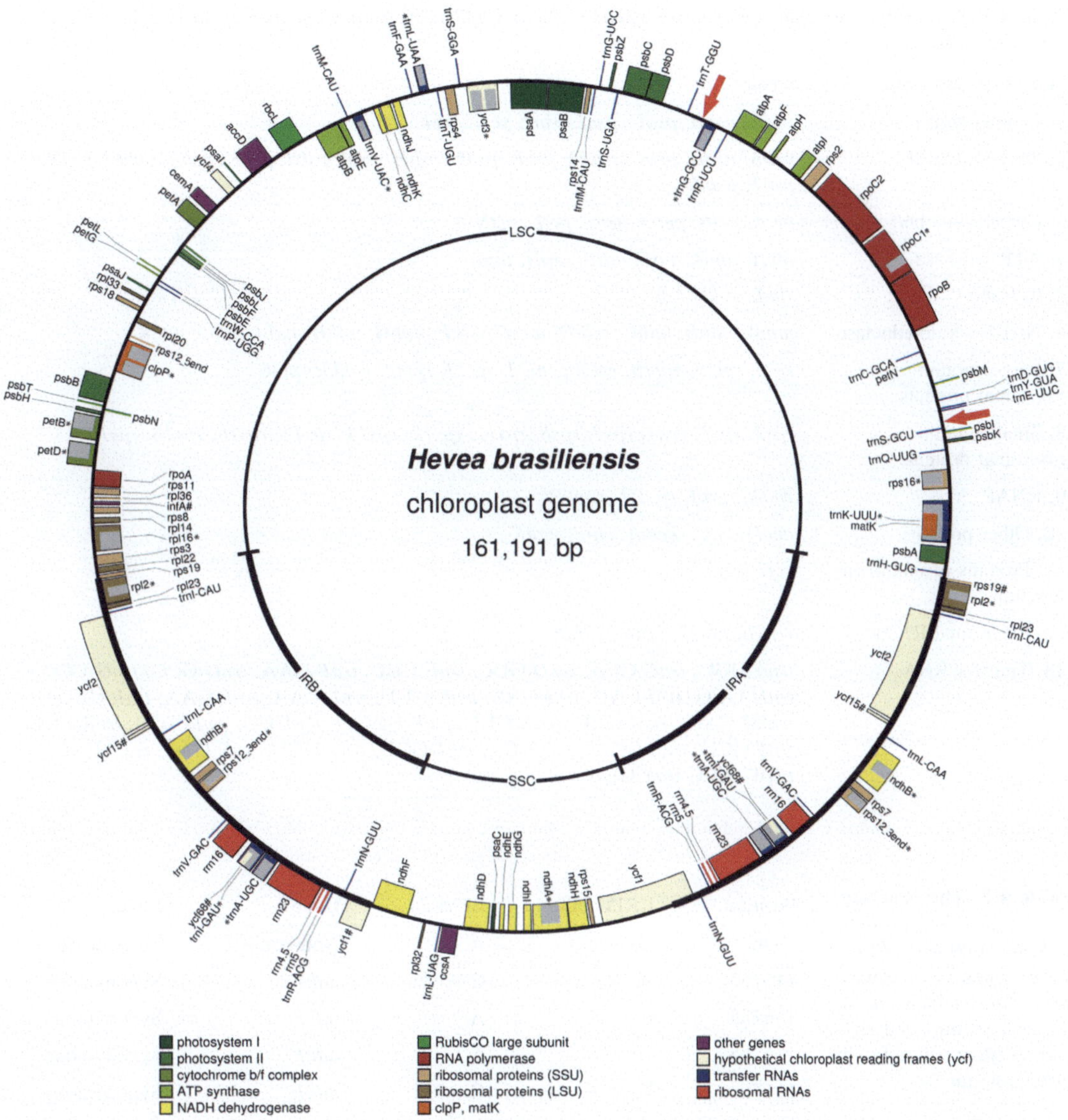

Fig. 4.1 The rubber tree chloroplast genome (RRIM 600). The inner circle delineates the single-copy regions flanked by the inverted repeats. The outer circle shows the exons of each gene with plus/minus strand coding indicated by the exon box appearing outside or inside the circle, respectively. Introns are represented by a grey box and * in the name. Pseudogenes are marked with an #. Arrows indicate the position of a 30 kb inversion compared to the cassava chloroplast genome (Tangphatsornruang et al. 2011)

identified cassava (*Manihot esculenta*) as the closest relative. The rubber tree chloroplast genome has a 30 kb inversion compared to the cassava genome. Comparing the chloroplast genomic sequence to chloroplast transcript sequences revealed 51 RNA editing sites that occur in 26 protein coding genes and 3 introns. The BPM 24 chloroplast sequence is identical to the RRIM 600 published sequence, with the exception of four synonymous coding SNPs in the gene *ndhF* and one non-coding SNP (Table 4.2).

Table 4.1 The rubber tree chloroplast genes (RRIM 600 and BPM 24) grouped according to function (Tangphatsornruang et al. 2011)

Functional grouping	Gene
1. Photosystem I	*psaA*, *psaB*, *psaC*, *psaI*, *psaJ*, *ycf3*, *ycf4*
2. Photosystem II	*psbA*, *psbB*, *psbC*, *psbD*, *psbE*, *psbF*, *psbH*, *psbI*, *psbJ*, *psbK*, *psbL*, *psbM*, *psbN*, *psbT*, *psbZ*
3. Cytochrome b6/f	*petA*, *petB*, *petD*, *petG*, *petL*, *petN*
4. ATP synthase	*atpA*, *atpB*, *atpE*, *atpF*, *atpH*, *atpI*
5. Rubisco	*rbcL*
6. NADH oxidoreductase	*ndhA*, *ndhB*, *ndhC*, *ndhD*, *ndhE*, *ndhF*, *ndhG*, *ndhH*, *ndhI*, *ndhJ*, *ndhK*
7. Large subunit ribosomal proteins	*rpl2*, *rpl14*, *rpl16*, *rpl20*, *rpl22*, *rpl23*, *rpl32*, *rpl33*, *rpl36*
8. Small subunit ribosomal proteins	*rps2*, *rps3*, *rps4*, *rps7*, *rps8*, *rps11*, *rps12*, *rps14*, *rps15*, *rps16*, *rps18*, *rps19*
9. RNAP	*rpoA*, *rpoB*, *rpoC1*, *rpoC2*
10. Other proteins	*accD*, *ccsA*, *cemA*, *clpP*, *matK*
11. Proteins of unknown function	*ycf1*, *ycf2*
12. Ribosomal RNAs	*rrn16*, *rrn23*, *rrn4.5*, *rrn5*
13. Transfer RNAs	*trnA*-UGC, *trnC*-GCA, *trnD*-GUC, *trnE*-UUC, *trnF*-GAA, *trnG*-GCC, *trnG*-UCC, *trnH*-GUG, *trnI*-CAU, *trnI*-GAU, *trnK*-UUU, *trnL*-CAA, *trnL*-UAA, *trnL*-UAG, *trnfM*-CAU, *trnM*-CAU, *trnN*-GUU, *trnP*-UGG, *trnQ*-UUG, *trnR*-ACG, *trnR*-UCU, *trnS*-GCU, *trnS*-GGA, *trnS*-UGA, *trnT*-GGU, *trnT*-UGU, *trnV*-GAC, *trnV*-UAC, *trnW*-CCA, *trnY*-GUA

Table 4.2 The sequence variants and their positions in RRIM 600 and BPM 24 chloroplasts. Gene names for coding variants are in the gene column and any coding effect is listed in the effect column

Position	RRIM 600	BPM 24	Gene	Effect
9946	G	A	Non-coding	Non-coding
116753	A	C	*ndhF*	Synonymous
116755	C	A	*ndhF*	Synonymous
116765	G	A	*ndhF*	Synonymous
116766	T	G	*ndhF*	Synonymous

4.2.2 Mitochondrial Genome

Plant mitochondrial genomes range in size from 200 kb in *Brassica hirta* (Palmer and Herbon 1987) to 11.3 Mb in *Silene conica* (Sloan et al. 2012). Mitochondrial genomes have expanded in land plants and accumulated large numbers of repeat sequences which cause frequent recombination events and dynamic genome rearrangements within a species, leading to the generation of multiple circular DNA strands with overlapping sequence and different copy number (Turmel et al. 2003; Bullerwell and Gray 2004; Allen et al. 2007; Chang et al. 2011).

The mitochondria of BPM 24 was investigated in order to obtain a reference mitochondrial genome and also to investigate the cause of male sterility (Shearman et al. 2014). Maternal inheritance of male sterility in BPM 24 and the lack of causative mutations in the chloroplast genome suggested that the cause of male sterility in BPM 24 could be within the mitochondrial genome.

The mitochondrial genome was assembled into 37 contigs using 454 sequence data, including 8 kb paired-end libraries and Illumina paired-end sequencing data. The scaffold graph, which shows joins between contigs, was complex with many contigs joining to multiple other contigs. This complex scaffold graph suggests that the genome exists not as a single circle, but as many subgenomic circles with partial sequence overlap. A master circle mitochondrial sequence was generated by traversing the contig graph in such a way that all contigs were used at least once with as little duplication as possible. This resulted in a mitochondrial sequence that was 1,325,823 bp in size with approximately 350 kb of duplicated sequence (Fig. 4.2).

The mitochondrial genome was annotated using Illumina RNA-seq data which allowed for 65 open reading frames to be identified (Table 4.3; Shearman et al. 2014). There was group II trans-splicing in three genes, *nad1*, *nad2* and *nad5*, as has been found in multiple plant species (Bonen 2008). Group II trans-splicing is the splicing of introns from two or more separate mRNA molecules that are encoded on opposite strands. Each of the trans-spliced *nad* genes identified in rubber tree are also trans-spliced in other species (Bonen 2008). There were 19 tRNA genes identified and five of them occurred twice in the assembled mitochondrial master circle (Table 4.3). Seven tRNA genes are from gene transfer events, largely consistent with previously identified chloroplast gene transfer events (Wang et al. 2007). A notable difference is that BPM 24 lacks a copy of *trnS*-GGA which is common in species closely related to rubber tree, including *Ricinus communis* (castor bean), suggesting that the chloroplast-derived *trnS*-GGA was lost in the rubber tree after the split from *R. communis*.

The BPM 24 mitochondrial contigs have 11 rearrangements compared to RRIM 600, with five of these occurring within 1 kb of a gene. The mitochondrial genome was found to contain multiple structural arrangements within a single individual for each of the six varieties that were sequenced (BPM 24, RRIM 600, RRII 105, RRIC 110, RRIT 251, PB 235), consistent with the existence of many subgenomic circles. One of these 11 rearrangements was unique to BPM 24 and includes 240 bp of sequence not shared by any other clone, making it a good candidate for the cause of male sterility. This additional sequence is expressed, according to RNA-seq data, and encodes an mRNA of 51 amino acids, 33 of which are identical to the tail end of *ATPase subunit 9* (*atp9*). An entire transmembrane domain is included in this additional sequence which likely allows it to compete with the functional copy of *atp9* in the ATP synthase complex, and reduce the efficiency of ATP production sufficiently to cause apoptosis in the high-energy demanding anthers. A novel or fusion transcript is a common occurrence in cytoplasmic male sterile plants and often involves a portion of, or is near an ATP synthase subunit gene (Carlsson et al. 2008). To conclude, the cause of cytoplasmic male sterility in BPM 24 should be the expression of this additional mRNA that resulted from a genomic rearrangement.

4.2.3 Nuclear Genome

The cost and difficulty of obtaining a de novo whole genome sequence depend on the type of sequencing technology used to interrogate each nucleotide. Short read shotgun sequencing using platforms such as Illumina, BGI and Ion Torrent, alone can generate a fragmented genome assembly that contains the majority of unique sequence. Such a genome assembly is heavily fragmented because the short read lengths cannot span repeat sequences. Long read shotgun sequencing platforms such as PacBio and Nanopore now allow for genomes to be less fragmented, but still far from complete. A method currently gaining traction is the use of short read sequencing with high-throughput chromosome conformation capture (HiC) to explore DNA linkage in chromatin (Putnam et al. 2016). Segments of DNA that are in close proximity are more likely to ligate than distant segments and this information can be used to scaffold contigs generated through shotgun

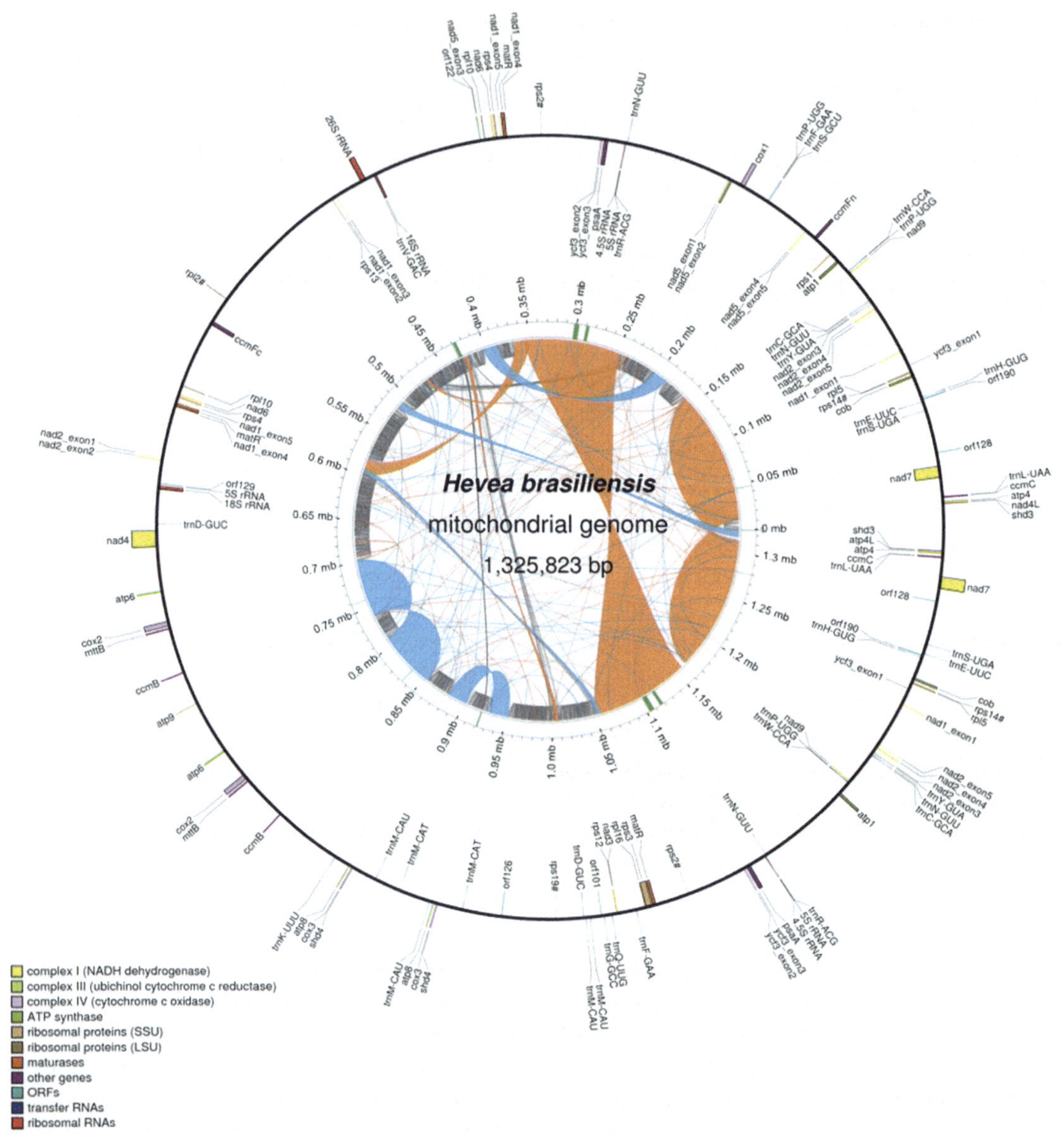

Fig. 4.2 The BPM 24 mitochondria master circle. Grey arches represent Illumina paired-end sequence mapping data, blue arches show direct repeats, and orange arches show inverted repeats (inner circle). The outer circle shows the exons of each gene with plus/minus strand coding indicated by the exon box appearing outside or inside the circle, respectively (Shearman et al. 2014)

sequencing. There are two methods employed; one involves using high molecular weight DNA to reconstitute chromatin in vitro followed by proximity ligation and shotgun sequencing, and the other uses in vivo chromatin to generate proximity ligation allowing for longer range capture than the in vitro method.

The rubber tree nuclear genome consists of 18 chromosomes and is roughly 2.15 Gb in size. The rubber tree genome is one of the more difficult genomes to sequence because of its size and high repeat content, with repeat sequences accounting for around 78% of the genome (Rahman et al. 2013). A draft sequence of the rubber tree genome,

Table 4.3 BPM 24 mitochondrial genes grouped according to function. Copy number is indicated in brackets and exon count is indicated in square brackets. Chloroplast derived tRNAs have -cp appended to them (Shearman et al. 2014)

Gene function	Gene name
Complex I	(2x)*nad1*[5], (2x)*nad2*[5], *nad3*, *nad4*[4], (2x)*nad4L*, *nad5*[5], (2x)*nad6*, (2x)*nad7*[5], (2x)*nad9*
Complex II	(2x)*shd3*, (2x)*shd4*
Complex III	(2x)*cob*
Complex IV	*cox1*[2], (2x)*cox2*[2], *cox3*
Complex V	(2x)*atp1*, (2x)*atp4*, (2x)*atp6*, (2x)*atp8*, *atp9*
Cytochrome-c biogenesis	(2x)*ccmB*, (2x)*ccmC*, *ccmFc*[2], *ccmFn*
SecY-independent transport	(2x)*mttB*
Ribosomal RNAs	*5S rRNA*, *18S rRNA*, *26S rRNA*
Ribosomal protein small subunit	*rps1*, *rps3*[2], (2x)*rps4*, *rps12*, *rps13*
Ribosomal protein large subunit	(2x)*rpl5*, (2x)*rpl10*, *rpl16*
Intron maturase	(3x)*matR*
Chloroplast transferred complete genes	(2x)*4.5S rRNA*, (2x)*5S rRNA*, *16S rRNA*, (2x)*psaA*, (2x)*ycf3*[3]
Conserved Hypothetical genes	*orf101*, *orf122*, *orf126*, (2x)*orf128*, *orf129*, (2x)*orf190*
Transfer RNA	(2x)*trnC*-GCA, *trnD*-GUC, *trnD*-GUC-cp, (2x)*trnE*-UUC, (2x)*trnF*-GAA, *trnG*-GCC, (2x)*trnH*-GUG-cp, *trnK*-UUU, (2x)*trnL*-UAA[2], (4x)*trnM*-CAU, (2x)*trnM*-CAU-cp, (4x)*trnN*-GUU-cp, *trnP*-UGG, (2x)*trnP*-UGG-*cp*, *trnQ*-UUG, (2x)*trnR*-ACG-cp, *trnS*-GCT, (2x)*trnS*-TGA, *trnV*-GAC-cp, (2x)*trnW*-CCA-*cp*, (2x)*trnY*-GTA
Pseudogenes	*rpl2*, (2x)*rps2*, (2x)*rps14*, *rps19*
cp-derived gene fragment transfer	*16S rRNA*, *23S rRNA*, *atpE*, *ndhF*, (2x)*psaB*, (2x)*psbC*, (2x)*rpoA*, *rps12*_3end, (2x)*ycf1*, (2x)*ycf15*, (2x)*ycf2*, *ycf68*

utilising a whole genome shotgun sequencing approach, became available in 2013 (Rahman et al. 2013). The large number of repeat sequences caused a heavily fragmented genome assembly, despite the genome assembly using data from three different sequencing platforms (454, Illumina and SOLiD), and multiple library types (single end, and 200 bp, 500 bp, 8 kb and 20 kb paired-end reads). Prior to the scaffolding step, the library consisted of 1,223,366 contigs ranging from 200 to 46,694 bp. Scaffolding reduced the total assembly to 1,150,326 scaffolds and contigs ranging from 200 to 531,465 bp. Linkage map information (Le Guen et al. 2011) was incorporated into the rubber tree genome allowing for 143 scaffolds to be anchored (Rahman et al. 2013). Additional attempts to use linkage maps to place scaffolds made incremental improvements, but for the most part, these only joined a small number of the larger contigs (Shearman et al. 2015; Pootakham et al. 2015).

The BPM 24 nuclear genome was assembled as an ongoing process over several years, combining data from multiple platforms as sequencing technology continued to advance (Pootakham et al. 2017). In total, there were ten separate libraries with insert sizes ranging from 350 bp to 12 kb consisting of: 6 Gb of single end and 1 Gb of 8 to 20 kb mate-pair libraries from 454 GS FLX + runs, 79 Gb of Illumina HiSeq 2000

Table 4.4 Comparison of assembly methods and data types for the whole genome assembly of BPM 24 (Pootakham et al. 2017)

Assembly statistics	454 + Illumina	454 + Illumina + PacBio	454 + Illumina + PacBio + HiC
Number of contigs/scaffolds	989,097	658,583	592,580
Total length of assembly (bp)	868,915,424	1,249,467,938	1,256,269,486
Longest contig/scaffold (bp)	76,143	1,224,126	2,026,921
Contig/scaffold N50 (bp)	1,316	51,412	96,825
Contig/scaffold L50	161,440	4,695	2,520
% N in scaffolds	0.002	0.00321	0.05463
Number of gaps	13,567	13,066	81,022
Non-gap bases	868,901,484	1,249,454,514	1,249,459,262

paired-end reads, and 56 Gb of 10 to 12 kb insert PacBio reads. In addition to this, scaffolding was performed by Dovetail Genomics using Capture Hi-C Analysis of Genome Organization (Chicago) technology to construct long-range linkage libraries based on proximity ligation of in vitro reconstituted chromatin (Putnam et al. 2016). A total of six Chicago libraries were constructed and sequenced on an Illumina HiSeq system which produced 769 million paired-end reads of DNA fragments between 1 and 50 kb, yielding approximately 44 fold coverage. As a result of this approach, multiple assemblies of the genome were produced allowing for comparison of the different assembly methods and data types (Table 4.4). The first assembly used SOAPdenovo to assemble the Illumina reads into contigs, followed by assembly with the 454 data using Newbler which resulted in 989,097 contigs totalling 869 Mb with an N50 of 1,316 bp. Adding the long read PacBio data to this for scaffolding using DBG2OLC software reduced the number of contigs to 658,583 and increased the total sequence to 1,249 Mb with an N50 of 51,412 bp. This assembly was then used as input for Chicago Hi-C, which reduced the number of contigs to 592,580 and increased the total sequence to 1,256 Mb and the N50 to 96,825 bp. The long reads of the PacBio platform allowed many repeat sequences to be traversed in a single read which greatly improved the assembly quality compared to short read sequences. The addition of Chicago Hi-C technology roughly doubled the N50 size of the assembly, thus representing a significant increase in assembly quality.

Adding linkage map information to the assembly allowed for 1,568 scaffolds totalling 363 Mb to be anchored onto 18 linkage groups. These anchored scaffolds included 68.4% of the predicted protein-coding genes and thus represents a large portion of the unique sequences in the rubber tree genome. The markers used to anchor these scaffolds in a genetic map were from genotype-by-sequencing which utilized methylation-sensitive enzymes, giving the resulting SNPs a high chance to be in euchromatic regions. The sequences which remain unanchored contain roughly 80% repetitive sequence, making it difficult to anchor.

Repeat sequences account for 69.2% of the assembled genome and are composed of 3.5 Mb (0.28%) of low complexity elements, 1.3 Mb (0.1%) of small RNA structures, 13 Mb (1.11%) of simple sequence repeats, 21 Mb (1.72%) of DNA transposons, 160 Mb (12.73%) of unclassified elements and nearly 670 Mb (53.31%) of retrotransposons (Table 4.5). The majority of retrotransposons are long terminal repeats (LTR; 51.55%), and include Copia-like (10.23%) and Gypsy-like (37.75%) elements, which are the most abundant LTR subfamilies in other sequenced *Euphorbia* genomes (Chan et al. 2010; Sato et al. 2011; Wang et al. 2014).

The protein-coding sequences were annotated using multiple ab initio gene prediction tools,

Table 4.5 Repeat sequences in the BPM 24 rubber tree genome (Pootakham et al. 2017)

Repeat element	Count	Total length (bp)	Total length (%)
DNA transposons	58,210	21,004,842	1.67
LINE	43,421	21,703,411	1.73
SINE	1,402	404,943	0.03
LTR: Gypsy	792,203	474,271,360	37.75
LTR: Copia	194,673	128,587,832	10.24
LTR: others	99,248	44,783,869	3.56
Simple repeats	289,782	13,970,828	1.11
Low complexity	62,563	3,522,906	0.28
Unclassified	578,157	159,991,742	12.74

homology searches against sequence databases and the transcript alignment program, PASA. Sequencing data included PacBio Iso-seq libraries (46,129 non-redundant transcripts) and published RNA-seq data obtained from leaf, bark, latex and root (Rahman et al. 2013; Chao et al. 2015; Li et al. 2016; Tang et al. 2016). The RNA-seq data was incorporated with outputs from gene prediction programs and homology searches using the program EvidenceModeler, which produced a consensus set of 43,868 protein-coding genes. In the rubber tree, the average gene length (2,747 bp), average exon length (223 bp), and average number of exons per gene (4) were comparable to those in cassava (Wang et al. 2014) and castor bean (Chan et al. 2010).

Of the 43,868 predicted genes, 30,232 (68.92%) were assigned InterPro motifs, 20,107 received GO annotations, and 33,718 (76.9%) and 16,794 (38.3%) had significant BLASTP matches to proteins in the Genbank non-redundant protein and the SwissProt databases, respectively. The long PacBio reads facilitated the identification of alternative splicing in several genes, consistent with Maize 58 and Sorghum 59. In addition to protein-coding genes, there were 623 tRNA, 274 (rRNA), 282 small nucleolar RNA (snoRNA), 164 small nuclear RNA (snRNA) and 193 micro RNA (miRNA) genes in the BPM 24 assembly. A large number of resistance-related genes, 1,275 putative pattern-recognition receptor genes, were identified in the rubber tree genome. Resistance genes tended to be grouped in clusters indicative of tandem duplication followed by function diversification.

A major evolution mechanism that is common in eudicots that can give rise to new genes is whole genome duplication followed by gene loss and diploidization. The rubber tree appears to have experienced a paleotetraploidization event based on 164 syntenic blocks, of ten or more genes each, containing 2,951 paralogous gene pairs distributed across the 18 linkage groups.

4.3 Synteny with Euphorbiaceae

Comparison of the BPM 24 rubber tree genome with other Euphorbiaceae species reaffirms that cassava is the closest ancestor (Pootakham et al. 2017). The evolutionary divergence of the rubber tree from other Euphorbiaceae was analysed by comparing the transversion rate at fourfold degenerate sites (4DTv) of orthologous gene pairs identified in syntenic blocks between the rubber tree and its close relatives. A whole genome duplication event took place after the ancestor of castor bean split from the other lineages and this is reflected in the 4DTv values from the transversion rate comparison. Both the rubber tree and cassava also have 18 chromosomes while physic nut (*Jatropha curcas*) has 11 and castor bean has 10, also supporting a whole genome duplication event before the rubber tree and cassava diverged (Fig. 4.3a). Speciation times were estimated based on orthologues of

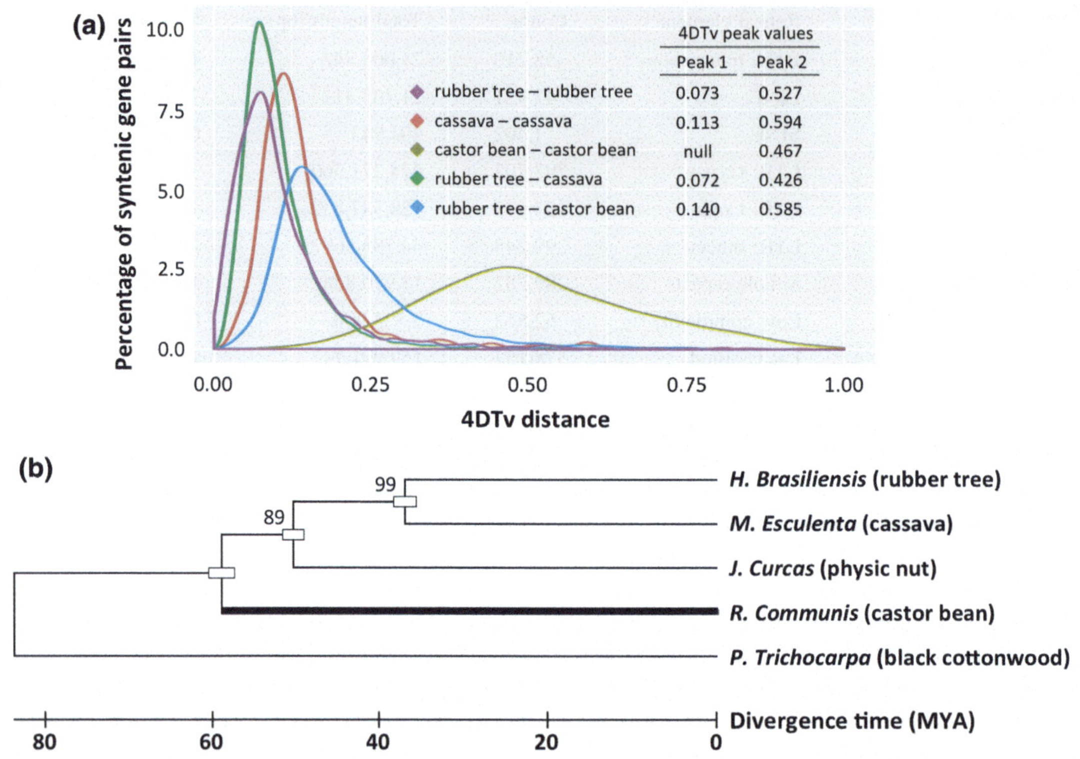

Fig. 4.3 **a** Distribution of 4DTv distances between syntenic gene pairs among rubber tree, cassava and castor bean. **b** A maximum likelihood phylogenetic tree of Euphorbiaceae species, with black cottonwood as an outgroup (Pootakham et al. 2017)

single gene families in each of the species, with black cottonwood as an outgroup (Fig. 4.3b). The origin of the Euphorbiaceae family has been estimated to be at 58 million years ago (mya) (Magallon et al. 1999). The maximum likelihood tree suggested a speciation time of around 36 mya for *Hevea* and *Manihot*, and a divergence time of the *Ricinus* lineage from other *Euphorbia* around 60 mya, this in agreement with the estimates from previous phylogenetic analyses (Bredeson et al. 2016).

Comparison of the gene sets of the rubber tree with other sequenced Euphorbiaceae species (*M. esculenta*, *J. curcas* and *R. communis*) identified 15,605 orthologous gene clusters within the rubber tree. The comparison again confirms that the rubber tree and cassava shared the most recent ancestor, with 13,680 gene clusters in common between the two species. The rubber tree shares 13,040 gene clusters with *J. curcas* and 13,028 with *R. communis* and a total of 11,759 gene clusters were common to all four Euphorbiaceae species considered. The rubber tree also had 934 gene families that were not found in the other species, where these were significantly enriched with genes related to molecular function categories such as catalytic and binding activities.

4.4 Conclusion

As we obtain whole genome sequence data from multiple individuals of each species, it has become increasingly clear that significant structural variation can exist between different individuals of the same species, including genome rearrangement and the presence/absence variations of entire genes (Hu et al. 2018; Sun et al. 2018). Thus, sequencing only a single individual results in a high chance of missing a large portion of the variation that exists within that species and limits the potential of identifying variants for QTLs of interest. Having whole genome assemblies of multiple rubber tree clones is important since marker-assisted selection for QTLs is desirable for rubber tree. We now have whole genome assemblies for BPM 24, RRIM 600 and Reyan 7-33-97, including chloroplast and mitochondrial genomes. These will serve as valuable resources for marker development, QTL mapping, and functional variant discovery. As the cost of long-read sequencing becomes more affordable, it is possible to obtain high-quality whole genome assemblies from additional rubber tree clones to investigate existing structural variations that may be associated with agronomic traits of interest.

References

Allen JO, Fauron CM, Minx P, Roark L, Oddiraju S, Lin GN, Meyer L, Sun H, Kim K, Wang C, Du F, Xu D, Gibson M, Cifrese J, Clifton SW, Newton KJ (2007) Comparisons among two fertile and three male-sterile mitochondrial genomes of maize. Genetics 177:1173–1192

Bonen L (2008) Cis- and trans-splicing of group II introns in plant mitochondria. Mitochondrion 8:26–34

Bredeson JV, Lyons JB, Prochnik SE, Wu GA, Ha CM, Edsinger-Gonzales E, Grimwood J, Schmutz J, Rabbi IY, Egesi C, Nauluvula P, Lebot V, Ndunguru J, Mkamilo G, Bart RS, Setter TL, Gleadow RM, Kulakow P, Ferguson ME, Rounsley S, Rokhsar DS (2016) Sequencing wild and cultivated cassava and related species reveals extensive interspecific hybridization and genetic diversity. Nat Biotechnol 34:562–570

Bullerwell CE, Gray MW (2004) Evolution of the mitochondrial genome: protist connections to animals, fungi and plants. Curr Opin Microbiol 7:528–534

Carlsson J, Leino M, Sohlberg J, Sundström JF, Glimelius K (2008) Mitochondrial regulation of flower development. Mitochondrion 8:74–86

Chan AP, Crabtree J, Zhao Q, Lorenzi H, Orvis J, Puiu D, Melake-Berhan A, Jones KM, Redman J, Chen G, Cahoon EB, Gedil M, Stanke M, Haas BJ, Wortman JR, Fraser-Liggett CM, Ravel J, Rabinowicz PD (2010) Draft genome sequence of the oilseed species *Ricinus communis*. Nat. Biotechnol. 28:951–956

Chang S, Yang T, Du T, Huang Y, Chen J, Yan J, He J, Guan R (2011) Mitochondrial genome sequencing helps show the evolutionary mechanism of mitochondrial genome formation in Brassica. BMC Genom 12:497

Chao J, Chen Y, Wu S, Tian WM (2015) Comparative transcriptome analysis of latex from rubber tree clone CATAS8-79 and PR107. Genom Data 5:120–121

Hu Z, Wang W, Wu Z, Sun C, Li M, Lu J, Fu B, Shi J, Xu J, Ruan J, Wei C, Li Z (2018) Novel sequences, structural variations and gene presence variations of Asian cultivated rice. Sci Data 5:180079

Le Guen V, Garcia D, Doaré F, Mattos CRR, Condina V, Couturier C, Chambon A, Weber C, Espéout S, Seguin M (2011) A rubber tree's durable resistance to *Microcyclus ulei* is conferred by a qualitative gene and a major quantitative resistance factor. Tree Genet Genomes 7:877–889

Li D, Wang X, Deng Z, Liu H, Yang H, He G (2016) Transcriptome analyses reveal molecular mechanism underlying tapping panel dryness of rubber tree (*Hevea brasiliensis*). Sci Rep 6:23540

Lloyd AH, Timmis JN (2011) The origin and characterization of new nuclear genes originating from a cytoplasmic organellar genome. Mol Biol Evol 28:2019–2028

Magallon S, Crane PR, Herendeen PS (1999) Phylogenetic pattern, diversity, and diversification of eudicots. Ann Mo Bot Gard 86:297–372

Palmer JD, Herbon LA (1987) Unicircular structure of the *Brassica hirta* mitochondrial genome. Curr Genet 11:565–570

Pootakham W, Ruang-Areerate P, Jomchai N, Sonthirod C, Sangsrakru D, Yoocha T, Theerawattanasuk K, Nirapathpongporn K, Romruensukharom P, Tragoonrung S, Tangphatsornruang S (2015) Construction of a high-density integrated genetic linkage map of rubber tree (*Hevea brasiliensis*) using genotyping-by-sequencing (GBS). Front Plant Sci 6:367

Pootakham W, Sonthirod C, Naktang C, Ruang-Areerate P, Yoocha T, Sangsrakru D, Theerawattanasuk K, Rattanawong R, Lekawipat N, Tangphatsornruang S (2017) De novo hybrid assembly of the rubber tree genome reveals evidence of paleotetraploidy in *Hevea* species. Sci Rep 7:41457

Priyadarshan PM, Clément-Demange A (2004) Breeding *Hevea* rubber: formal and molecular genetics. Adv Genet 52:51–115

Putnam NH, O'Connell BL, Stites JC, Rice BJ, Blanchette M, Calef R, Troll CJ, Fields A, Hartley PD, Sugnet CW, Haussler D, Rokhsar DS, Green RE (2016) Chromosome-scale shotgun assembly using an in vitro method for long-range linkage. Genome Res 26:342–350

Rahman AY, Usharraj AO, Misra BB, Thottathil GP, Jayasekaran K, Feng Y, Hou S, Ong SY, Ng FL,

Lee LS, Tan HS, Sakaff MK, Teh BS, Khoo BF, Badai SS, Aziz NA, Yuryev A, Knudsen B, Dionne-Laporte A, Mchunu NP, Yu Q, Langston BJ, Freitas TA, Young AG, Chen R, Wang L, Najimudin N, Saito JA, Alam M (2013) Draft genome sequence of the rubber tree *Hevea brasiliensis*. BMC Genom 14:75

Sato S, Hirakawa H, Isobe S, Fukai E, Watanabe A, Kato M, Kawashima K, Minami C, Muraki A, Nakazaki N, Takahashi C, Nakayama S, Kishida Y, Kohara M, Yamada M, Tsuruoka H, Sasamoto S, Tabata S, Aizu T, Toyoda A, Shin-i T, Minakuchi Y, Kohara Y, Fujiyama A, Tsuchimoto S, Kajiyama S, Makigano E, Ohmido N, Shibagaki N, Cartagena JA, Wada N, Kohinata T, Atefeh A, Yuasa S, Matsunaga S, Fukui K (2011) Sequence analysis of the genome of an oil-bearing tree, *Jatropha curcas* L. DNA Res 18:65–76

Shearman JR, Sangsrakru D, Ruang-Areerate P, Sonthirod C, Uthaipaisanwong P, Yoocha T, Poopear S, Theerawattanasuk K, Tragoonrung S, Tangphatsornruang S (2014) Assembly and analysis of a male sterile rubber tree mitochondrial genome reveals DNA rearrangement events and a novel transcript. BMC Plant Biol 14:45

Shearman JR, Sangsrakru D, Jomchai N, Ruang-Areerate P, Sonthirod C, Naktang C, Theerawattanasuk K, Tragoonrung S, Tangphatsornruang S (2015) SNP identification from RNA sequencing and linkage map construction of rubber tree for anchoring the draft genome. PLoS One 10:e0121961

Sloan DB, Alverson AJ, Chuckalovcak JP, Wu M, McCauley DE, Palmer JD, Taylor DR (2012) Rapid evolution of enormous, multichromosomal genomes in flowering plant mitochondria with exceptionally high mutation rates. PLoS Biol 10:e1001241

Stifel SD (1973) The growth of the rubber economy of Southern Thailand. J Southeast Asian Stud 4:107–132

Sun S, Zhou Y, Chen J, Shi J, Zhao H, Zhao H, Song W, Zhang M, Cui Y, Dong X, Liu H, Ma X, Jiao Y, Wang B, Wei X, Stein JC, Glaubitz JC, Lu F, Yu G, Liang C, Fengler K, Li B, Rafalski A, Schnable PS, Ware DH, Buckler ES, Lai J (2018) Extensive intraspecific gene order and gene structural variations between Mo17 and other maize genomes. Nat Genet 50:1289–1295

Tang C, Yang M, Fang Y, Luo Y, Gao S, Xiao X, An Z, Zhou B, Zhang B, Tan X, Yeang HY, Qin Y, Yang J, Lin Q, Mei H, Montoro P, Long X, Qi J, Hua Y, He Z, Sun M, Li W, Zeng X, Cheng H, Liu Y, Yang J, Tian W, Zhuang N, Zeng R, Li D, He P, Li Z, Zou Z, Li S, Li C, Wang J, Wei D, Lai CQ, Luo W, Yu J, Hu S, Huang H (2016) The rubber tree genome reveals new insights into rubber production and species adaptation. Nat Plants 2:16073

Tangphatsornruang S, Uthaipaisanwong P, Sangsrakru D, Chanprasert J, Yoocha T, Jomchai N, Tragoonrung S (2011) Characterization of the complete chloroplast genome of *Hevea brasiliensis* reveals genome rearrangement, RNA editing sites and phylogenetic relationships. Gene 475:104–112

Turmel M, Otis C, Lemieux C (2003) The mitochondrial genome of *Chara vulgaris*: insights into the mitochondrial DNA architecture of the last common ancestor of green algae and land plants. Plant Cell 15:1888–1903

Wang D, Wu YW, Shih AC, Wu CS, Wang YN, Chaw SM (2007) Transfer of chloroplast genomic DNA to mitochondrial genome occurred at least 300 MYA. Mol Biol Evol 24:2040–2048

Wang W, Feng B, Xiao J, Xia Z, Zhou X, Li P, Zhang W, Wang Y, Møller BL, Zhang P, Luo M-C, Xiao G, Liu J, Yang J, Chen S, Rabinowicz PD, Chen X, Zhang H-B, Ceballos H, Lou Q, Zou M, Carvalho LJCB, Zeng C, Xia J, Sun S, Fu Y, Wang H, Lu C, Ruan M, Zhou S, Wu Z, Liu H, Kannangara RM, Jørgensen K, Neale RL, Bonde M, Heinz N, Zhu W, Wang S, Zhang Y, Pan K, Wen M, Ma P-A, Li Z, Hu M, Liao W, Hu W, Zhang S, Pei J, Guo A, Guo J, Zhang J, Zhang Z, Ye J, Ou W, Ma Y, Liu X, Tallon LJ, Galens K, Ott S, Huang J, Xue J, An F, Yao Q, Lu X, Fregene M, López-Lavalle LAB, Wu J, You FM, Che, M, Hu S, Wu G, Zhong S, Ling P, Chen Y, Wang Q, Liu G, Liu B, Li K, Peng M (2014) Cassava genome from a wild ancestor to cultivated varieties. Nat Commun 5:5110

Wicke S, Schneeweiss GM, dePamphilis CW, Müller KF, Quandt D (2011) The evolution of the plastid chromosome in land plants: gene content, gene order, gene function. Plant Mol Biol 76(3–5):273–297

Zimorski V, Ku C, Martin WF, Gould SB (2014) Endosymbiotic theory for organelle origins. Curr Opin Microbiol 22:38–48

Development of Molecular Markers in *Hevea brasiliensis* for Marker-Assisted Breeding

5

Wirulda Pootakham, Jeremy R. Shearman, and Sithichoke Tangphatsornruang

Abstract

The para rubber tree (*Hevea brasiliensis*) is a major producer of high-quality natural rubber, accounting for more than 98% of the total production worldwide. Conventional breeding through recurrent selection has been used to obtain high-yielding clones in the past several decades. Although traditional breeding based on phenotypic selection has been effective, it is very laborious and time-consuming. The use of molecular markers in the rubber tree began in the 1980s, with isozymes being the first set of genetic markers developed followed by restriction fragment length polymorphism (RFLP), random amplified polymorphic DNA (RAPD) and the more commonly adopted simple sequence repeat (SSR) markers. At the turn of the millennium, the first genetic linkage map for the rubber tree was constructed based on anonymous markers such as RFLP and AFLP. In the past decade, several linkage maps have been reported, most of which were based on SSR or microsatellite markers. More recently, researchers have shifted from SSR markers to direct analyses of sequence variations such as single nucleotide polymorphisms (SNPs). Even though SNPs are generally less polymorphic than multi-allelic SSRs, their ubiquity in eukaryotic genomes makes them valuable for clone identification and marker-trait association analyses. Rapid advances in next-generation sequencing technologies have facilitated initial efforts in large-scale SNP discovery for genetic map construction in the rubber tree, leading to the first reports of SNP-based linkage maps in 2015. Since then, several high-density maps have been generated in order to identify markers and QTLs associated with traits of interest such as growth and latex yield. As the cost of SNP genotyping becomes more affordable, breeders can take full advantage of marker-assisted selection to expedite the development of elite rubber tree clones with multiple desirable agronomic traits.

5.1 Introduction

Commercial breeding of *Hevea brasiliensis*, the para rubber tree, began in the early nineteenth century through mass selection, shortly after the introduction of rubber tree seeds to Southeast Asia (Dijkman 1951). Since then, considerable

W. Pootakham · J. R. Shearman · S. Tangphatsornruang (✉)
National Omics Center, National Center for Genetic Engineering and Biotechnology (BIOTEC), National Science and Technology Development Agency (NSTDA), Science Park, Thailand, Phahonyothin Road, Khlong Luang, Khlong Nueng, Pathumthani 12120, Thailand
e-mail: sithichoke.tan@biotec.or.th

M. Matsui and K.-S. Chow (eds.), *The Rubber Tree Genome*, Compendium of Plant Genomes,
https://doi.org/10.1007/978-3-030-42258-5_5

improvements have been achieved in both yield and quality characters. Breeding efforts have focused on obtaining clones with multiple desirable traits, with improvement in latex yield and growth being a primary goal for most breeders (Priyadarshan 2003). During the 1920s, commercial planting of unselected seedlings obtained from the Upper Amazon (through Kew Gardens) in Southeast Asia yielded 496 kg/ha of latex per year (Priyadarshan 2017). With recurrent breeding and selection of primary clones, the latex production of the best-yielding clones gradually improved to 1,600 kg/ha/year during the 1950s. (Baulkwill and Webster 1989; Priyadarshan 2017). A few decades later, the yield potential was further increased to 2,500 kg/ha/year in PB, RRIM, RRII, RRIC, IRCA and RRIV clones (Priyadarshan 2017). During this period of 70 years (1920–1990), latex yield in Southeast Asian plantations had been enhanced roughly five times (from ~500 to 2,500 kg/ha/year) through intensive recurrent selection, which led to the development of improved clones such as BPM 24, PB 217, PB 235, PB 260, RRIC 100, RRII 105, RRIM 501, RRIM 600 and RRIM 712 (Tan 1987; Priyadarshan 2003; Priyadarshan et al. 2009). Besides latex yield, agronomic traits involving resistance to abiotic stresses such as wind damage, low temperature, drought and tapping panel dryness have also been of interest. In Latin America, rubber tree breeding effort is focused exclusively on obtaining clones with acceptable yield and resistance to South American Leaf Blight (SALB) (Dean 1987). Although traditional breeding continues to play an important role in enhancing yield and other economically desirable qualities, it is inherently impeded by the long selection cycle, and enormous resources required for evaluating and maintaining rubber trees in breeding programs. A typical rubber breeding cycle lasts approximately 35–40 years (Priyadarshan 2017). The ability to select for commercially valuable traits at an early stage will have a tremendous impact on reducing the time and resources needed to develop superior *H. brasiliensis* clones.

5.2 Development of Molecular Markers in the Rubber Tree

The advent of DNA-based genetic markers allowed plant breeders to move from phenotype-based towards genotype-based selection of agronomical traits, accelerating the breeding programs. Marker-assisted selection employs DNA markers that are tightly linked to the target loci (associated with traits of interest) to assist in phenotypic screenings (Lande and Thompson 1990). In recent years, marker-assisted selection has played a pivotal role in agricultural breeding programs as it has several advantages over traditional breeding approaches. Genotypic screening is often simpler, faster and much less laborious than phenotypic screening. In addition, marker-based selection can be performed early at the seedling or juvenile stage. The ability to identify clones that are likely to exhibit desirable traits prior to field trials would considerably reduce the duration and size of breeding programs. This is especially true for rubber trees as they have a relatively long immature period (up to 5 years). Most importantly, many agriculturally valuable traits such as yield and growth are frequently controlled by more than one gene. The identification of quantitative trait loci (QTLs) based solely on conventional phenotypic evaluation is not possible, and the use of molecular markers creates opportunities for breeders to select for progenies with important quantitative traits.

5.2.1 Isozyme, RFLP, AFLP and RAPD Markers

The dawn of molecular marker development in the rubber tree began in the early 1980s with the first set of genetic markers developed by CIRAD, which consisted of 14 polymorphic biochemical markers (Seguin et al. 2001). These isozyme markers were used primarily for clone identification, parentage identification and genetic diversity analysis (Chevallier 1988; Leconte et al. 1994). Because of the instability of isozyme molecules at room temperature, the analyses had to be performed at field sites, rendering them inconvenient to use. The small number of markers available and the low degree of polymorphism at these loci also limited their usage. The utilization of RFLP (Restriction Fragment Length Polymorphism) and RAPD (Random Amplified Polymorphic DNA) markers for clonal identification (Besse et al. 1993) and genetic diversity assessment (Besse et al. 1994; Seguin et al. 1995; Varghese et al. 1997) in *H. brasiliensis* began in the 1990s. A decade later, Lespinasse et al. (2000) constructed the first genetic linkage map for the rubber tree based on 106 F_1 progeny from a cross between clones PB 260 and RO 38 primarily using RFLP and AFLP (Amplified Fragment Length Polymorphism) markers. The linkage map consisted of 18 linkage groups, corresponding to the haploid chromosome number. The map contained 10 isozymes, 301 RFLPs, 388 AFLPs and 18 microsatellite markers, and spanned a total distance of 2,144 centiMorgan (cM) with an average marker density of one marker per 3 cM (Table 5.1) (Lespinasse et al. 2000).

RFLPs, AFLPs and RAPDs are usually among the first types of markers developed because they are relatively easy to generate, and they do not require prior sequence information. However, they are generally not highly polymorphic, and are often inherited in a dominant manner. In addition, screening a large number of hybridization-based RFLP markers can be laborious and time-consuming. RAPD analysis offers a cost- and time-effective alternative to RFLP;

Table 5.1 Details of the published genetic linkage maps of *Hevea brasiliensis*

References	Map length (cM)	LG number	Markers							Average marker distance (cM)
			Isozyme	RFLP	AFLP	RAPD	SSR	SNP	Total	
Lespinasse et al. (2000)	2,144	18	10	301	388	–	18	–	717	3
Feng et al. (2010)	1,937.06	18	–	–	–	–	91	–	91	21.29
Souza et al. (2011)	2,471.20	23	–	–	–	–	225	–	225	11
Le Guen et al. (2011)	2,441	18	–	–	187	–	195	–	383	6.37
Triwitayakorn et al. (2011)	842.9	23	–	–	–	–	97	–	97	11.9
Souza et al. (2013)	2,688.80	23	–	–	–	–	284	–	284	10
Pootakham et al. (2015a, b)	2,052	18	–	–	–	–	–	2,321	2,321	0.89
Shearman et al. (2015)	4,160	18	–	–	–	–	–	2,186	2,186	1.9
Conson et al. (2018)	3,779.70	18	–	–	–	–	224	855	1,079	3.5
Xia et al. (2018)	2,094.10	18	–	–	–	–	–	6,940	6,940	0.3
An et al. (2019)	2,670.27	18	–	–	–	–	–	4,543	4,543	0.59

however, the low degree of reproducibility of RAPD markers due to non-specific amplification of DNA fragments limits the transferability of these markers from one laboratory to another (Penner et al. 1993).

5.2.2 Development of Microsatellite or Simple Sequence Repeat Markers

The next class of markers generated for rubber tree is the simple sequence repeats (SSRs) or microsatellites. The transition from RFLP and other anonymous markers to site-specific markers (e.g. SSRs) was first introduced in barley (Tragoonrung et al. 1992). Shortly after their introduction, SSRs emerged as the markers of choice in many areas of molecular genetics since they exhibit a remarkable degree of allelic diversity, even among closely related lines. Microsatellites typically behave as co-dominant markers, making them particularly useful for population genetic studies and mapping. They are also relatively abundant and randomly distributed in the genome. The development of microsatellite markers in *H. brasiliensis* were initially achieved by screening small-insert genomic libraries for the presence of AC/GT or CT/GA tandem repeats (Lekawipat et al. 2003; Bindu-Roy et al. 2004; Le Guen et al. 2009; Souza et al. 2009). These microsatellites were subsequently employed in population genetic studies and the evaluation of genetic diversity in *Hevea* taxa (Bindu-Roy et al. 2004; Feng et al. 2009; Souza et al. 2009; Le Guen et al. 2010; de Souza et al. 2015).

The development of SSR markers by Feng et al. (2009) was facilitated by an in silico search of available expressed sequence tags (ESTs) in the public databases. The authors analyzed a total of 10,018 ESTs and discovered 799 SSR loci located in 643 non-redundant ESTs (Feng et al. 2009). Of the 799 SSRs identified, 45.3% were mononucleotide repeats, 42.2% were dinucleotide repeats, 11.9% were trinucleotide repeats, and the remaining 0.6% were tetranucleotide and hexanucleotide repeats, with AG and AAG being the most abundant motifs (Feng et al. 2009). A total of 184 EST-SSRs were used for primer design, and 87 primer pairs exhibited polymorphisms when tested with 12 cultivated rubber tree accessions (Feng et al. 2009). The largest set of polymorphic SSR markers was screened from a genomic library enriched for CA- and GA-containing microsatellites by a research group at CIRAD (Le Guen et al. 2010). A total of 396 sequences were tested for polymorphisms in ten rubber tree genotypes, and 296 new polymorphic microsatellite markers were characterized and reported (Le Guen et al. 2010). Cubry et al. (2014) utilized disease resistance and defense-related ESTs retrieved from public databases, and sequences obtained from suppression subtractive hybridization libraries constructed specifically for disease resistance studies to develop a new set of EST-SSRs from disease-related genes (Cubry et al. 2014). The authors identified 673 microsatellite motifs from 10,499 ESTs. Two hundred and sixty-four SSR loci were tested, and 164 polymorphic loci were used to evaluate the degree of genetic diversity in 19 rubber tree accessions (Cubry et al. 2014).

With rapid advancement in sequencing throughput together with an overall decrease in sequencing cost, next-generation sequencing technologies enabled efficient identification of a large number of SSR markers at a fraction of the cost of the conventional approach. Next-generation sequencing has been applied to microsatellite identification in various plant species including sweet potato (Schafleitner et al. 2010; Wang et al. 2010), soybean (Li et al. 2010), mungbean (Tangphatsornruang et al. 2009), cucumber (Cavagnaro et al. 2010) and pine (Parchman et al. 2010). The first attempt to utilize next-generation sequencing data for the identification of SSR markers in *H. brasiliensis* was reported by Triwitayakorn et al. (2011). Pyrosequencing was used to obtain a transcriptome assembly consisting of 113,313 contigs, from which 17,819 EST-SSRs were identified. The average frequency of EST-SSRs was one in every 3.38 kb, which was lower than the number reported (one SSR in every 2.25 kb) by Feng et al. (2009). Similar to the previous discovery, the most common dinucleotide and trinucleotide repeats were AG/CT and AAG/CTT, respectively (Feng et al. 2009). From a set of 323 EST-SSR primer pairs amplifiable in *H. brasiliensis*, 59 primer pairs were polymorphic between RRIM 600 and RRII 105 parental clones. These, together with 98 previously published SSR markers, were used to construct a genetic linkage map using 81 F_1 individuals. The map consisted of 97 SSR loci distributed over 23 linkage groups, with an average inter-marker distance of 11.9 cM (Table 5.1) (Triwitayakorn et al. 2011). The number of linkage groups exceeded the haploid chromosome number, suggesting that more markers were needed to fill the gap between adjacent markers. Another SSR-based linkage map was constructed using 270 individuals derived from a cross between PB 217 and PR 255 (Souza et al. 2013). This genetic map spanned the length of 2,688 cM and contained 284 markers distributed over 23 linkage groups, with an average inter-marker distance of 9.4 cM (Table 5.1) (Souza et al. 2013). Because the number of SSR markers placed on these maps was fairly small, the average distances between adjacent markers were quite large, rendering them less practical for downstream association analyses.

5.3 Single Nucleotide Polymorphism (SNP) Discovery in Rubber Tree

To construct high-density linkage maps, researchers have shifted from SSR markers to direct analyses of sequence variations such as SNPs (Pootakham et al. 2013; Zhang et al. 2013; Huang et al. 2014; Pootakham et al. 2014). The ubiquity of SNPs in eukaryotic genomes and their usefulness as genetic markers has been well established over the last decade. Rapid advances in genotyping technologies make SNP markers ideal for high-throughput applications in plant genetics and breeding. Recent studies of sequence diversity have shown that SNP frequencies in plants are one in every 100–300 bases (Edwards et al. 2007). Even though SNPs are generally less polymorphic than multi-allelic SSRs, their abundance in the genome makes them valuable for population-based genetic analyses, especially for the construction of high-density linkage maps (Rafalski 2002). The development of next-generation sequencing technologies has facilitated rapid and inexpensive analysis of genome sequences that have not been extensively characterized. Several research groups have successfully applied next-generation sequencing for SNP discovery in non-model organisms (Novaes et al. 2008; Leonforte et al. 2013; Pootakham et al. 2015a). Traditionally, SNP markers have been discovered through the process of EST sequence mining, which occasionally did not provide sufficient coverage to reliably identify the polymorphisms (Labate and Baldo 2005).

Advances in high-throughput, next-generation sequencing technologies have enabled initial efforts in SNP discovery in the rubber tree. Pootakham et al. (2011) reported the first attempt to identify SNPs in *Hevea* species using 454-derived EST sequences. A set of 21,870 contigs (spanning 9.2 Mb) assembled from quality-filtered reads from RRIM 600 was used as a reference sequence to which individual reads from clones RRIM 600 and RRII 105 were aligned (Pootakham et al. 2011). The authors identified 5,883 bi-allelic SNPs, most of which (56.4%) represented transition events. A total of 27 putative SNPs were evaluated using the conventional Sanger sequencing approach, and 10 positions were shown to be polymorphic among 28 accessions used (Pootakham et al. 2011). The accuracy of SNP prediction/identification based on EST sequence data was not very high, and that was mainly due to the low depth coverage at the variant positions. Even with an overall decrease in sequencing cost, it remains highly expensive to employ whole-genome sequencing or high-coverage EST sequencing to identify SNPs in organisms with large genomes such as the rubber tree. Recently, next-generation sequencing has been coupled with reduced representation methods to identify SNPs, not only because of the cost effectiveness but also because many research questions can be answered with a set of markers without requiring every base of a genome to be sequenced. Elshire et al. (2011) proposed a method for constructing and sequencing multiplexed reduced representation libraries called genotyping-by-sequencing (GBS). This method utilizes restriction enzymes to digest the genome into short fragments, with a high degree of consistency between samples, that are subsequently sequenced on high-throughput platforms (Elshire et al. 2011). GBS has successfully been used to generate high-density genetic maps in several non-model species with limited genomic resources, including raspberry (Ward et al. 2013), oat (Huang et al. 2014) and apple (Gardner et al. 2014).

Pootakham et al. (2015a, b) applied the GBS approach to construct the first high-density, SNP-based linkage map in *H. brasiliensis* from two full-sib populations. GBS-based genetic maps of populations P (BPM 24 × RRIM 600) and C (BPM 24 × RRIC110) consisted of 1,704 and 1,719 SNP markers encompassing 2,041 and 1,874 cM, respectively (Pootakham et al. 2015b). SNP markers shared between the two component maps (1,114 SNPs) were used to merge them, yielding an integrated linkage map containing 2,321 markers with a cumulative length of 2,052 cM (Table 5.1) (Pootakham et al. 2015b). Substantial improvement in marker density was observed in the composite map (one SNP in every 0.89 cM) compared to those of the component maps (one SNP in every 1.23–1.25 cM). This integrated map was used to anchor 28,965 contigs, covering 135 Mb of the published rubber tree genome (Rahman et al. 2013; Pootakham et al. 2015b). de Souza et al. (2016) reported the development of SNP markers from genomic sequences obtained from ~75.9 million short reads from GT 1 and RRIM 701 genotypes (de Souza et al. 2016). A total of 3,779 putative SNPs were identified, and 143 loci were selected for validation using eight *H. brasiliensis* accessions. Thirty SNP markers were shown to be polymorphic while the remaining were monomorphic or failed to amplify (de Souza et al. 2016). Those monomorphic markers may have been incorrectly identified because of sequencing errors despite the fact that the average depth coverage (35×) for mapped reads was relatively high (de Souza et al. 2016). More recently, SNP markers identified through GBS were used for the analysis of genetic structure and the estimation of linkage disequilibrium in germplasm accessions (de Souza et al. 2018).

RNA sequencing is another approach often employed to reduce genome complexity for SNP discovery in organisms with large genome sizes. RNA sequencing enables the identification of SNPs located almost exclusively in coding and untranslated regions. Mantello et al. (2014) assembled RNA sequencing data generated from *H. brasiliensis* bark tissue and identified 17,927 SSRs along with 404,114 SNPs from 50,384 contigs. The authors validated 78 SNPs present in sequences related to the rubber biosynthesis pathway in 36 rubber tree genotypes (Mantello et al. 2014). A similar approach was also adopted by Salgado et al. (2014). The authors obtained RNA sequencing data from 33 organs and open-pollinated seedlings of the RRIM 600 clone and assembled them into 17,166 contigs. Subsequently, 2,191 SNPs and 1,397 SSRs were identified. A total of 191 SNPs were validated in 23 *Hevea* genotypes, and 172 loci exhibited polymorphisms (Salgado et al. 2014).

Shearman et al. (2015) performed large-scale SNP discovery using short-read RNA sequencing data from clones RRIM 600, BPM 24, RRIT 110, PB 235, RRIT 251 and RRII 105 (Shearman et al. 2015). The authors identified a total of 368,594 variants, of which 340,631 were SNPs, and 8,293 and 17,272 were deletions and insertions, respectively (Shearman et al. 2015). The ratio of transition:transversion events was similar to previously reported figures (Pootakham et al. 2011, 2015b). RNA sequencing-based SNP discovery revealed that as much as 91% of the variants identified fell within 1 kb of the predicted gene regions—these genic markers are very useful for the identification of putative causative mutations associated with desirable agronomic traits. A total of 2,186 non-redundant SNPs were used to construct a genetic map using 149 F_1 progenies from a cross between RRIM

600 and RRII 105 (Shearman et al. 2015). The total map length was 4,160 cM encompassing 18 linkage groups, with a median distance between markers of 1.36–2.72 cM (Table 5.1). Even though the length of this linkage map appeared to be higher than some of the previously reported maps, it was similar to the estimated length of a PB 260 map reported by Le Guen et al. (2011), which was calculated using the Hulbert et al. method (Le Guen et al. 2011).

More recently, the amplified-fragment SNP and methylation (AFSM) approach was applied to identify a large number of SNPs in the rubber tree for the construction of an ultra-high density linkage map (Xia et al. 2018). Xia et al. (2018) discovered 571,267 SNPs and 134,184 indels in an F_1 population consisting of 187 progenies from a cross between Yunyan 277-5 (fast-growing, high latex yield) and IAN 873 (fast-growing and medium latex yield). Of those markers, 20,066 SNPs and 2,672 indels were used for map construction. The final genetic map comprised 6,940 SNP markers distributed over 18 linkage groups, covering a map distance of 2,094.1 cM (Table 5.1) (Xia et al. 2018). This is currently the densest genetic map available for *Hevea* species with an impressive average marker interval of 0.3 cM, which was significantly smaller than the inter-marker distances reported in previously published maps (Lespinasse et al. 2000; Souza et al. 2013; Pootakham et al. 2015b; Shearman et al. 2015; Conson et al. 2018). While the first two SNP-based linkage maps reported by Pootakham et al. (2015a, b) and Shearman et al. (2015) were able to anchor 12% (135 Mb) and 10% (115 Mb) of the published genome sequence, respectively, the ultra high-density map was able to anchor over 70% (962 Mb) of the sequenced genome (Pootakham et al. 2015b; Shearman et al. 2015; Xia et al. 2018). Molecular markers are essential for the construction of a linkage map, the first crucial step towards marker-assisted selection. Genetic maps generated from full-sib populations are valuable for the identification of genes/QTLs regulating agronomic traits of interest such as growth rate, latex yield and disease tolerance. The following section discusses QTL mapping and population-based genome-wide association studies (GWAS) in the rubber tree.

5.4 Marker-Trait Association Analyses: QTL Mapping, GWAS and Genomic Selection

Pedigree-based QTL mapping and association mapping, also known as linkage disequilibrium (LD) mapping, have been applied to identify genomic regions containing major genes and QTLs regulating agronomic traits in *Hevea.* LD is the non-random association of alleles at different loci (Weir 1979). Both linkage and association mapping approaches are based on the LD between molecular markers and functional loci. While linkage mapping uses LD generated by the mating design, association mapping takes advantage of both LD and historical recombination present within the gene pool of an organism, thereby utilizing a broader population to overcome the limitations of family-based QTL mapping (Breseghello and Sorrells 2006; Ersoz et al. 2007; Myles et al. 2009). GWAS offers high-resolution power where there is large genotypic diversity and low LD for the germplasm investigated. It is also less time-consuming since mapping populations need not be generated (Saba Rahim et al. 2018).

A few of the most important traits for rubber tree farming are tree growth rate, girth and latex yield, which take several years to collect phenotypic data. Therefore, development of markers associated with these traits has been the focus of several research groups. Prior to a draft genome being available, there were limited studies on rubber tree trait loci. Early research into rubber tree QTLs was based on SSR markers that were also used to generate a linkage map. One such study used 603 SSR markers to generate a linkage map containing 23 linkage groups and performed QTL analysis on young plant height and girth measured over the course of 2 years (Souza et al. 2013). The authors identified seven QTLs for height measured during summer which explained approximately 30% of the variation; and two different QTLs for height measured during winter, each explaining approximately 6% of the variation. Girth was also affected by different QTLs based on whether the measurements were taken in winter or summer; where one QTL was identified from summer and three from winter measurements (Souza et al. 2013). Without a physical map, it was difficult to then proceed from such findings to identifying the underlying genetic implication of each QTL. One way around this problem is to use markers developed from transcript variants, as performed by Chanroj et al. (2017) in a genome-wide association study using 170 rubber tree accessions and 14,155 SNP markers identified from transcriptome sequence data (Shearman et al. 2015). The authors investigated latex yield and tree girth phenotypes in dry and wet seasons in North East Thailand, an effort which reflected the commitment involved in studying a species with such a long generation time. GWAS performed on these phenotypes using a MLM + Q + K model (mixed linear model including cluster membership probability Q and kinship K) identified a single QTL for each trait with girth having the same QTL for both dry and wet seasons (Table 5.2). At the time this study was carried out, the only reference genome available consisted of 1.1 million contigs. This made it difficult to proceed unless the GWAS markers are the underlying causative variants, though this is unlikely considering the relatively low number of SNPs investigated. Since only a small number of contigs were anchored into linkage groups, there was no readily accessible way to identify the genes surrounding the QTL peaks.

Two studies that began before reference genomes became available used both SSR and SNP markers to generate a linkage map and perform QTL mapping for tree height and circumference traits. Conson et al. (2018) used 225 SSR and 186 SNPs in an F1 population of 146 individuals from a cross between GT 1 and RRIM 701 to identify 38 QTLs (Conson et al. 2018). The authors first generated a genetic map using these markers followed by subsequent composite interval mapping at 1-cM intervals with significance thresholds calculated from 1,000 permutations. The phenotypes studied were tree height, number of whorls and stem diameter at 50 cm above ground level measured in plants aged 12, 17, 22, 28, 35, 47, and 59 months between 2013 and 2017. They identified 24 QTLs for stem diameter, seven for tree height, and seven for whorl number. Rosa et al. (2018) used 354 SSR and 151 SNP markers in 251 offsprings from a cross between PR 255 and PB 217, and phenotyped height and circumference over the course of 6 years, taking measurements in summer and winter (Rosa et al. 2018). They generated 23 linkage groups and performed composite interval mapping for height, circumference, and total latex production with significance thresholds calculated by permutation to identify 19 QTLs. In both cases, the authors did not incorporate reference genome information into the mapping results to identify candidate genes.

One study incorporated reference genome information from the start of the genotyping portion of the study. Xia et al. (2018) used 6940 SNPs and indels (insertions/deletions) that they identified and genotyped using a form of genotyping-by-sequencing to perform QTL analysis for dry latex yield, subsequently identifying 17 QTLs (Xia et al. 2018). The mapping population consisted of 187 F1 progenies from a cross between a female Yunyan 277-5 and a male IAN 873. Dry latex yield was measured by performing ten tappings per month in the months of

Table 5.2 QTLs for yield and girth identified in a GWAS of 170 accessions (Chanroj et al. 2017)

Trait	Contig accession	Best blast match
Yield, dry season	AJJZ010195800.1	Chlorophyll a–b binding protein 13, Chloroplastic
Yield, wet season	AJJZ010265910.1	Splicing factor U2af small subunit B-like
Girth	AJJZ011022278.1	Putative sodium-coupled neutral amino acid transporter 7

May and October 2013 and multiple QTL model mapping was performed. The markers were identified using a genotype-by-sequencing approach where the sequence reads were mapped against the Reyan 7-33-97 reference genome (Tang et al. 2016). This reference genome consists of 7,453 scaffolds and Xia et al. (2018) were able to anchor approximately 70% of the total sequence, allowing them to also investigate genes that were within the QTL regions with a fairly low chance of missing any gene. Two major QTL regions were identified (LVXX01000148.1__1091479 and LVXX010 00430.1__338729) that explained 38.3% and 33.3% of the phenotypic variance, respectively. The authors identified a list of candidate genes from within the QTL regions, but did not sequence the parental clones which would enable causative variant discovery. Recently, An et al. (2019) employed the specific-locus amplified fragment sequencing (SLAF-seq) method to genotype 206 F_1 progenies derived from a cross between CATAS 8-79 and MT/C/11 9/67 and constructed a high-density genetic map consisting of 4,543 SNPs with an average inter-marker distance of 0.59 cM (An et al. 2019). This was the first study to perform conditional QTL mapping in the rubber tree that explored QTLs associated with different growth stages. The authors identified a total of 11 and 12 QTLs regulating stem growth and latex yield, respectively. These QTLs explained 3.15–18.4% of the phenotypic variations observed (An et al. 2019). Although there has been a large number of reports on several locations of QTLs associated with latex yield and growth rate, these markers were identified by genotype-phenotype association in different environments. With the genome sequences available, candidate genes closely linked with markers have been annotated. However, there is no evidence from further experiments that can demonstrate the function of those candidate genes and how they interact and contribute to latex yield or growth rate in response to different environments.

The cost of genotyping a large number of individuals is decreasing rapidly while the cost of carrying out phenotypic evaluation, especially in the field, is getting more expensive. Genomic selection has arisen as an alternative genome-wide SNP-based approach to combine genotypic and phenotypic data to increase the accuracy of predicting breeding values and selecting traits, especially those with low degrees of heritability (Meuwissen et al. 2001). Recent reports on testing genomic selection models on rubber tree breeding have shown promising results (Cros et al. 2019; Souza et al. 2019). Souza et al. (2019) reported the first attempt to apply genomic selection to the rubber tree. The authors evaluated the efficiency of the genomic selection approach using 435 rubber tree individuals based on 30,546 GBS-derived SNP loci (Souza et al. 2019). Prediction models were estimated for diameter and height traits at different ages. Multi-environment models were shown to be superior to single-environment models regardless of the kernel methods employed (linear or non-linear), which suggested that the inclusion of interactions between markers and environmental conditions increased the percentage of variance explained by the model and the prediction accuracy (Souza et al. 2019). Genomic selection has potential to be implemented in rubber tree breeding programs in the future; however, additional work is still required to carefully evaluate genomic selection models using rubber tree populations with different genetic backgrounds and structure than that reported by Souza et al. (2019).

5.5 Conclusion

Although conventional plant breeding based on phenotypic selection is very effective, it is time-consuming and has suffered from the limitations of complex traits. The basic requirement for marker-assisted selection is the identification of a large number of polymorphic markers for the construction of high-resolution linkage maps, which are subsequently used to perform marker-trait association analysis. Over the past few decades, molecular marker analyses in the rubber tree has shifted from RFLPs and RAPDs to the highly effective SNPs. Rapid advances in sequencing technologies together with the available rubber genome sequences (Rahman et al. 2013; Lau et al. 2016; Tang et al. 2016; Pootakham et al. 2017) have made SNP discovery both technically and economically feasible. In addition, the increasing quality of reference genomes provides the necessary information to proceed from identifying QTLs to discovering the underlying genetic causes. As the cost of SNP genotyping becomes more affordable, rubber tree breeders worldwide can take full advantage of marker-assisted selection to expedite the development of elite clones with improved latex yield as well as resistance to biotic stress (e.g. pests and fungal pathogens) and abiotic stresses (e.g. drought, low temperature and tapping panel dryness).

References

An Z, Zhao Y, Zhang X, Huang X, Hu Y, Cheng H, Li X, Huang H (2019) A high-density genetic map and QTL mapping on growth and latex yield-related traits in *Hevea brasiliensis* Müll. Arg. Industrial Crops and Products 132:440–448. https://doi.org/10.1016/j.indcrop.2019.03.002

Baulkwill WJ, Webster CC (1989) Rubber. Longman Scientific and Technical, Burnt Mill, Harlow, Essex, England and Wiley, Wiley, New York

Besse P, Lebrun P, Seguin M, Lanaud C (1993) DNA fingerprints in *Hevea brasiliensis* (rubber tree) using human minisatellite probes. Heredity 70:237–244

Besse P, Seguin M, Lebrun P, Chevallier MH, Nicolas D, Lanaud C (1994) Genetic diversity among wild and cultivated populations of *Hevea brasiliensis* assessed by nuclear RFLP analysis. Theor Appl Genet 88:199–207

Bindu-Roy C, Nazeer MA, Saha T (2004) Identification of simple sequence repeats in rubber (*Hevea brasiliensis*). Curr Sci 87:807–811

Breseghello F, Sorrells ME (2006) Association analysis as a strategy for improvement of quantitative traits in plants. Crop Sci 46:1323–1330. https://doi.org/10.2135/cropsci2005.09-0305

Cavagnaro PF, Senalik DA, Yang L, Simon PW, Harkins TT, Kodira CD, Huang S, Weng Y (2010) Genome-wide characterization of simple sequence repeats in cucumber (*Cucumis sativus* L.). BMC Genomics 11, 569. https://doi.org/10.1186/1471-2164-11-569

Chanroj V, Rattanawong R, Phumichai T, Tangphatsornruang S, Ukoskit K (2017) Genome-wide association mapping of latex yield and girth in Amazonian accessions of *Hevea brasiliensis* grown in a suboptimal climate zone. Genomics 109:475–484

Chevallier MH (1988) Genetic variability of *Hevea brasiliensis* germplasm using isozyme markers. J Nat Rubber Res 3:42–53

Conson ARO, Taniguti CH, Amadeu RR, Andreotti IA, De Souza LM, Dos Santos LHB, Rosa JRBF, Mantello CC, Da Silva CC, José Scaloppi Junior E, Ribeiro RV, Le Guen V, Garcia AAF, Gonçalves PDS, De Souza AP (2018). High-resolution genetic map and QTL analysis of growth-related traits of *Hevea brasiliensis* cultivated under suboptimal temperature and humidity conditions. Front Plant Sci 9. https://doi.org/10.3389/fpls.2018.01255

Cros D, Mbo-Nkoulou L, Bell JM, Oum J, Masson A, Soumahoro M, Tran DM, Achour Z, Le Guen V, Clement-Demange A (2019) Within-family genomic selection in rubber tree (*Hevea brasiliensis*) increases genetic gain for rubber production. Ind Crops Prod 138:111464

Cubry P, Pujade-Renaud V, Garcia D, Espeout S, Le Guen V, Granet F, Seguin M (2014) Development and characterization of a new set of 164 polymorphic EST-SSR markers for diversity and breeding studies in rubber tree (*Hevea brasiliensis* Müll. Arg.). Plant Breed 133:419–426. https://doi.org/10.1111/pbr.12158

De Souza LM, Dos Santos LHB, Rosa JRBF, Da Silva CC, Mantello CC, Conson ARO, Scaloppi EJ, Fialho JDF, De Moraes MLT, Gonçalves PDS, Margarido GRA, Garcia AAF, Le Guen V, De Souza AP (2018) Linkage disequilibrium and population structure in wild and cultivated populations of rubber tree (*Hevea brasiliensis*). Front Plant Sci 9. 10.3389/fpls.2018.00815

De Souza LM, Le Guen V, Cerqueira-Silva CBM, Silva CC, Mantello CC, Conson ARO, Vianna JPG, Zucchi MI, Scaloppi Junior EJ, Fialho JDF, De Moraes MLT, Gonçalves PDS, Souza APD (2015) Genetic diversity strategy for the management and use of rubber genetic resources: more than 1,000 wild and cultivated accessions in a 100-genotype core collection. PLoS ONE 10:e0134607. https://doi.org/10.1371/journal.pone.0134607

De Souza LM, Toledo-Silva G, Cardoso-Silva CB, Da Silva CC, Andreotti IAA, Conson ARO, Mantello CC, Le Guen V, Souza AP (2016). Development of single nucleotide polymorphism markers in the large and complex rubber tree genome using next-generation sequence data. Mol Breed 36:115

Dean W (1987) Brazil and the struggle for rubber. Cambridge University Press, UK

Dijkman MJ (1951) *Hevea*. Thirty years of research in the far east. Chronica Botanica. Waltham, Mass

Edwards D, Batley J, Cogan NOI, Forster JW, Chagne D (2007) Single nucleotide polymorphism discovery. In: Association mapping in plants. Springer, NY, pp 53–76

Elshire RJ, Glaubitz JC, Sun Q, Poland JA, Kawamoto K, Buckler ES, Mitchell SE (2011) A robust, simple genotyping-by-sequencing (GBS) approach for high diversity species. PLoS ONE 6:e19379. https://doi.org/10.1371/journal.pone.0019379

Ersoz ES, Yu J, Buckler ES (2007) Applications of linkage disequilibrium and association mapping in crop plants. In: Varshney RK, Tuberosa R (eds) Genomics-assisted crop improvement: Vol. 1: Genomics approaches and platforms. Springer Netherlands, Dordrecht, pp 97–119

Feng SP, Li WG, Huang HS, Wang JY, Wu YT (2009) Development, characterization and cross-species/genera transferability of EST-SSR markers for rubber tree (*Hevea brasiliensis*). Mol Breed 23: 85–97

Feng SP, Li WG, Wang JY and Wu YT (2010) Construction of genetic linkage map for rubber tree (*Hevea brasiliensis*) based on SSR markers. Yi Chuan 32:857–863

Gardner KM, Brown P, Cooke TF, Cann S, Costa F, Bustamante C, Velasco R, Troggio M, Myles S (2014) Fast and cost-effective genetic mapping in apple using next-generation sequencing. G3 (Bethesda) 4, 1681–1687. https://doi.org/10.1534/g3.114.011023

Huang YF, Poland JA, Wight CP, Jackson EW, Tinker NA (2014) Using genotyping-by-sequencing (GBS) for genomic discovery in cultivated oat. PLoS ONE 9:e102448. https://doi.org/10.1371/journal.pone.0102448

Labate JA, Baldo A (2005) Tomato SNP discovery by EST mining and resequencing. Mol Breed 16:343–349. https://doi.org/10.1007/s11032-005-1911-5

Lande R, Thompson R (1990) Efficiency of marker-assisted selection in the improvement of quantitative traits. Genetics 124:743–756

Lau NS, Makita Y, Kawashima M, Taylor TD, Kondo S, Othman AS, Shu-Chien AC, Matsui M (2016) The rubber tree genome shows expansion of gene family associated with rubber biosynthesis. Sci Rep 6:28594. https://doi.org/10.1038/srep28594

Le Guen V, Doare F, Weber C, Seguin M (2009) Genetic structure of Amazonian populations of *Hevea brasiliensis* is shaped by hydrographical network and isolation by distance. Tree Genetics and Genomes 5:673–683

Le Guen V, Garcia D, Doaré F, Mattos CR, Condina V, Couturier C, Chambon A, Weber C, Espéout S, Seguin M (2011) A rubber tree's durable resistance to *Microcyclus ulei* is conferred by a qualitative gene and a major quantitative resistance factor. Tree Genet Genomes 7:877–889. https://doi.org/10.1007/s11295-011-0381-7

Le Guen V, Gay C, Xiong TC, Souza LM, Rodier-Goud M, Seguin M (2010) Development and characterization of 296 new polymorphic microsatellite markers for rubber tree (*Hevea brasiliensis*). Plant Breed 130:294–296

Leconte A, Lebrun P, Nicolas D, Seguin M (1994) Electrophoresis application to *Hevea* clone identification. Plant Res Dev 1:28–36

Lekawipat N, Teerawatanasuk K, Rodier-Goud M, Seguin M, Vanavichit A, Toojinda T, Tragoonrung S (2003) Genetic diversity analysis of wild germplasm and cultivated clones of *Hevea brasiliensis* Muell. Arg. by using microsatellite markers. J Rubber Res 6:36–47

Leonforte A, Sudheesh S, Cogan N, Salisbury P, Nicolas M, Materne M, Forster J, Kaur S (2013) SNP marker discovery, linkage map construction and identification of QTLs for enhanced salinity tolerance in field pea (*Pisum sativum* L.). BMC Plant Biol 13:1–14. https://doi.org/10.1186/1471-2229-13-161

Lespinasse D, Rodier-Goud M, Grivet L, Leconte A, Legnate H, Seguin M (2000) A saturated genetic linkage map of rubber tree (*Hevea* spp.) based on RFLP, AFLP, microsatellite and isozyme markers. Theor Appl Genet 100:975–984

Li AQ, Zhao CZ, Wang XJ, Liu ZJ, Zhang LF, Song GQ, Yin J, Li CS, Xia H, Bi YP (2010) Identification of SSR markers using soybean (*Glycine max*) ESTs from globular stage embryos. Electron J Biotechn 13:6–7

Mantello CC, Cardoso-Silva CB, Da Silva CC, De Souza LM, Scaloppi Junior EJ, De Souza Gonçalves P, Vicentini R, De Souza AP (2014) *De Novo* assembly and transcriptome analysis of the rubber tree (*Hevea brasiliensis*) and SNP markers development for rubber biosynthesis pathways. PLoS ONE 9: e102665. https://doi.org/10.1371/journal.pone.0102665

Meuwissen TH, Hayes BJ, Goddard ME (2001) Prediction of total genetic value using genome-wide dense marker maps. Genetics 157:1819–1829

Myles S, Peiffer J, Brown PJ, Ersoz ES, Zhang Z, Costich DE, Buckler ES (2009) Association mapping: critical considerations shift from genotyping to experimental design. Plant Cell 21:2194–2202. https://doi.org/10.1105/tpc.109.068437

Novaes E, Drost DR, Farmerie WG, Pappas GJ Jr, Grattapaglia D, Sederoff RR, Kirst M (2008) High-throughput gene and SNP discovery in *Eucalyptus grandis*, an uncharacterized genome. BMC Genom 9:312. https://doi.org/10.1186/1471-2164-9-312

Parchman TL, Geist KS, Grahnen JA, Benkman CW, Buerkle CA (2010) Transcriptome sequencing in an ecologically important tree species: assembly, annotation, and marker discovery. BMC Genom 11:180. https://doi.org/10.1186/1471-2164-11-180

Penner GA, Bush A, Wise R, Kim W, Domier L, Kasha K, Laroche A, Scoles G, Molnar SJ, Fedak G (1993) Reproducibility of random amplified polymorphic DNA (RAPD) analysis among laboratories. PCR Methods Appl 2:341–345

Pootakham W, Chanprasert J, Jomchai N, Sangsrakru D, Yoocha T, Therawattanasuk K, Tangphatsornruang S (2011) Single nucleotide polymorphism marker development in the rubber tree, *Hevea brasiliensis* (Euphorbiaceae). Am J Bot 98:e337–e338

Pootakham W, Jomchai N, Ruang-Areerate P, Shearman JR, Sonthirod C, Sangsrakru D, Tragoonrung S, Tangphatsornruang S (2015a) Genome-wide SNP discovery and identification of QTL associated with agronomic traits in oil palm using genotyping-by-sequencing (GBS). Genomics 105:288–295. https://doi.org/10.1016/j.ygeno.2015.02.002

Pootakham W, Ruang-Areerate P, Jomchai N, Sonthirod C, Sangsrakru D, Yoocha T, Theerawattanasuk K, Nirapathpongporn K, Romruensukharom P, Tragoonrung S, Tangphatsornruang S (2015b) Construction of a high-density integrated genetic linkage map of rubber tree (*Hevea brasiliensis*) using genotyping-by-sequencing (GBS). Front Plant Sci 6:367. https://doi.org/10.3389/fpls.2015.00367

Pootakham W, Shearman JR, Ruang-Areerate P, Sonthirod C, Sangsrakru D, Jomchai N, Yoocha T, Triwitayakorn K, Tragoonrung S, Tangphatsornruang S (2014) Large-scale SNP discovery through RNA sequencing and SNP genotyping by targeted enrichment sequencing in cassava (*Manihot esculenta* Crantz). PLoS ONE 9:e116028. https://doi.org/10.1371/journal.pone.0116028

Pootakham W, Sonthirod C, Naktang C, Ruang-Areerate P, Yoocha T, Sangsrakru D, Theerawattanasuk K, Rattanawong R, Lekawipat N, Tangphatsornruang S (2017) De novo hybrid assembly of the rubber tree genome reveals evidence of paleotetraploidy in *Hevea* species. Sci Rep 7:41457. https://doi.org/10.1038/srep41457

Pootakham W, Uthaipaisanwong P, Sangsrakru D, Yoocha T, Tragoonrung S, Tangphatsornruang S (2013) Development and characterization of single-nucleotide polymorphism markers from 454 transcriptome sequences in oil palm (*Elaeis guineensis*). Plant Breed 132:711–717

Priyadarshan PM (2003) Breeding *Hevea brasiliensis* for environmental constraints. In: Advances in agronomy. Academic Press, pp 351–400

Priyadarshan PM (2017) Refinements to *Hevea* rubber breeding. Tree Genet Genomes 13:20

Priyadarshan PM, Gonçalves PS, Omokhafe KO (2009) Breeding *Hevea* rubber. In: Jain SM, Priyadarshan PM (eds) Breeding plantation tree crops: tropical species. Springer New York, New York, NY, pp 469–522

Rafalski A (2002) Applications of single nucleotide polymorphisms in crop genetics. Curr Opin Plant Biol 5:94–100

Rahman AYA, Usharraj A, Misra B, Thottathil G, Jayasekaran K, Feng Y, Hou S, Ong SY, Ng FL, Lee LS, Tan HS, Sakaff MKLM, Teh BS, Khoo B, Badai SS, Aziz NA, Yuryev A, Knudsen B, Dionne-Laporte A, Mchunu N, Yu Q, Langston B, Freitas TA, Young A, Chen R, Wang L, Najimudin N, Saito J, Alam M (2013) Draft genome sequence of the rubber tree *Hevea brasiliensis*. BMC Genom 14:75

Rosa JRBF, Mantello CC, Garcia D, De Souza LM, Da Silva CC, Gazaffi R, Da Silva CC, Toledo-Silva G, Cubry P, Garcia AAF, De Souza AP, Le Guen V (2018) QTL detection for growth and latex production in a full-sib rubber tree population cultivated under suboptimal climate conditions. BMC Plant Biol 18:223. https://doi.org/10.1186/s12870-018-1450-y

Saba Rahim M, Sharma H, Parveen A, Roy JK (2018) Trait mapping approaches through association analysis in plants. In: Varshney RK, Pandey MK, Chitikineni A (eds) Plant genetics and molecular biology. Springer International Publishing, Cham, pp 83–108

Salgado LR, Koop DM, Pinheiro DG, Rivallan R, Le Guen V, Nicolás MF, De Almeida LGP, Rocha VR, Magalhães M, Gerber AL, Figueira A, Cascardo JCDM, De Vasconcelos AR, Silva WA, Coutinho LL, Garcia D (2014) De novo transcriptome analysis of *Hevea brasiliensis* tissues by RNA-seq and screening for molecular markers. BMC Genom 15:236. https://doi.org/10.1186/1471-2164-15-236

Schafleitner R, Tincopa LR, Palomino O, Rossel G, Robles RF, Alagon R, Rivera C, Quispe C, Rojas L, Pacheco JA, Solis J, Cerna D, Young Kim J, Hou J, Simon R (2010) A sweetpotato gene index established by de novo assembly of pyrosequencing and Sanger sequences and mining for gene-based microsatellite markers. BMC Genom 11:604. https://doi.org/10.1186/1471-2164-11-604

Seguin M, Besse P, Lebrun P, Chevallier M (1995) *Hevea* germplasm characterization using isozymes and RFLP markers. SPB Academic Publishing, Amsterdam

Seguin M, Gay C, Xiong TC, Rodier-Goud M (2001) Microsatellite markers for genome analysis of rubber tree (*Hevea* spp.). In: Jérôme S-B (ed) Biotechnology and rubber tree. Proceedings of IRRDB symposium. IRRDB, CIRAD-CP-HEVEA: IRRDB, CIRAD-CP-HEVEA

Shearman JR, Sangsrakru D, Jomchai N, Ruang-Areerate P, Sonthirod C, Naktang C, Theerawattanasuk K, Tragoonrung S, Tangphatsornruang S (2015) SNP identification from RNA sequencing and linkage map construction of rubber tree for anchoring the draft genome. PLoS ONE 10:e0121961. https://doi.org/10.1371/journal.pone.0121961

Souza LM, Francisco FR, Gonçalves PS, Scaloppi Junior EJ, Le Guen V, Fritsche-Neto R, Souza AP (2019) Genomic selection in rubber tree breeding: a comparison of models and methods for dealing with G × *E*. *bioRxiv*, 603662. https://doi.org/10.1101/603662

Souza LM, Gazaffi R, Mantello CC, Silva CC, Garcia D, Le Guen V, Cardoso SEA, Garcia AaF, Souza, AP (2013) QTL mapping of growth-related traits in a full-Sib family of rubber tree (*Hevea brasiliensis*) evaluated in a sub-tropical climate. PLoS ONE 8, e61238. https://doi.org/10.1371/journal.pone.0061238

Souza LM, Mantello CC, Suzuki F, Gazaffi R, Garcia D, Le Guen V, Garcia AAF, Souza AP (2011) Development of a genetic linkage map of rubber tree (*Hevea braziliensis*) based on microsatellite markers. BMC Proc 5:39

Souza LM, Mantello CC, Santos MO, De Souza Gonçalves P, Souza AP (2009) Microsatellites from rubber tree (*Hevea brasiliensis*) for genetic diversity analysis and cross-amplification in six *Hevea* wild species. Conserv Genet Resour 1:75. https://doi.org/10.1007/s12686-009-9018-7

Tan H (1987) Strategies in rubber tree breeding. In: Abbot AJ, Atkin RK (eds) Improving vegetatively propagated crops. Academic Press, London, pp 28–54

Tang C, Yang M, Fang Y, Luo Y, Gao S, Xiao X, An Z, Zhou B, Zhang B, Tan X, Yeang H-Y, Qin Y, Yang J, Lin Q, Mei H, Montoro P, Long X, Qi J, Hua Y, He Z, Sun M, Li W, Zeng X, Cheng H, Liu Y, Yang J, Tian W, Zhuang N, Zeng R, Li D, He P, Li Z, Zou Z, Li S, Li C, Wang J, Wei D, Lai C-Q, Luo W, Yu J, Hu S, Huang H (2016) The rubber tree genome reveals new insights into rubber production and species adaptation. Nat Plants 2:16073. https://doi.org/10.1038/nplants.2016.73

Tangphatsornruang S, Somta P, Uthaipaisanwong P, Chanprasert J, Sangsrakru D, Seehalak W, Sommanas W, Tragoonrung S, Srinives P (2009) Characterization of microsatellites and gene contents from genome shotgun sequences of mungbean (*Vigna radiata* (L.) Wilczek). BMC Plant Biol 9:137

Tragoonrung S, Kanazin V, Hayes PM, Blake TK (1992) Sequence-tagged-site-facilitated PCR for barley genome mapping. Theor Appl Genet 84:1002–1008

Triwitayakorn K, Chatkulkawin P, Kanjanawattanawong S, Sraphet S, Yoocha T, Sangsrakru D, Chanprasert J, Ngamphiw C, Jomchai N, Therawattanasuk K, Tangphatsornruang S (2011) Transcriptome sequencing of *Hevea brasiliensis* for development of microsatellite markers and construction of a genetic linkage map. DNA Res 18. https://doi.org/10.1093/dnares/dsr034

Varghese YA, Knaak C, Sethuraj MR, Ecke W (1997) Evaluation of random amplified polymorphic DNA (RAPD) markers in *Hevea brasiliensis*. Plant Breed 116:47–52. https://doi.org/10.1111/j.1439-0523.1997.tb00973.x

Wang Z, Fang B, Chen J, Zhang X, Luo Z, Huang L, Chen X, Li Y (2010) De novo assembly and characterization of root transcriptome using Illumina paired-end sequencing and development of cSSR markers in sweet potato (*Ipomoea batatas*). BMC Genom 11:726. https://doi.org/10.1186/1471-2164-11-726

Ward J, Bhangoo J, Fernandez-Fernandez F, Moore P, Swanson J, Viola R, Velasco R, Bassil N, Weber C, Sargent D (2013) Saturated linkage map construction in *Rubus idaeus* using genotyping by sequencing and genome-independent imputation. BMC Genom 14:2

Weir BS (1979) Inferences about linkage disequilibrium. Biometrics 35:235–254. https://doi.org/10.2307/2529947

Xia Z, Liu K, Zhang S, Yu W, Zou M, He L, Wang W (2018) An ultra-high density map allowed for mapping QTL and candidate genes controlling dry latex yield in rubber tree. Ind Crops Prod 120:351–356

Zhang Y, Wang L, Xin H, Li D, Ma C, Ding X, Hong W, Zhang X (2013) Construction of a high-density genetic map for sesame based on large scale marker development by specific length amplified fragment (SLAF) sequencing. BMC Plant Biol 13:141

6 Genome-Wide Analysis of Transcription Start Sites and Core Promoter Elements in *Hevea brasiliensis*

Yuko Makita, Yukio Kurihara, Nyok-Sean Lau, Mika Kawashima, Ahmad Sofiman Othman, and Minami Matsui

Abstract

Next-generation sequencing (NGS) technologies have enabled genome analysis of numerous economically important plants including the natural rubber tree, *Hevea brasiliensis*. As a result, genomic sequence data of three rubber tree clones and various types of transcriptome data such as RNA-seq, ESTs, full-length cDNAs, and isoform sequencing (Iso-seq) are currently available in the public domain. A combination of high-throughput omics approaches has led us to a deeper understanding of gene regulation. Information on the precise transcription start sites (TSSs) of genes is essential to reveal transcriptional regulation. In this chapter, we introduce the application of the CAGE (cap analysis gene expression) method for genome-wide identification of TSSs in *H. brasiliensis*. CAGE is a technique to obtain accurate and comprehensive TSSs throughout the genome. The data obtained from CAGE can be used to update annotation of 5′ UTRs and tissue-specific TSSs. Accurate information about TSSs also enables elucidation of gene regulatory elements in core promoters, such as TATA boxes, initiator elements (Inrs) and transcription factor-binding sites (TFBSs). By including RNA-seq and full-length cDNA data, it is also possible to generate statistics in order to discover novel genes, non-coding RNAs (ncRNAs), and antisense transcripts. In this chapter, we discuss CAGE analysis in combination with other transcriptomic data in *H. brasiliensis* with the ultimate aim of controlling rubber biosynthetic pathways. The CAGE data are accessible at http://rubber.riken.jp.

6.1 Introduction

Transcriptome analysis using RNA-seq has become a commonly used technology. In *Hevea brasiliensis*, RNA-seq analyses of rubber tissues (Tang et al. 2016) and clone specificity (Chao et al. 2015) as well as responses to abiotic/biotic stresses such as wounding and jasmonic acid (Liu et al. 2018), and ethylene treatment and infection by *Microcyclus ulei* (Hurtado Páez et al. 2015) have been published. RNA-seq data assist in identifying differentially expressed genes in different groups of tissue samples or

Y. Makita · Y. Kurihara · N.-S. Lau · M. Kawashima · M. Matsui (✉)
Synthetic Genomics Research Group, RIKEN Center for Sustainable Resource Science (CSRS), 1-7-22 Suehiro-cho, Tsurumi-ku, Yokohama, Kanagawa 230-0045, Japan
e-mail: minami@riken.jp

N.-S. Lau · A. S. Othman
Centre for Chemical Biology, Universiti Sains Malaysia, 11900 Bayan Lepas, Penang, Malaysia

A. S. Othman
School of Biological Sciences, Universiti Sains Malaysia, 11800 Minden, Penang, Malaysia

M. Matsui and K.-S. Chow (eds.), *The Rubber Tree Genome*, Compendium of Plant Genomes,
https://doi.org/10.1007/978-3-030-42258-5_6

treatments. RNA-seq also provides evidence for novel protein-coding genes and novel isoforms (Campbell et al. 2014), rare and novel transcripts (Yin et al. 2019), simple sequence repeats (SSRs), single nucleotide polymorphisms (SNPs) (Mantello et al. 2014) and post-transcriptional modifications (Zhang et al. 2019). Although RNA-seq is a powerful method, it cannot accurately identify the positions of transcription start sites (TSSs). For this purpose, an alternative method is required. Precise TSS determination is vital for understanding transcriptional and post-transcriptional regulation.

Computational prediction of transcription factor (TF)-binding sites is difficult since their binding motifs are often short (6–10 bp) and contain nucleotide variation. One such factor, the TATA box, is among the most studied motifs in the core promoter region of genes in eukaryotes and archaea. The motif sequence of the TATA box often contains "TATA" bases but this motif is not strongly conserved. Similar sequences can be found throughout the genome. To determine the correct TATA box for a gene, the distance from the TSS (around −35 bp) is required for analysis. Not only are core promoter elements located around TSSs, TF-binding sites are also often found near them. Determining the precise TSS location helps in finding TF-binding motifs and core promoter elements (Morton et al. 2014).

Genome-wide transcription start site detection technologies also identify tissue- or condition-specific TSSs. It has been reported that transcripts from alternative TSSs generate different protein isoforms, especially at their N-terminus regions, and these isoforms can localize in different subcellular components depending on light conditions (Ushijima et al. 2017). It has been reported that upstream open reading frames (uORFs) in the 5′ leader sequence of mRNAs often inhibit translation and cause mRNA degradation by nonsense-mediated decay (NMD). We have reported that light causes TSSs to change from upstream to downstream sites of uORFs to ensure expression of genes by preventing uORF inhibition (Kurihara et al. 2018).

Several methods have been used to determine TSSs. We chose to use the cap analysis gene expression (CAGE) method, whereby TSS accuracy may be supported by other genomic techniques (Adiconis et al. 2018). CAGE has also been applied to predict enhancer RNAs in mammals (Andersson et al. 2014).

In the RRIM 600 draft genome, 5′ UTRs were predicted in 30% of rubber genes (Lau et al. 2016). TSS positions of these genes were estimated using EST and/or RNA-seq data. Since more recent RNA-seq technology is capable of capturing rare transcripts, predicted 5′ UTRs tend to be longer (Makita et al. 2017, 2018). Full-length cDNAs (fl-cDNAs) that have been prepared using the same technology as in the CAGE method can also be used for TSS determination. Using such data, we have improved TSS annotation in the *H. brasiliensis* RRIM 600 clone (Makita et al. 2017). PacBio single-molecule real-time (SMRT) sequencing is useful to obtain long reads (>10 kb on average), and Iso-Seq technology employs such reads to produce transcript isoform lengths ranging from the 5′ end to their poly-A tails (Minoche et al. 2015). In *H. brasiliensis*, 46,129 non-redundant transcripts from leaves were obtained with Iso-Seq in the clone, BPM 24 (Pootakham et al. 2017). Recently, 37,224 unique transcripts from latex were determined in the RRIM 600 clone (Chow et al. 2019). Fl-cDNA libraries and Iso-Seq are the best approaches to obtain transcript isoforms throughout the genome. However, the cost is higher if the focus is only on obtaining TSS information. CAGE is a specialized method for identifying TSSs and obtains sequences that are only about 50 bp from the 5′-end of each transcript. The frequency of TSS reads is correlated to the level of gene expression. Genes that have more than one TSS are also revealed. Tissue- or condition-dependent alternative TSSs help to shed light on transcriptional and post-transcriptional regulation. Information about TSSs on a genome-wide scale is required for better understanding of transcriptional regulation of rubber biosynthesis. In this chapter, we introduce the TSS landscape of the *H. brasiliensis* RRIM 600 clone, particularly in relation to rubber biosynthesis genes and promoter sequences.

6.2 How to Obtain TSS Locations and Expression Levels with CAGE Technology

CAGE produces a snapshot of the 5′-ends of a mRNA population in a biological sample. Since it captures the cap structure at the 5′-end of transcripts, it enables identification of TSSs of not only mRNAs but also non-coding RNAs (Murata et al. 2014). Theoretically, CAGE cannot detect cap-less transcripts such as pre-tRNA, rRNA, and chloroplast and mitochondrial RNAs. However, these sequences, especially rRNAs, are often included in CAGE results and need to be removed beforehand. In the construction of a CAGE library, RNAs are reverse transcribed to cDNAs and biotinylated 5′-capped structures of RNA-cDNA hybrids are trapped with streptavidin beads (Fig. 6.1a). Following RNA digestion, linkers are ligated to single-stranded cDNAs and second strand synthesis is carried out. The 5′-ends of the second strand cDNAs, which correspond to TSS positions, are sent for next-generation sequencing (NGS) (Kodzius et al. 2006). The newest CAGE protocol is optimized for the Illumina sequencer. The 5′-end short nucleotide sequences (CAGE tags) are mapped onto the genome (Fig. 6.1b). Similar to RNA-seq, the number of CAGE tags for each promoter gives the level of RNA expression. More important, CAGE provides simultaneous information on various alternative TSSs and their expression levels. We obtained tissue RNAs from latex, bark and leaf tissues of clone RRIM 600 and constructed CAGE libraries for sequencing.

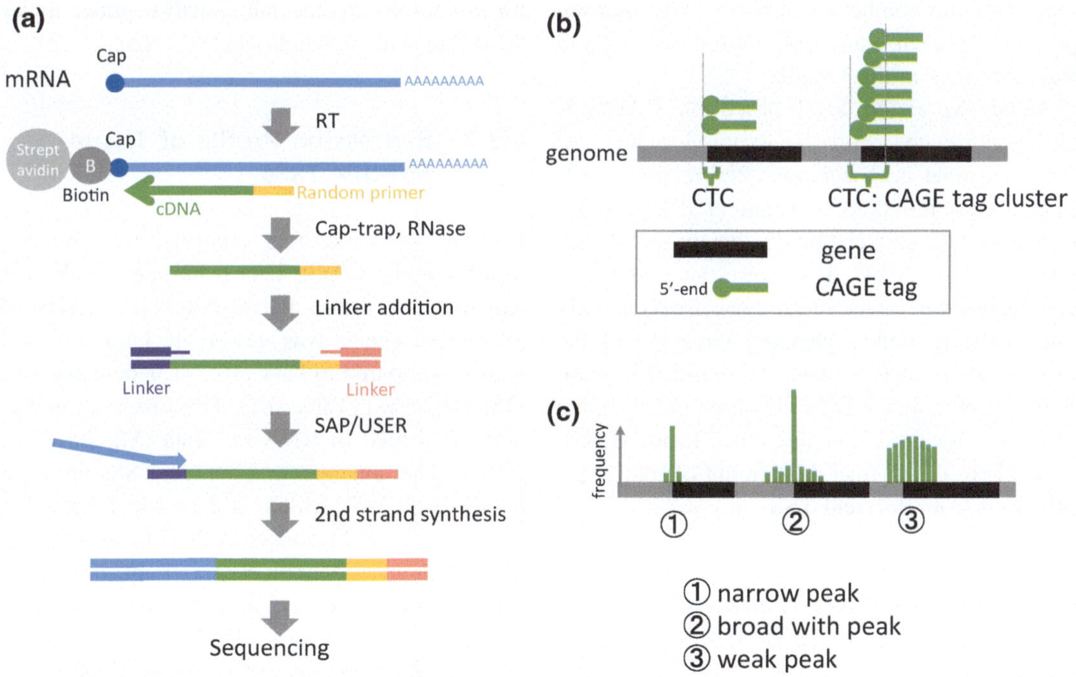

Fig. 6.1 Flow of CAGE analysis and clustering of each TSS. **a** Representation of the CAGE preparation protocol. The short fragments from the 5′ end of capped transcripts are captured by biotinylation of the cap structures. This method specifically detects capped and not cap-less RNAs. **b** Illustration of CAGE tags and CAGE tag clusters (CTCs). The short fragment from the 5′ end of a capped transcript is called a CAGE tag or CAGE read. After sequencing, CAGE tags are mapped onto the reference genome. To count the number of TSSs, we make groups of CAGE tags and each representative TSS group is called a CAGE tag cluster (CTC). The number of CAGE tags represents the expression level for each CTC

For our computational analysis in *H. brasiliensis*, CAGE tags were mapped onto the RRIM 600 genome (Lau et al. 2016) using the mapping software BWA (Li and Durbin 2009). Around half of TSSs were localized on a single nucleotide residue while the rest were not only found on a single nucleotide residue but also at several neighboring positions. Morton et al. (2014) categorized TSS tags into three patterns: narrow peak, broad with peak, and weak peak (Fig. 6.1c). To make clusters of such neighboring TSSs, we first applied the Paraclu software, which reports genomic intervals containing many CAGE tags compared to the surrounding regions (Ohmiya et al. 2014). Paraclu tends to make relatively wide clusters and often merges two closely located clusters into an incorrect one. To obtain detailed and separate CAGE tag clusters (CTCs), we clustered CAGE tags when neighboring tags were within 20 bp of each other (Nepal et al. 2013). These CAGE tags were normalized to TPM (tags per million). TPM represents the number of transcript tags that are present for each transcript within one million transcript tags in the sample.

Gene expression level based on CAGE is highly correlated with the expression level of RNA-seq data in *Arabidopsis* (correlation coefficients were 0.9–0.98, Kurihara et al. 2018) and in human cells (correlation coefficient was 0.998, Kawaji et al. 2014). However, the correlation coefficient was lower in *H. brasiliensis* mainly due to the limited sequencing coverage of the large genome of this plant. To avoid this problem, we set a higher TPM threshold (>1.0) in the following analysis. On the other hand, CAGE was able to distinguish the amount of transcripts produced from different TSSs.

6.3 Genome-Wide Transcription Start Sites in *H. brasiliensis*

6.3.1 Overview of Tissue-Specific CAGE Data

In model organisms, results from CAGE data were evaluated with TSS gene annotations that were supported by other experimental evidence, such as primer extension. However, as there is still room for improvement in TSS annotation of the *H. brasiliensis* RRIM 600 genome, we compared the location of CAGE-detected TSSs (in other words, CTCs) with the current gene models (Lau et al. 2016). Around 65% of CTCs were located in 5′ UTRs or upstream regions (<2,000 bp) of the current gene models (Fig. 6.2a). This ratio decreased if we used the lower expression threshold (TPM). Nearly 10% of CTCs were located in CDSs and of these, we often found CAGE tags at the beginning of exons. This phenomenon is called "exon painting" (Fejes-Toth et al. 2009). The biological function of exon painting is not fully understood but such CAGE features are considered as degradation products. CAGE analyses can help to identify novel ORFs and non-coding (ncRNAs) in combination with expression data of RNA-seq from various conditions and tissue samples. More than 10% of the CTCs were found in intergenic regions. Thus, these are candidates for novel ORFs (especially small peptides under 50 amino acids in length) and ncRNAs.

6.3.2 Expression Profile of Tissue-Specific TSSs

For genome-wide TSS analysis, we obtained tissue-specific CAGE data from latex, bark, and leaf tissues of clone RRIM 600. The number of expressed genes was fewer in latex (10,442 genes) compared to bark (16,771 genes) and leaf (15,567 genes) (Table 6.1). This observation was also confirmed in RNA-seq data (Makita et al. 2017). The average number of TSSs in each tissue was 3–5, which is larger compared to *Arabidopsis* (Tokizawa et al. 2017) and sorghum (2-3 TSSs/gene, Makita et al. unpublished data). Latex is produced and accumulated in specialized cells of laticifers. The highly expressed genes in latex are important for elucidating the rubber biosynthetic pathway in *H. brasiliensis*. We investigated gene expression in latex, bark, and leaf tissues (Fig. 6.2b) and found that a number of the expressed genes were present in all three tissues; however, there were also tissue-

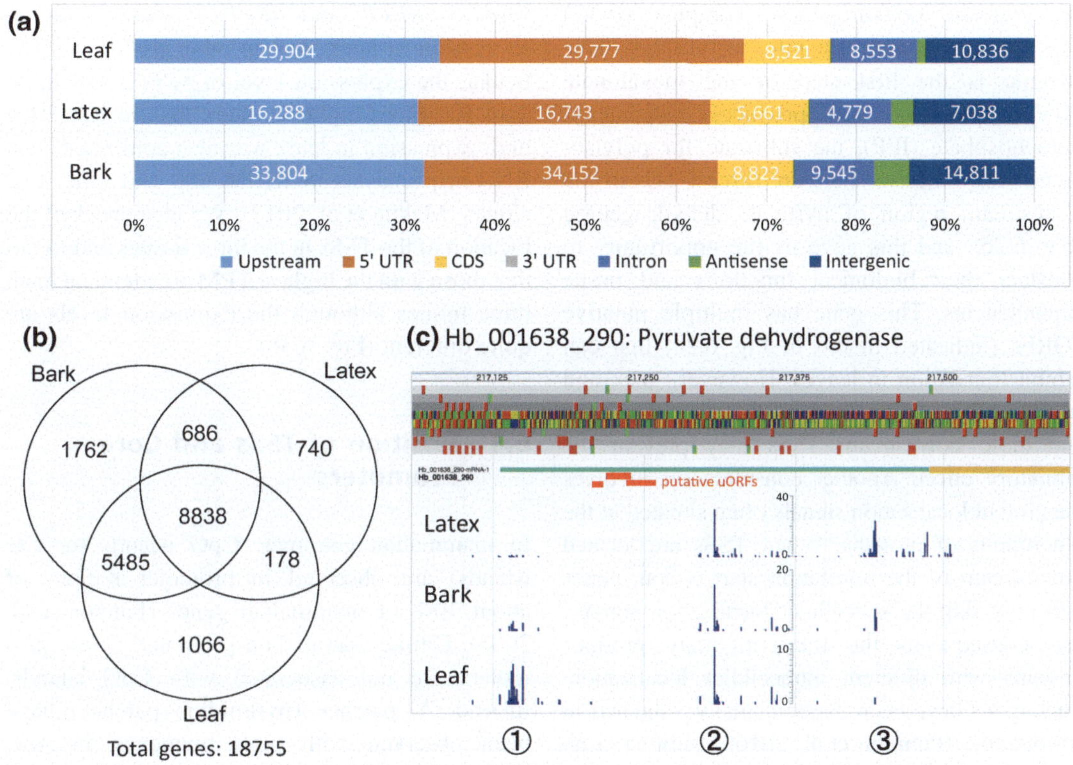

Fig. 6.2 **a** Distribution of gene locations of CTCs in leaf, latex, and bark. **b** Venn diagram showing the numbers of genes detected by CAGE tags in bark, latex, and leaf. **c** Example of a genome browser view of tissue-specific TSSs. The Hb_0001638_290 gene encodes pyruvate dehydrogenase. This gene shows different TSS profiles in latex, bark, and leaf. Putative uORFs are indicated in red

Table 6.1 Statistics of tissue-specific CAGE data

Rubber tissue	Number of CTCs	Number of CTCs with gene annotation	Number of genes detected CTCs	Average number of TSSs per gene
Bark	106,950	92,424	16,771	5.5
Latex	40,397	35,273	10,442	3.4
Leaf	93,447	82,447	15,567	5.3

specific genes that were only expressed in one sample.

We focused on the locations of TSSs in each sample. Of the 15,187 genes that were expressed in at least two samples, TSSs of 9,462 genes (62%) were localized in the same genomic position, and TSSs of 2,464 genes (16%) were candidates for alternative TSSs (these genomic positions were more than 50 bp apart). Among the alternative TSS candidates, 2,039 genes had multiple TSSs only in the promoter region and the rest had TSSs in both promoter and non-promoter (CDS/intron) regions. These alternative TSSs are considered to be under different transcriptional regulation in each sample. The CAGE data also revealed novel TSSs on the antisense strand in 405 genes.

An example of a tissue-specific alternative TSS is shown in Fig. 6.2c. Hb_001638_290 encodes pyruvate dehydrogenase, an enzyme of the pyruvate dehydrogenase complex (PDC) which converts pyruvate into acetyl-CoA. This enzyme is the first step of the mevalonate (MVA) pathway that produces isopentenyl pyrophosphate (IPP), the substrate for polyisoprene biosynthesis. There are three CTCs in the 5′-upstream region of pyruvate dehydrogenase (Fig. 6.2c), and this gave us the opportunity to consider their biological functions and tissue dependencies. This gene has multiple putative uORFs (indicated in red in Fig. 6.2c) that can inhibit translation in leaves. However, latex and bark have new TSSs downstream of these uORFs and these downstream TSSs may prevent the inhibitory effect. Another consideration involves the protein localization signals often situated at the N-terminus of proteins. When TSSs are located downstream of the translation start codon, genes will only skip the subcellular localization signals, and consequently, the transcript may produce proteins with different intracellular localization. Such cases have been experimentally validated in *Arabidopsis* (Pumplin et al. 2016; Ushijima et al. 2017) and maize (Mejía-Guerra et al. 2015).

6.3.3 Transcriptome Analysis on Latex Biosynthetic Genes

Latex contains rubber serum and rubber particles (Ferreira et al. 2009). The rubber elongation factor (REF) and the small rubber particle protein (SRPP) are localized on the surface of rubber particles and are possibly involved in rubber biosynthesis. *Cis*-prenyltransferase (CPT) is suggested to be involved in *cis*-polyisoprene formation. A rubber transferase activator (CPTL) which has homology to the human Nogo-B receptor is suggested to be essential for rubber biosynthesis by interacting with CPTs on the surface of rubber particles (Epping et al. 2015). From both genome and fl-cDNA analyses, we identified 6 *CPTs*, 1 *CPTL*, 8 *SRPP*, and 9 *REF* genes in the RRIM 600 genome (Lau et al. 2016; Makita et al. 2017). When their expression levels were investigated in clone RRIM 600 (Fig. 6.3), *CPT1*, *CPT2*, *CPTL*, *REF1-3*, *REF8*, and *SRPP1-2* showed higher expression in latex than in other tissues. In particular, the expression level of *SRPP1* was more than 100 times higher in latex than in leaf. This high expression in latex was also confirmed with RNA-seq analysis in RRIM 600 and other tree clones (Makita et al. 2017). We also checked the location of the TSSs in the three tissues and found that those with the highest TPM are identical in all three tissues although the expression levels are quite different (Fig. 6.3b).

6.4 Position of TSSs and Core Promoters

In mammalian genomes, CpG islands (or CG islands) are observed in promoter regions of about 40% of mammalian genes (Fatemi et al. 2005). Unlike mammalian promoters, plant promoters are not associated with CpG islands. Instead, Y patches (pyrimidine patches) have been observed with high frequency in *Arabidopsis* and rice promoters (Yamamoto et al. 2007). We checked the frequency of CpG (CG) islands and Y patches (CT/TC) around TSSs in *H. brasiliensis* and confirmed similar results to those in *Arabidopsis* and rice (Fig. 6.4). In mammals, TATA-type and CpG-associated promoters have complementary roles: TATA-type promoters show tissue-specific expression and TSS fluctuation is limited, while CpG-associated promoters show ubiquitous expression and broad TSS clusters (Sakakibara et al. 2007). In *H. brasiliensis*, TATA-type promoters show limited TSS fluctuation, as in mammals. Genes possessing TATA boxes and initiator elements (Inrs) show higher expression levels, this being the same as in *Arabidopsis* (Yamamoto et al. 2007).

In addition to the core promoter elements, such as TATA boxes and Inrs, transcription factor-binding sites (TFBSs) are often located in promoter regions. A systematic deletion assay of 45 randomly selected promoters in human suggests that the region spanning −300 to −50 bp upstream of a TSS positively regulates promoter

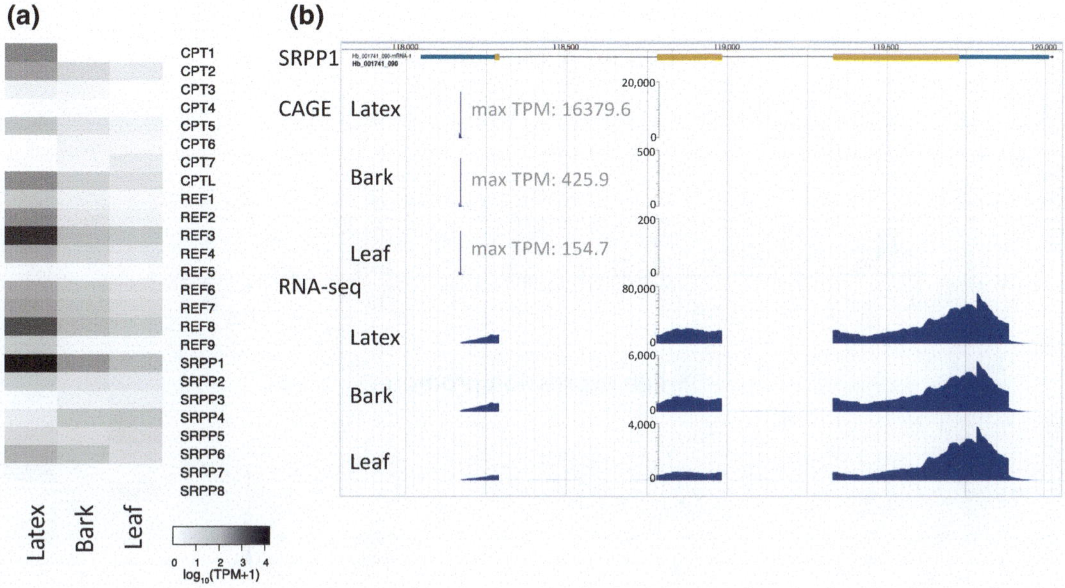

Fig. 6.3 **a** Heatmap of the transcription levels of *cis*-prenyltransferase (CPT), rubber transferase activator (CPT-like, CPTL), and two rubber particle proteins, rubber elongation factor (REF) and small rubber particle protein (SRPP). The expression values in TPM are normalized values of CAGE. Darker shades indicate higher expression. **b** Genome browser view of the SRPP1 gene. The upper three rows show TSSs from CAGE and the lower three show TSSs from RNA-seq. In this case, we can fix the 5′ UTR annotation with CAGE and RNA-seq data

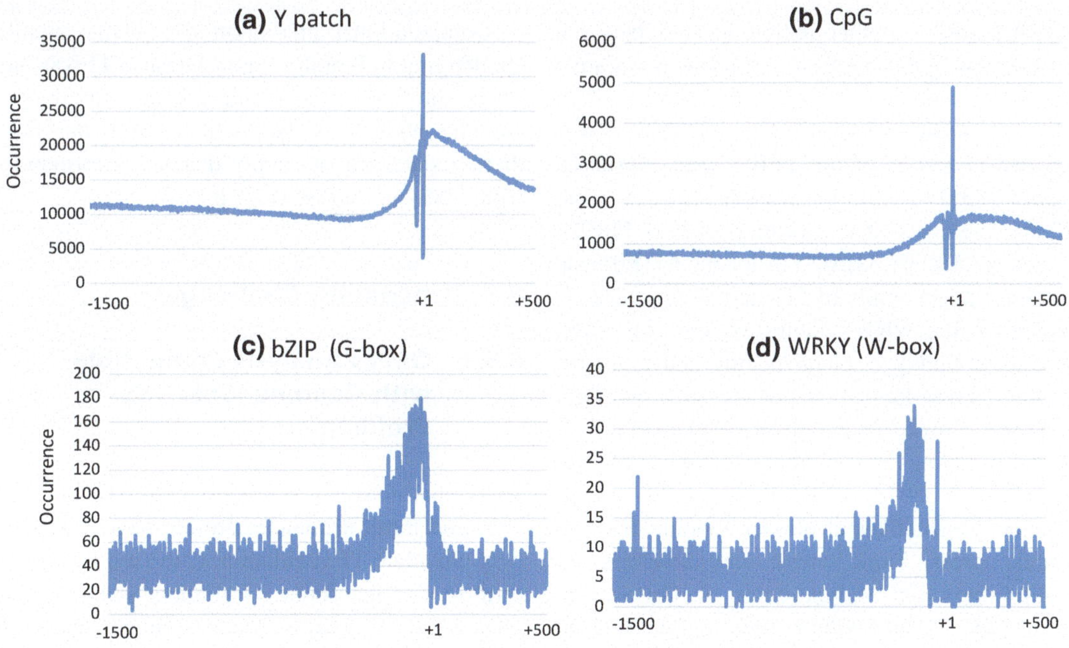

Fig. 6.4 Motif occurrence around the TSS (−1,500 bp to +500 bp) of Y patch (**a**), CpG skew (**b**), G-box motif of bZIP TFs recognition sequences (**c**) and W-box motif for WRKY TFs recognition sequences (**d**)

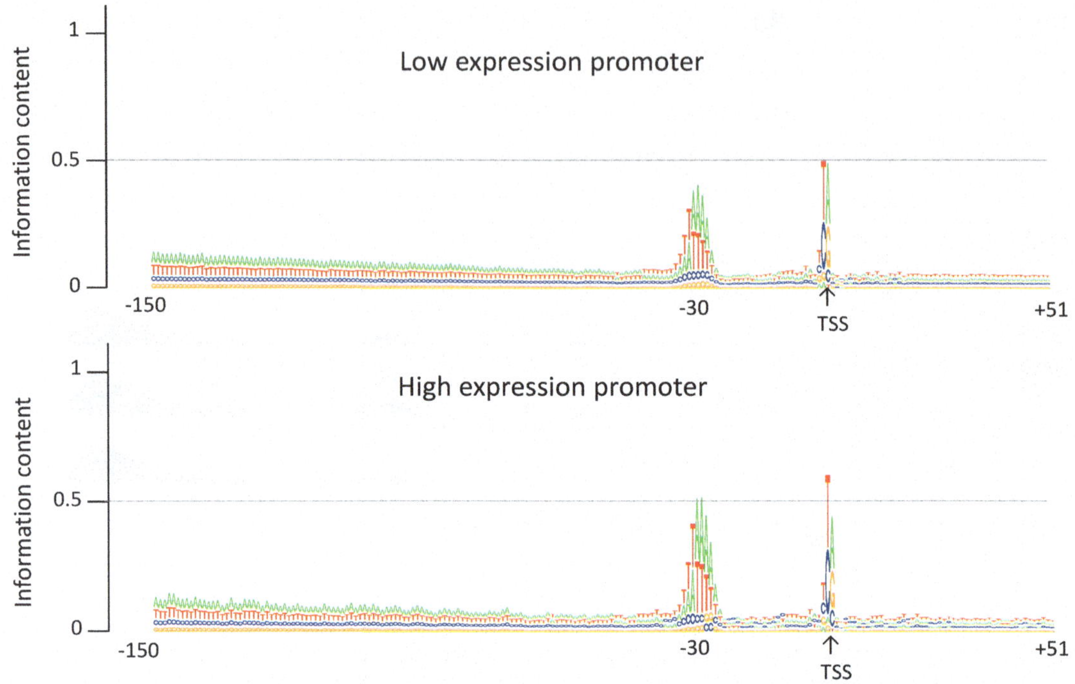

Fig. 6.5 Sequence conservation around TSS (−150 to +50 bp) for lowly expressed (TPM < 10) and highly expressed (TPM ≧ 10) genes. A TATA box is confirmed around −30 bp from the TSS. Inrs sequences were observed −1 to +1 bp from the TSS

activity, while another region located between −1,000 and −500 bp upstream forms a negative regulatory domain (Cooper et al. 2006). In *Arabidopsis*, positional distribution of TFBSs was verified and 86% of the TFBSs were observed from −1,000 to +200 bp of each TSS (Yu et al. 2016). We checked the location of two TFBSs: G-box ([A/C/T][C/G]ACGT) for the bZIP (basic leucine zipper) family of TFs and W-box (TGAC [C/T]) for the WRKY family of TFs (Fig. 6.4). The bZIP family is plant-specific and regulates various physiological responses such as light response, flower development, seed maturation, stress signaling and pathogen defense (Dröge-Laser et al. 2018). WRKY is another plant-specific TF family and plays important roles in abiotic stress, biotic stress, and development. Both G-box and W-box types of TFBS were frequently found within the region located between −50 and −250 bp in promoter sequences.

Computational prediction of TFBSs is still challenging in bioinformatics because TFBSs are often short (6–10 bp) and contain ambiguity. Genome-wide TSS data will assist better prediction since we can apply distance information from TSSs to TFBSs (Fig. 6.5).

6.5 Upcoming Challenges

6.5.1 Omics Analysis Compatible with Genome-Wide TSS Profiling

After obtaining genome-wide TSSs, TF-binding sites and motif sequences are the next targets to help elucidate TF regulation. ChIP-seq is a powerful tool to determine TF-binding sites. On the other hand, its analysis requires genetic transformation to obtain in vivo genomic DNA-binding sites. Taking into consideration the generation

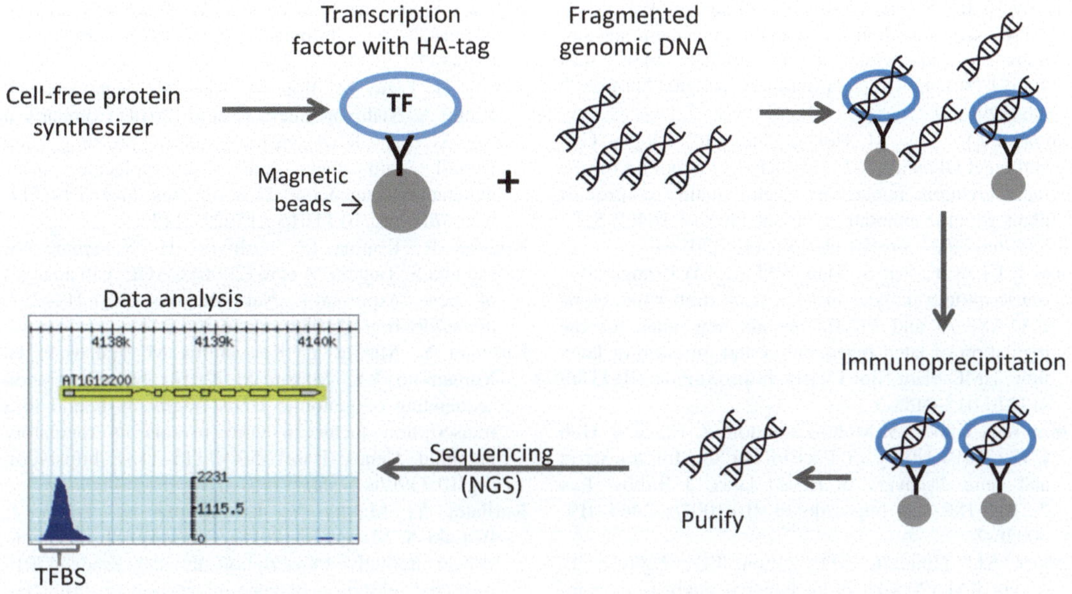

Fig. 6.6 Representation of gDB-seq that detects transcription factor-binding sites (TFBSs) on a genome-wide scale. In gDB-seq, we used an in vitro genomic DNA-binding assay coupled with immunoprecipitation and sequencing

time of *H. brasiliensis*, more conventional methods will be required. We have successfully developed gDB-seq, an in vitro genome DNA-binding assay coupled with immunoprecipitation (Fig. 6.6) (Kurihara et al. 2014). A similar method of DNA affinity purification sequencing (DAP-seq) was used to define 529 TF-binding sites in *Arabidopsis* (O'Malley et al. 2016). Transcriptional regulation is also associated with chromatin levels. Long distance gene control is now a new trend in biology (Dong et al. 2018). The assay for transposase-accessible chromatin (ATAC-seq) is a powerful technique to assess genome-wide chromatin accessibility. ATAC-seq is a fast and sensitive analysis of the epigenome compared to DNase-seq or MNase-seq (Buenrostro et al. 2015) and can be applied to the analysis of genome integrity (Maher et al. 2018).

6.5.2 Future Perspective of CAGE and Gene Regulation for Rubber Research

In this chapter, we have described genome-wide analyses of TSSs to understand transcriptional regulation, especially for proteins associated with rubber particles. In *H. brasiliensis*, analysis of genome-wide TSSs and their regulation has just begun and the data needed to reveal the effects of the environment on TSS utilization are completely lacking. To improve latex productivity and tree resistance to environmental stresses and diseases, much data collection is required. Accurate genome sequence of rubber is vital to enable the use of modern technologies to help bring about this improvement. At the same time, field performance data of a broad range of *H. brasiliensis* clones will be essential for understanding differences in latex production and disease resistance.

References

Adiconis X, Haber AL, Simmons SK, Levy Moonshine A, Ji Z, Busby MA et al (2018) Comprehensive comparative analysis of 5'-end RNA-sequencing methods. Nat Methods 15(7):505–511. https://doi.org/10.1038/s41592-018-0014-2

Andersson R, Gebhard C, Miguel-Escalada I, Hoof I, Bornholdt J, Boyd M et al (2014) An atlas of active enhancers across human cell types and tissues. Nature 507(7493):455–461. https://doi.org/10.1038/nature12787

Buenrostro JD, Wu B, Chang HY, Greenleaf WJ (2015) ATAC-seq: a method for assaying chromatin accessibility genome-wide. Curr Protoc Mol Biol 109:21.29.1-9. https://doi.org/10.1002/0471142727.mb2129s109

Campbell MS, Law M, Holt C, Stein JC, Moghe GD, Hufnagel DE et al (2014) MAKER-P: a tool kit for the rapid creation, management, and quality control of plant genome annotations. Plant Physiol 164(2):513–524. https://doi.org/10.1104/pp.113.230144

Chao J, Chen Y, Wu S, Tian WM (2015) Comparative transcriptome analysis of latex from rubber tree clone CATAS8-79 and PR107 reveals new cues for the regulation of latex regeneration and duration of latex flow. BMC Plant Biol 15:104. https://doi.org/10.1186/s12870-015-0488-3

Chow K-S, Khoo J-S, Mohd-Zainuddin Z, Ng S-M, Hoh C-C (2019) Utility of PacBio Iso-Seq for transcript and gene discovery in *Hevea* latex. J Rubber Res 22:169–186. https://doi.org/10.1007/s42464-019-00026-7

Cooper SJ, Trinklein ND, Anton ED, Nguyen L, Myers RM (2006) Comprehensive analysis of transcriptional promoter structure and function in 1% of the human genome. Genome Res 16(1):1–10. https://doi.org/10.1101/gr.4222606

Dong Q, Li N, Li X, Yuan Z, Xie D, Wang X et al (2018) Genome-wide Hi-C analysis reveals extensive hierarchical chromatin interactions in rice. Plant J 94 (6):1141–1156. https://doi.org/10.1111/tpj.13925

Dröge-Laser W, Snoek BL, Snel B, Weiste C (2018) The *Arabidopsis* bZIP transcription factor family—an update. Curr Opin Plant Biol 45(Pt A):36–49. https://doi.org/10.1016/j.pbi.2018.05.001

Epping J, Deenen NV, Niephaus E, StolzeA, Fricke J, Huber C et al (2015) A rubber transferase activator is necessary for natural rubber biosynthesis in dandelion. Nat Plants 1:15048. https://doi.org/10.1038/nplants.2015.48

Fatemi M, Pao MM, Jeong S, Gal-Yam EN, Egger G, Weisenberger DJ et al (2005) Footprinting of mammalian promoters: use of a CpG DNA methyltransferase revealing nucleosome positions at a single molecule level. Nucl Acids Res 33(20):e176. https://doi.org/10.1093/nar/gni180

Fejes-Toth K, Sotirova V, Sachidanandam R, Assaf G, Hannon GJ, Kapranov P, Foissac S, Willingham AT, Duttagupta R, Dumais E et al (2009) Post-transcriptional processing generates a diversity of 50-modified long and short RNAs. Nature 57:1028–1032

Ferreira M, Mendonça RJ, Coutinho-Netto J, Mulato M (2009) Angiogenic properties of natural rubber latex biomembranes and the serum fraction of *Hevea brasiliensis*. Braz J Phys 39(3). https://doi.org/10.1590/s0103-97332009000500010

Hurtado Páez UA, García Romero IA, Restrepo Restrepo S, Aristizábal Gutiérrez FA, Montoya Castaño D (2015) Assembly and analysis of differential transcriptome responses of *Hevea brasiliensis* on interaction with *Microcyclus ulei*. PLoS ONE 10(8): e0134837. https://doi.org/10.1371/journal.pone.0134837

Kawaji H, Lizio M, Itoh M, Kanamori-Katayama M, Kaiho A, Nishiyori-Sueki H et al (2014) Comparison of CAGE and RNA-seq transcriptome profiling using clonally amplified and single-molecule next-generation sequencing. Genome Res 24(4):708–717. https://doi.org/10.1101/gr.156232.113

Kodzius R, Kojima M, Nishiyori H, Nakamura M, Fukuda S, Tagami M et al (2006) CAGE: cap analysis of gene expression. Nat Methods 3(3):211–222. https://doi.org/10.1038/nmeth0306-211

Kurihara Y, Makita Y, Kawashima M, Hamasaki H, Yamamoto YY, Matsui M (2014) Next-generation sequencing of genomic DNA fragments bound to a transcription factor in vitro reveals its regulatory potential. Genes (Basel) 5(4):1115–1131. https://doi.org/10.3390/genes5041115

Kurihara Y, Makita Y, Kawashima M, Fujita T, Iwasaki S, Matsui M (2018) Transcripts from downstream alternative transcription start sites evade uORF-mediated inhibition of gene expression in. Proc Natl Acad Sci U S A 115(30):7831–7836. https://doi.org/10.1073/pnas.1804971115

Lau NS, Makita Y, Kawashima M, Taylor TD, Kondo S, Othman AS et al (2016) The rubber tree genome shows expansion of gene family associated with rubber biosynthesis. Sci Rep 6:28594. https://doi.org/10.1038/srep28594

Li H, Durbin R (2009) Fast and accurate short read alignment with Burrows-Wheeler transform. Bioinformatics 25(14):1754–1760. https://doi.org/10.1093/bioinformatics/btp324

Liu JP, Hu J, Liu YH, Yang CP, Zhuang YF, Guo XL et al (2018) Transcriptome analysis of *Hevea brasiliensis* in response to exogenous methyl jasmonate provides novel insights into regulation of jasmonate-elicited rubber biosynthesis. Physiol Mol Biol Plants 24(3):349–358. https://doi.org/10.1007/s12298-018-0529-0

Maher KA, Bajic M, Kajala K, Reynoso M, Pauluzzi G, West DA et al (2018) Profiling of accessible chromatin regions across multiple plant species and cell types reveals common gene regulatory principles and new control modules. Plant Cell 30(1):15–36. https://doi.org/10.1105/tpc.17.00581

Makita Y, Ng KK, Veera Singham G, Kawashima M, Hirakawa H, Sato S et al (2017) Large-scale collection of full-length cDNA and transcriptome analysis in *Hevea brasiliensis*. DNA Res. https://doi.org/10.1093/dnares/dsw056

Makita Y, Kawashima M, Lau NS, Othman AS, Matsui M (2018) Construction of Pará rubber tree genome and multi-transcriptome database accelerates rubber researches. BMC Genom 19(Suppl 1):922. https://doi.org/10.1186/s12864-017-4333-y

Mantello CC, Cardoso-Silva CB, da Silva CC, de Souza LM, Scaloppi Junior EJ, de Souza Gonçalves P et al (2014) *De novo* assembly and transcriptome analysis of the rubber tree (*Hevea brasiliensis*) and SNP markers development for rubber biosynthesis pathways. PLoS ONE 9(7):e102665. https://doi.org/10.1371/journal.pone.0102665

Mejía-Guerra MK, Li W, Galeano NF, Vidal M, Gray J, Doseff AI et al (2015) Core promoter plasticity between maize tissues and genotypes contrasts with predominance of sharp transcription initiation sites. Plant Cell 27(12):3309–3320. https://doi.org/10.1105/tpc.15.00630

Minoche AE, Dohm JC, Schneider J, Holtgräwe D, Viehöver P, Montfort M et al (2015) Exploiting single-molecule transcript sequencing for eukaryotic gene prediction. Genome Biol 16:184. https://doi.org/10.1186/s13059-015-0729-7

Morton T, Petricka J, Corcoran DL, Li S, Winter CM, Carda A et al (2014) Paired-end analysis of transcription start sites in *Arabidopsis* reveals plant-specific promoter signatures. Plant Cell 26(7):2746–2760. https://doi.org/10.1105/tpc.114.125617

Murata M, Nishiyori-Sueki H, Kojima-Ishiyama M, Carninci P, Hayashizaki Y, Itoh M (2014) Detecting expressed genes using CAGE. Methods Mol Biol 1164:67–85. https://doi.org/10.1007/978-1-4939-0805-9_7

Nepal C, Hadzhiev Y, Previti C, Haberle V, Li N, Takahashi H et al (2013) Dynamic regulation of the transcription initiation landscape at single nucleotide resolution during vertebrate embryogenesis. Genome Res 23(11):1938–1950. https://doi.org/10.1101/gr.153692.112

Ohmiya H, Vitezic M, Frith MC, Itoh M, Carninci P, Forrest AR et al (2014) RECLU: a pipeline to discover reproducible transcriptional start sites and their alternative regulation using capped analysis of gene expression (CAGE). BMC Genom 15:269. https://doi.org/10.1186/1471-2164-15-269

O'Malley RC, Huang SC, Song L, Lewsey MG, Bartlett A, Nery JR et al (2016) Cistrome and epicistrome features shape the regulatory DNA landscape. Cell 165(5):1280–1292. https://doi.org/10.1016/j.cell.2016.04.038

Pootakham W, Sonthirod C, Naktang C, Ruang-Areerate P, Yoocha T, Sangsrakru D et al (2017) *De novo* hybrid assembly of the rubber tree genome reveals evidence of paleotetraploidy in *Hevea* species. Sci Rep 7:41457. https://doi.org/10.1038/srep41457

Pumplin N, Sarazin A, Jullien PE, Bologna NG, Oberlin S, Voinnet O (2016) DNA methylation influences the expression of DICER-LIKE4 isoforms, which encode proteins of alternative localization and function. Plant Cell 28(11):2786–2804. https://doi.org/10.1105/tpc.16.00554

Sakakibara Y, Irie T, Suzuki Y, Yamashita R, Wakaguri H, Kanai A et al (2007) Intrinsic promoter activities of primary DNA sequences in the human genome. DNA Res 14(2):71–77. https://doi.org/10.1093/dnares/dsm006

Tang C, Yang M, Fang Y, Luo Y, Gao S, Xiao X et al (2016) The rubber tree genome reveals new insights into rubber production and species adaptation. Nat Plants 2(6):16073. https://doi.org/10.1038/nplants.2016.73

Tokizawa M, Kusunoki K, Koyama H, Kurotani A, Sakurai T, Suzuki Y et al (2017) Identification of *Arabidopsis* genic and non-genic promoters by paired-end sequencing of TSS tags. Plant J 90(3):587–605. https://doi.org/10.1111/tpj.13511

Ushijima T, Hanada K, Gotoh E, Yamori W, Kodama Y, Tanaka H et al (2017) Light controls protein localization through phytochrome-mediated alternative promoter selection. Cell 171(6):1316–1325.e12. https://doi.org/10.1016/j.cell.2017.10.018

Yamamoto YY, Ichida H, Abe T, Suzuki Y, Sugano S, Obokata J (2007) Differentiation of core promoter architecture between plants and mammals revealed by LDSS analysis. Nucl Acids Res 35(18):6219–6226. https://doi.org/10.1093/nar/gkm685

Yin H, Zhang X, Zhang B, Luo H, He C (2019) Revealing the dominant long noncoding RNAs responding to the infection with *Colletotrichum gloeosporioides* in *Hevea brasiliensis*. Biol Direct 14(1):7. https://doi.org/10.1186/s13062-019-0235-z

Yu CP, Lin JJ, Li WH (2016) Positional distribution of transcription factor binding sites in *Arabidopsis thaliana*. Sci Rep 6:25164. https://doi.org/10.1038/srep25164

Zhang Y, Leclercq J, Wu S, Ortega-Abboud E, Pointet S, Tang C et al (2019) Genome-wide analysis in *Hevea brasiliensis* laticifers revealed species-specific post-transcriptional regulations of several redox-related genes. Sci Rep 9(1):5701. https://doi.org/10.1038/s41598-019-42197-8

Genomics of Rubber Biosynthesis in *Hevea brasiliensis*

7

Keng-See Chow, Azlina Bahari,
Mark A. Taylor, and David F. Marshall

Abstract

The *Hevea brasiliensis* rubber tree is the largest commercial producer of natural rubber. The ability to synthesize a naturally occurring polymer is one of the main attractions of studying latex rubber biosynthesis. Natural rubber consists of high molecular weight *cis*-polyisoprene, a branch product of the isoprenoid network, which is synthesized by sequential condensations of isopentenyl diphosphate (IPP). To date, genomic technologies have enabled the generation of *Hevea* transcriptomes, draft genomes and rubber particle proteomes for a number of key tree clones (or cultivars). These have contributed significantly to the molecular genetics of genes encoding the mevalonate (MVA) and methylerythritol 4-phosphate (MEP) pathways of IPP synthesis, including rubber transferase candidates that participate in the final steps of *cis*-polyisoprene formation. Rapidly developing sequencing technologies will continue to provide opportunities to yield new insights into rubber biosynthesis. This chapter discusses the prospects of furthering genomics analysis of rubber biosynthesis in the following areas: regulation of pathways supplying and utilising IPP for rubber and non-rubber isoprenoids, characterization of rubber particle proteins in relation to molecular weight distribution and lipid droplet membrane proteins, and the potential benefits of haplotype analysis of rubber biosynthesis pathway genes for genomics-driven breeding.

Keng-See Chow—Formerly Malaysian Rubber Board

K.-S. Chow (✉)
Academy of Sciences Malaysia, Selangor, Malaysia
e-mail: kschow88@gmail.com

A. Bahari
Genetic Resources and Improvement Unit, Production Development Division, Malaysian Rubber Board, 47000 Sungai Buloh, Selangor, Malaysia

M. A. Taylor
The James Hutton Institute, Invergowrie, Dundee DD2 5DA, Scotland, UK

D. F. Marshall
Scotland's Rural College (SRUC), Peter Wilson Building, West Mains Road, Edinburgh EH9 3JG, Scotland, UK

7.1 Introduction

Natural rubber or *cis*-polyisoprene is a product of the isoprenoid pathway synthesized in the latex of the *Hevea brasiliensis* rubber tree. Notwithstanding its status as a world crop, the domestication of the *Hevea brasiliensis* tree is relatively recent in comparison with other major crops such as wheat, rice, barley, maize, potatoes, cassava, lentils and chickpeas (Murphy 2007). The success of *Hevea* as a major world crop is because latex can be harvested on a commercial scale,

M. Matsui and K.-S. Chow (eds.), *The Rubber Tree Genome*, Compendium of Plant Genomes,
https://doi.org/10.1007/978-3-030-42258-5_7

combined with the continued demand for the unparalleled mechanical and chemical properties of natural rubber.

From a biological perspective, latex is the cytoplasmic content of the laticiferous network that is ubiquitously distributed throughout the *Hevea* rubber tree (Gomez and Moir 1979, de Faÿ et al. 1989). Latex is by far the most economically valuable part of the tree. Therefore, research on improving rubber production is focused on latex biochemistry and physiology. Other plant systems are often limited by either the amount of, or access to biological material for investigation. In contrast in *Hevea*, the ease with which latex may be harvested afforded biochemists and molecular biologists copious amounts of laboratory material. Tapping rubber trees for ex vivo laboratory analysis may even be likened to having a "cytoplasmic tap". Another unique aspect of latex analysis is its amenability for separation into multiple fractions by centrifugation, a method refined by Moir (1959). Thus, 'subcellular fractionation' takes on a slightly different perspective when latex is separated into semi-purified compartments. Moir's method laid the foundation for the development of methods for the genomic analysis of latex genes and proteins by many groups (Kush et al. 1990; Tang et al. 2007; Wang et al. 2010, 2018; Thulaseedharan et al. 2012; An et al. 2012, 2013; Dai et al. 2013b).

The mechanism of rubber biosynthesis has intrigued generations of scientists. The ability of the *Hevea* tree to synthesize a natural polymer of molecular weight exceeding 1 million Daltons is a biological enigma. Evidence from biochemical experimentations has established rubber particles as the site of rubber polymerization (Archer and Audley 1987). In latex, rubber is compartmentalized within these particles and may constitute 30–50% of latex weight depending on tree genotype (Kekwick 1989). Figure 7.1 shows an overview of the general isoprenoid network. The pathway is generally similar in plants, with the exception of the branch leading to *cis*-polyisoprene synthesis which is found only in rubber-producing species. Isopentenyl diphosphate (IPP) is the basic chemical precursor of all classes of isoprenoids. Early studies on rubber biosynthesis focused on sucrose as the initial substrate, its entry into the mevalonate (MVA) pathway of IPP synthesis and the analysis of limiting steps. The cytosolic MVA pathway was accepted as the key pathway until the discovery of the methylerythritol 4-phosphate (MEP) route for IPP synthesis in plants (Rodríguez-Concepción and Boronat 2015). In *Hevea*, the presence of the MEP pathway in latex was first indicated by the discovery of ESTs encoding 1-Deoxy-D-xylulose 5-phosphate synthase (DXS) (Han et al. 2000; Chow et al. 2007). The presence of an alternative supply pathway of the IPP substrate adds a potential new level of complexity in the identification of biosynthetic pathway genes and the impact of their expression on *cis*-polyisoprene synthesis.

The stages of *cis*-polyisoprene synthesis are well-studied and amply reviewed (Cornish 2014; Cherian et al. 2019). Following the isomerization of IPP to dimethylallyl diphosphate (DMAPP) (Fig. 7.1), sequential condensations of IPP by cytosolic *trans*-prenyltransferases produce the initiator molecules, geranyl diphosphate (GPP), farnesyl diphosphate (FPP) and geranyl geranyl diphosphate (GGPP). These steps are followed by the addition of IPP units in the *cis*-configuration through the activity of rubber transferase. A rubber transferase enzyme for the synthesis of *cis*-rubber has never been purified or a gene encoding that activity sequenced. The prevailing thinking is that the *Hevea* transferase is actually a complex of proteins working in concert both spatially and enzymatically to make high-molecular weight rubber. The identification of candidate genes and proteins underpinning rubber transferase activity in *Hevea* is one of the most important contributions of genomics and bioinformatics in providing insights into the mechanism of *cis*-polyisoprene formation.

Since the late 1990s, building upon findings from conventional latex biochemistry and physiology, DNA sequencing and omics approaches are increasingly changing the landscape of rubber biosynthesis research. This chapter will discuss the contribution of *Hevea* genome and transcriptome sequencing in elucidating the molecular genetics of rubber biosynthesis-related genes, and how combined with laboratory

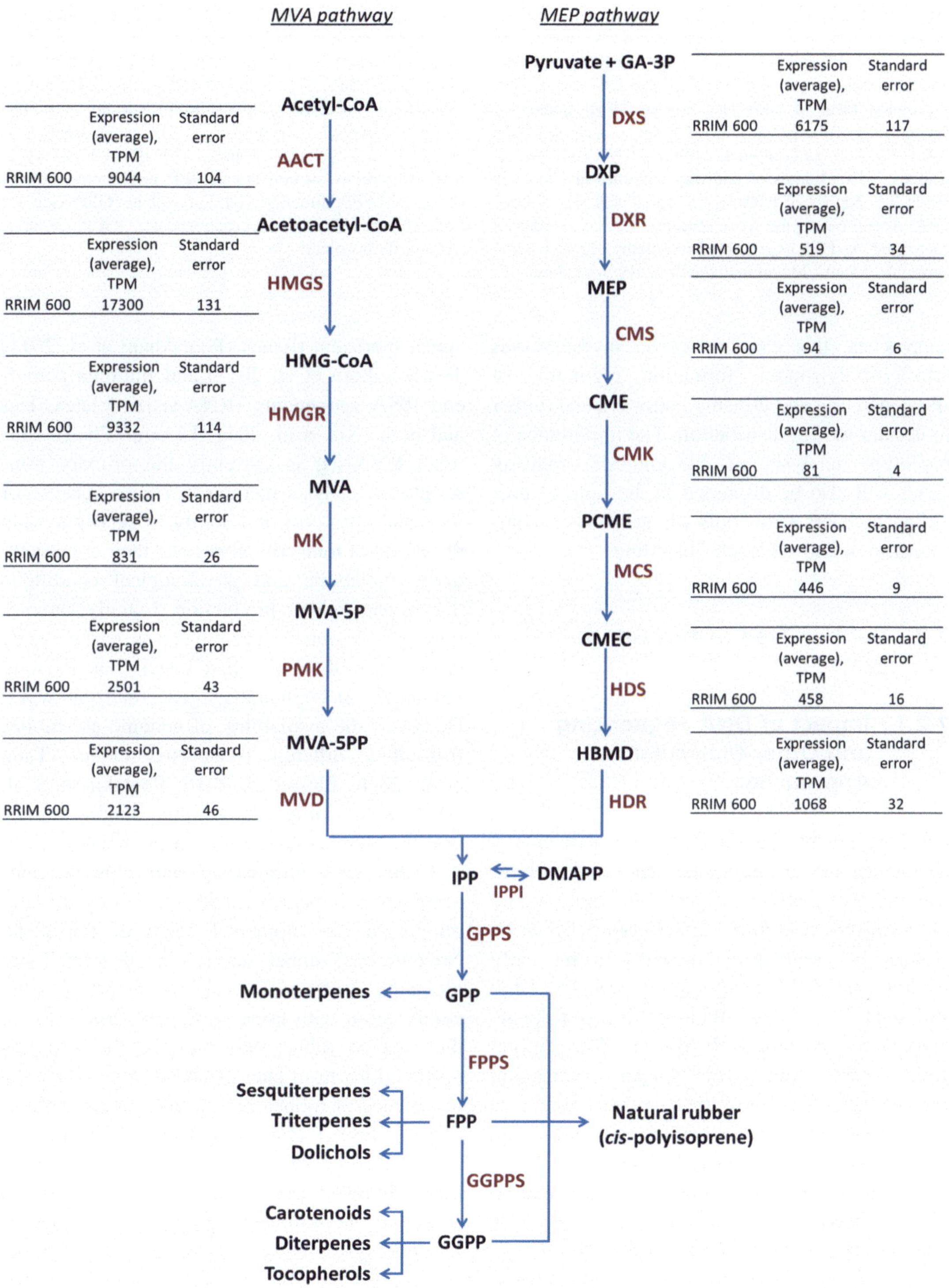
MVA pathway
MEP pathway
Acetyl-CoA
AACT
Expression (average), TPM Standard error
RRIM 600 9044 104
Acetoacetyl-CoA
HMGS
Expression (average), TPM Standard error
RRIM 600 17300 131
HMG-CoA
HMGR
Expression (average), TPM Standard error
RRIM 600 9332 114
MVA
MK
Expression (average), TPM Standard error
RRIM 600 831 26
MVA-5P
PMK
Expression (average), TPM Standard error
RRIM 600 2501 43
MVA-5PP
MVD
Expression (average), TPM Standard error
RRIM 600 2123 46
Pyruvate + GA-3P
DXS
Expression (average), TPM Standard error
RRIM 600 6175 117
DXP
DXR
Expression (average), TPM Standard error
RRIM 600 519 34
MEP
CMS
Expression (average), TPM Standard error
RRIM 600 94 6
CME
CMK
Expression (average), TPM Standard error
RRIM 600 81 4
PCME
MCS
Expression (average), TPM Standard error
RRIM 600 446 9
CMEC
HDS
Expression (average), TPM Standard error
RRIM 600 458 16
HBMD
HDR
Expression (average), TPM Standard error
RRIM 600 1068 32
IPP
IPPI
DMAPP
GPPS
Monoterpenes
GPP
FPPS
Sesquiterpenes
Triterpenes
Dolichols
FPP
Natural rubber
(*cis*-polyisoprene)
GGPPS
Carotenoids
Diterpenes
Tocopherols
GGPP

Fig. 7.1 Overview of the general isoprenoid biosynthesis in plants showing the mevalonate (MVA) and MEP pathways of IPP synthesis, and routes of rubber and non-rubber product synthesis in the post-IPP steps. The expression level of each enzyme (in TPM, transcripts per million) of the MVA and MEP steps in clone RRIM 600 is the average computed from six RNA-seq datasets (Bahari 2019). Names of pathway enzymes are listed in Table 7.1. Names of pathway intermediates are: Acetyl-CoA, Acetyl coenzyme A; Acetoacetyl-CoA, Acetoacetyl coenzyme A; HMG-CoA, Hydroxymethylglutaryl coenzyme A; MVA, Mevalonate; MVA-5P, Mevalonate-5-phosphate; MVA-5PP, Mevalonate-5-diphosphate; GA-3P, D-Glyceraldehyde-3-phosphate; DXP, 1-Deoxy-D-xylulose-5-phosphate; MEP, 2-C-methyl-D-erythritol 4-phosphate; CME, 4-(cytidine 5′-diphospho)-2-C-methyl-D-erythritol; PCME, 2-phospho-4-(cytidine 5′-diphospho)-2-C-methyl-D-erythritol; CMEC, 2-C-methyl-D-erythritol 2,4-cyclodiphosphate, HBMD, 1-Hydroxy-2-methyl-2-butenyl 4-diphosphate; IPP, Isopentenyl diphosphate; DMAPP, dimethylallyl diphosphate; GPP, Geranyl diphosphate; FPP, Farnesyl diphosphate; GGPP, Geranylgeranyl diphosphate

approaches, this can impact on investigations into *cis*-polyisoprene formation, regulation of IPP generating and utilising pathways and rubber molecular weight distribution. The importance of haplotype analysis of biosynthesis pathway genes will also be discussed in the light of providing breeders with tools for germplasm characterization and molecular breeding.

7.2 Rubber Tree in the Genomics Era

7.2.1 Impact of DNA Sequencing and Other Molecular Approaches

Until the advent of automated DNA sequencing, characterization of individual genes and proteins was the staple of *Hevea* molecular biology and biochemistry. One of the first *Hevea* genes to be cloned and sequenced manually using early recombinant DNA methodology was the large sub-unit of ribulose-1,5-bisphosphate carboxylase/oxygenase (RuBisCo) (Cheong and Koh 1993). Automated Sanger sequencing marked the beginning of the Expressed Sequence Tags (EST) era in plants, beginning with the first 1,500 ESTs from the model plant, *Arabidopsis* (Newman et al. 1994). Latex was the natural primary choice of *Hevea* tissue type for EST analysis (Han et al. 2000; Ko et al. 2003; Chow et al. 2007). The development of second-generation sequencing platforms then led to a step change in the generation of *Hevea* transcript data, e.g. 454 pyrosequencing of leaf and shoot apical meristem tissues (Pootakham et al. 2011; Triwitayakorn et al. 2011) and Illumina paired-end RNA sequencing (RNA-seq) of latex, leaf and bark (Xia et al. 2011; Li et al. 2012). Illumina RNA-seq is currently the primary transcriptome analysis method in *Hevea* because of its scalability, cost and ability to survey a wide repertoire of transcript genes and their expression under metabolic and physiological conditions that impact rubber production. Initially, annotation of assembled transcripts was based on de novo assembly and relied heavily on existing ontologies of comparative genes and pathways. However, the availability of genome assemblies from three different rubber tree clones (Tang et al. 2016; Lau et al. 2016; Pootakham et al. 2017) is providing increasingly valuable resources for reference-based transcript analysis.

Unlike gene sequencing, transcriptome characterisation is dependent on the timing of sampling and developmental stage of the plant. Nevertheless, rubber research has benefitted significantly from gene expression profiling in combination with large scale gene discovery. In the area of rubber biosynthesis, the two key observations from latex RNA sequencing are the prevalence of rubber particle membrane proteins (i.e. the rubber elongation factor, REF and the small rubber particle protein, SRPP) among the most abundant transcripts, and evidence for a potential role of the MEP pathway providing IPP for *cis*-polyisoprene based on the detection of the enzyme for the first step, DXS (Ko et al. 2003; Chow et al. 2007). A combination of Sanger and RNA-seq transcript sets was able to survey the extent of rubber and non-rubber isoprenoid branch products as an alternative to metabolite analysis (Chow et al. 2012). Promoters of highly expressed Sanger ESTs were characterised for the purpose of understanding mechanisms of laticifer-specific gene expression (Aoki et al. 2014). When a high-yielding *Hevea* genotype (SHY) from China was identified, Tang et al. (2013) analysed its features of "super-productivity" using cDNA-AFLP. To explore the biosynthesis capacity of two high latex-yielding clones (GT 1 and PR 255), SNPs were identified and validated for MVA and MEP pathway genes found among RNA-seq transcripts (Mantello et al. 2014). Rubber clones displaying contrasting phenotypes relating to physiology of latex flow and regeneration, ethylene responses, laticifer morphology, elite performance and other tree productivity aspects were analysed by comparative transcriptomics and proteomics (Chao et al. 2015; Li et al. 2015; Dai et al. 2016; Li et al. 2016a, b; Tan et al. 2017).

Fully characterizing the members of a gene family is important because of the potential for sub-functionalisation of different transcript isoforms. In an effort to optimize transcript coverage, a study on RNA-seq read depth requirement suggested that a 5–8 Gb read amount is a reasonable compromise for achieving high transcript number and N50 length (Chow et al. 2014). To enhance annotation of rubber biosynthesis genes, Makita et al. (2017) generated 23,843 unigenes by a combination of Sanger EST sequencing and Illumina RNA-seq of full-length cDNA libraries from several *Hevea* tissues. In addition, this study reported preliminary statistics of transcription start sites, and analysis of laticifer genes and transcription factors (TFs) that are abundantly expressed in latex. Continuous effort is needed to validate genome annotations as transcript resources grow and improve. For *Hevea*, it is crucial to constantly update models for genes related to rubber biosynthesis pathways. In recent years, a third-generation sequencing platform, PacBio Iso-seq, has been widely adopted in plants. The key benefit of Iso-seq which outweighs its higher cost is longer read-length for elucidation of novel transcripts and alternatively spliced isoforms, although the high error rate is still an issue. Currently, *Hevea* Iso-seq resources are limited to a small dataset from BPM 24 leaf (Pootakham et al. 2017) and a larger one from RRIM 600 latex (Chow et al. 2019). The benefits and cost reductions of long read technologies such as Pac-Bio and Oxford Nanopore will inevitably result in more high-quality data in the near future for both genome and transcriptome analysis.

7.2.2 Molecular Genetics of Genes Related to Pathways of Rubber Biosynthesis

7.2.2.1 MVA and MEP Pathways

The MVA and MEP pathways consist of six and seven steps respectively (Fig. 7.1). Many genomic and cDNA sequences encoding enzymes controlling MVA and MEP pathway steps in *Hevea* have been cloned using conventional approaches. Concurrent detection of transcripts of all MVA and MEP pathway steps in the same experiment has been recorded by both Sanger EST sequencing (Sando et al. 2008a, b) and RNA-seq (Li et al. 2012; Mantello et al. 2014). Subsequently, the availability of rubber genome annotations made possible the characterization of complete gene families on a genome-wide scale (Tang et al. 2016; Lau et al. 2016; Pootakham et al. 2017). Recently, Bahari (2019), assembled a new set of *Hevea* reference transcripts and mapped them to the Reyan 7-33-97 genome assembly (Tang et al. 2016). A total of 41 gene models were identified, with 19 for the MVA and 22 for the MEP pathways (Table 7.1). This is marginally more than the number of gene family members found by Tang et al. (2016) i.e.18 for MVA and 22 for MEP pathways). The reason for this could be the incorporation of additional transcriptome sequences in the new reference transcriptome (Bahari 2019).

The same mapping analysis revealed the expression pattern of gene isoforms in the rubber clone, RRIM 600 (Fig. 7.1). Generally, transcription level of the MVA pathway genes appeared higher than that of the MEP pathway

Table 7.1 Gene models and isoforms for genes related to MVA and MEP pathway steps based on mapping of a *Hevea* reference trancriptome to the Reyan 7-33-97 genome

Pathway	Gene	Number of identified isoforms
MVA	Acetyl-CoA acetyltransferase (AACT)	4
	Hydroxymethylglutaryl-CoA synthase (HMGS)	2
	Hydroxymethylglutaryl-CoA reductase (HMGR)	6
	Mevalonate kinase (MK)	3
	Phosphomevalonate kinase (PMK)	2
	Diphosphomevalonate decarboxylase (MVD)	2
MEP	1-Deoxy-D-xylulose 5-phosphate synthase (DXS)	10
	1-Deoxy-D-xylulose 5-phosphate reductoisomerase (DXR)	2
	2-C-methyl-D-erythritol 4-phosphate cytidylyltransferase (CMS)	2
	4-(Cytidine 5′-diphospho)-2-C-methyl-D-erythritol kinase (CMK)	2
	2-C-methyl-D-erythritol 2,4-cyclodiphosphate synthase (MCS)	2
	4-Hydroxy-3-methylbut-2-enyl diphosphate synthase (HDS)	2
	4-Hydroxy-3-methylbut-2-enyl diphosphate reductase (HDR)	2
Formation of isoprenoid initiators	Isopentenyl diphosphate isomerase (IPPI)	2
	Geranyl diphosphate synthase (GPPS)	5
	Farnesyl diphosphate synthase (FPPS)	4
	Geranylgeranyl diphosphate synthase (GGPPS)	5
Rubber formation steps	Rubber elongation factor (REF)	8
	Small rubber particle protein (SRPP)	10
	cis-prenyl transferase (CPT)	6
	Rubber biosynthesis stimulator protein (RBSP)	5
	Rubber biosynthesis inhibitor protein (RBIP)	2

step transcripts. This trend of relative transcriptional activity appears to reflect observations in previous studies and was seen not just in clone RRIM 600 but also in PB 350 and RRIM 901 (Sando et al. 2008a, b; Chow et al. 2012; Makita et al. 2017). Current knowledge of the features and regulation of IPP-generating pathways in *Hevea* is very much modelled on what is known in other plants. Further conclusions about the MVA pathway being the main IPP source for rubber biosynthesis requires considerations of other aspects such as trafficking of pathway intermediates between subcellular compartments, post-transcriptional and post-translational regulation and quantification of downstream rubber and non-rubber products (see Sect. 7.3.1). This is because transcript levels are only one component in the determination of flux through metabolic pathways.

7.2.2.2 Final Steps of Rubber Polymerisation

In *Hevea*, the precise steps and mechanism of the post-IPP branch of *cis*-polyisoprene synthesis by the rubber transferase remain to be fully elucidated. Synthesis of initiators of the rubber molecule by *trans*-prenyltransferases is well understood but not the mechanism of *cis*-rubber formation. The rubber transferase is postulated to be a protein complex embedded in the rubber particle membrane (Cornish 2001; Cornish and Xie 2012). Currently, the mechanistic action of rubber transferase is thought to depend on a set of either cytosolic or membrane-bound proteins. Throughout decades of research, several prime candidates have been proposed as components of this complex: REF, SRPP, *cis*-prenyltransferase (CPT), a rubber biosynthesis stimulator protein (RBSP) and a rubber biosynthesis inhibitor protein (RBIP). More recently, the participation of binding proteins (Yamashita et al. 2016; Brown et al. 2017) and accessory peptides (Cornish et al. 2018) have been highlighted. The following sections summarise the molecular genetics and functional investigations by various groups in attempts to piece together the main reactions involved in rubber polymerisation.

REF and SRPP

In any Coomassie Blue stained gel separating total latex proteins, the two most abundant bands correspond to the REF and SRPP proteins (approximately 14 and 22 kDa respectively). Collectively, they constitute the most highly expressed transcript components in *Hevea* latex (Ko et al. 2003). These membrane-bound rubber particle proteins started to receive attention at the same time as the emergence of recombinant DNA technology applications in rubber and latex allergy research. Initially, a 14.6 kDa protein proposed to play a role in rubber formation was named REF and its amino acid sequence determined (Light and Dennis 1989; Dennis et al. 1989). Soon after, a REF cDNA was cloned and sequenced (Goyvaerts et al. 1991; Attanyaka et al. 1991). In latex allergy studies, Hev b 1 and Hev b 3 proteins originating from rubber particle membranes are the major allergens which spina bifida patients are sensitised to as a result of multiple surgical procedures (Yeang et al. 1996). Tryptic peptide sequences of the Hev b 1 allergen identified it as an REF (Czuppon et al. 1993). Subsequently, Hev b 3 was named as a 22.4 kDa SRPP through the discovery of an abundant EST encoding the allergen (Oh et al. 1999) and following resolution of variant allergen sizes of Hev b 3 (Yeang et al. 1996, 1998). Concurrently, SRPP was reported to be involved in the incorporation of IPP in biosynthesis assays containing washed rubber particles (Oh et al. 1999).

REF and SRPP have been known to share high levels of amino acid sequence similarity (Oh et al. 1999). From the latex EST data, the REF/SRPP family was extended to at least nine variant proteins (Chow et al. 2007). These included the initially cloned REF and SRPP cDNAs (Goyvaerts et al. 1991; Attanyaka et al. 1991; Oh et al. 1999) together with additional REF-like and SRPP-like sequences encoding proteins ranging from 11–27 kDa. In retrospect, this provided a genetic basis for the previous detection of Hev b 3 protein isoforms sizes ranging from 23–27 kDa by different latex allergy research groups, originally thought to be due to post-translational modification (Yeang et al. 1998) and possibly protein fragmentation or aggregation (Yeang et al. 1996). Subsequently, Western immunoblots using antibodies to REF and SRPP detected protein bands corresponding to several of the REF-like and SRPP-like ESTs (Chow et al. 2007). More recent genome analysis has revealed that the REF/SRPP gene family is even larger. Rahman et al. (2013) found 12 REFs and 10 SRPPs and Lau et al. (2016) reported 9 REFs and 8 SRPPs in their respective genome assemblies of clone RRIM 600. Using the Cap Analysis Gene Expression (CAGE) technique, Lau et al. (2016) found tissue-dependent transcription start sites for some of these isoforms, this indicating another regulatory level of gene expression. Based on a high-resolution genome assembly of Reyan 7-33-97, Tang et al. (2016) extended the gene family to 18 REF/SRPP members. This large scale expansion of the REF/SRPP gene family appears to be a unique feature of *Hevea*. Based on variation in the length of the carboxy-terminus (absent in REF), they were designated SRPP1-10 and REF1-8, whereby REF1 and SRPP1 coincided with the initially cloned 14.6 kDa REF and 22.4 kDa SRPP isoforms respectively. Four isoforms, REF1, SRPP1, REF3 and REF7, were found to account for 96.8% of REF/SRPP gene expression in latex and therefore regarded as laticifer-specific. Phylogenetic analysis of the 18 REF/SRPP sequences in *Hevea* indicates that they are descended from a common ancestor; however, REF1 emerged later than SRPP1 in the evolutionary timeline (Tang et al. 2016).

Parallel to genome sequencing, proteomic analysis of rubber particle membrane proteins also revealed size variants of REF and SRPP (Xiang et al. 2012; Dai et al. 2013a, 2017), only some of which were validated by additional characterization. Two novel REF isoforms, 23.6 kDa and 27.3 kDa in size, were identified by 16BAC/SDS-PAGE followed by in-gel trypsin digest and MALDI TOF/TOF MS, and subsequently validated by transcript sequence search and analysis, Western blotting and RACE amplification (Dai et al. 2017). The 27.3 kDa protein is an example of an REF isoform which was not elucidated from genome sequencing. Both proteins, also identified from clone Reyan 7-33-97, were found to be very highly similar to the REF8 protein reported by Tang et al. (2016). An explanation for this could be the occurrence of alternatively spliced forms of REF8 as shown from characterisation of a latex Iso-seq transcriptome (Chow et al. 2019).

The 18 Reyan 7-33-97 REF/SRPPs were distributed between seven scaffolds (Tang et al. 2016). However, one of the scaffolds (scaffold_1222) harboured most of the isoforms (Fig. 7.2). Comparison of 63 REF/SRPP proteins from 13 plant species showed that 13 REF/SRPP isoforms were phylogenetically clustered into a single clade. Twelve of them were co-located on scaffold_1222, these including the four laticifer-specific isoforms (REF1, SRPP1, REF3 and REF7) (Fig. 7.2). Taken together, these findings point to the significance of the genomic region associated with scaffold_1222, in the evolution of the REF/SRPP gene family towards enhancing rubber biosynthesis. REF and SRPP forms on the other six scaffolds may be related to stress tolerance (while also not being laticifer-specific). These isoforms clustered together with REF and SRPP homologs that were previously reported to be involved in other physiological functions. For example, high expression of REF and SRPP homologs from *Arabidopsis* (At1g67360, At2g47780 and At3g05500) showed higher survival rates in transgenic seedlings under water stress (Kim et al. 2016). Another aspect of the biological roles of REF and SRPP that is of increasing interest is the evolutionary and functional relationship between membrane proteins of rubber particles and plant lipid droplets, both of which share a similar membrane monolayer structure (Murphy 2012; Chapman et al. 2012; Huang 2018) (see Sect. 7.3.2).

Given the significance of scaffold_1222 in rubber biosynthesis, the organisational structure of REF and SRPP isoforms was investigated in the genome assemblies of three clones. Firstly, the same REF/SRPP gene family of 18 members in clone Reyan 7-33-97 as reported by Tang et al. (2016) were found in the rubber genome assemblies of RRIM 600 (Lau et al. 2016) and BPM 24 (Pootakham et al. 2017) (Table 7.2). Secondly, REF/SRPPs isoforms were located in two scaffolds in RRIM 600 and BPM 24, whereby all 12 REF and SRPP genes on scaffold_1222 were identified in the BPM 24 genome but only 10 were found in the RRIM 600 genome (Fig. 7.2). The general order of isoform location is similar in all three clones but there are indications of inversion of chromosomal segments associated with these scaffolds (Fig. 7.2). This may be indicative of a genetic recombination event or scaffolding errors during genome assembly. In-depth analysis of genomic organization of REF/SRPP gene family members will be beneficial for understanding its impact on the capacity of the *Hevea* tree to synthesize rubber (see Sect. 7.3.3).

In this chapter, generally the names REF and SRPP have been used to refer to the first proteins reported by Light and Dennis (1989) and Oh et al.

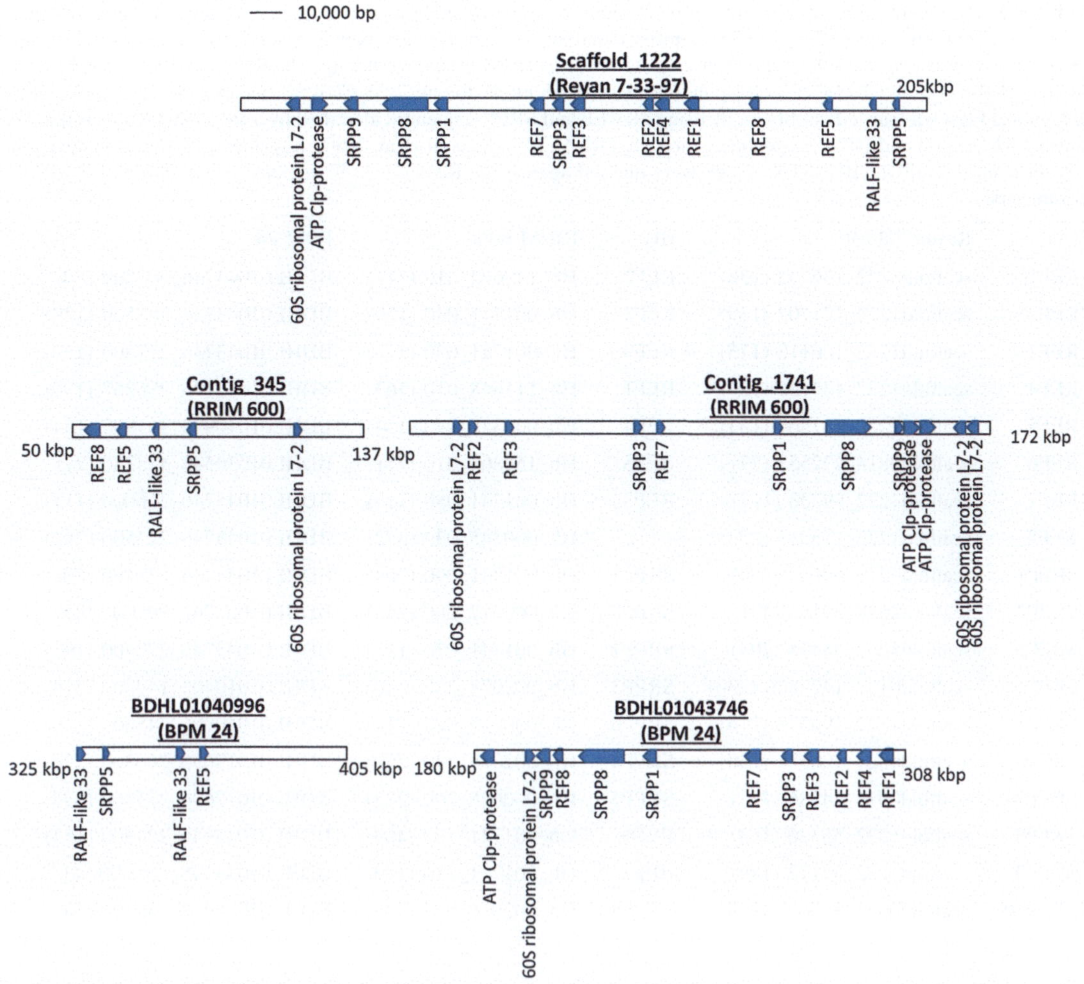

Fig. 7.2 Organisational structure of 12 REF and SRPP genes in the genome assemblies of tree clones Reyan 7-33-97, RRIM 600 and BPM 24. Only Reyan 7-33-97 has a single scaffold harbouring 12 REF and SRPP isoforms (scaffold_1222)

(1999), except while discussing their multiple forms. Currently, there is some confusion arising from arbitrary numbering of REF and SRPP isoforms by different genome sequencing projects (Tang et al. 2016; Lau et al. 2016; Pootakham et al. 2017). This issue also applies to descriptions of size variants of REFs and SRPPs from proteomics analysis (Dai et al. 2017; Tong et al. 2017). To resolve potential confusion, there is an immediate need to reconcile the different names assigned to REF/SRPP isoforms in research papers and to develop a nomenclature that can be shared by the global rubber community.

Cis-Prenyltransferase

Cis-prenyltransferases (CPTs) are a class of enzymes which catalyse the synthesis of polyisoprenoids by sequential addition of IPP. Unlike *trans*-prenyltransferases which generate products with chain lengths from C10-C50, CPTs make products ranging from C10 to C>10,000. *Cis*-prenyltransferases may be classified into subfamilies which make short (C15), medium (C50-55), long (C70-120) and extra-long polyisoprenes (C>10,000) (Grabińska et al. 2016). While other plant CPTs make shorter polyprenols such as dolichols (Cunillera et al. 2000; Brasher et al.

Table 7.2 REF and SRPP isoforms in three rubber tree clones (predicted protein length in parantheses). The scaffold in each row contains the same REF or SRPP isoform. Columns (A) and (B) show names assigned to each isoform by Tang et al. (2016) for Reyan 7-33-97 and by Lau et al. (2016) for RRIM 600, respectively. The REF and SRPP sequences in clones RRIM 600 and BPM 24 corresponding to those in clone Reyan 7-33-97 were obtained by reciprocal blast. Reciprocal blast was carried out firstly, by using 18 REF and SRPP sequences identified by Tang et al. (2016) as queries against RRIM 600 and BPM 24 genomes through TBLASTN (Altschul et al. 1997) (e-value < 1e-10). Secondly, the top hit from the initial TBLASTN results were queried against the Reyan 7-33-97 genome using similar TBLASTN parameters

(A)	Reyan 7-33-97	(B)	RRIM 600	BPM 24
REF1	scaffold1222_136753 (138)	REF2	Hb_149683_010 (117)	BDHL01043746_305800 (80)
REF2	scaffold1222_121702 (139)	REF3	Hb_001741_040 (129)	BDHL01043746_288500 (139)
REF3	scaffold1222_100110 (175)	REF9	Hb_001741_030 (175)	BDHL01043746_277500 (130)
REF4	scaffold1222_124165 (164)	REF1	Hb_111648_010 (163)	BDHL01043746_296267 (130)
REF5	scaffold1222_181260 (154)	REF4	Hb_000345_040 (154)	BDHL01040996_361900 (154)
REF6	scaffold0818_272553 (77)	REF5	Hb_152868_010 (77)	BDHL01030548_473280 (77)
REF7	scaffold1222_89338 (117)	REF8	Hb_001741_080 (117)	BDHL01043746_266000 (117)
REF8	scaffold1222_175215 (222)	-	Hb_000345_030 (372)	BDHL01043746_223500 (282)
SRPP1	scaffold1222_60641 (204)	SRPP1	Hb_001741_090 (204)	BDHL01043746_229400 (204)
SRPP2	scaffold2538_3915 (243)	SRPP2	Hb_000617_260 (243)	BDHL01002762_99500 (243)
SRPP3	scaffold1222_95474 (204)	SRPP7	Hb_001741_070 (152)	BDHL01043746_272400 (197)
SRPP4	scaffold0197_1429806 (230)	SRPP3	Hb_002078_240 (230)	BDHL01010403_149000 (214)
SRPP5	scaffold1222_196376 (216)	SRPP6	Hb_000345_070 (216)	BDHL01040996_345000 (216)
SRPP6	scaffold0824_400587 (230)	SRPP4	Hb_002078_240 (230)	BDHL01017696_766300 (230)
SRPP7	scaffold0916_24536 (223)	SRPP5	Hb_000635_200 (223)	BDHL01017696_275800 (223)
SRPP8	scaffold1222_55336 (153)	REF6	Hb_001741_100 (469)	BDHL01043746_224500 (153)
SRPP9	scaffold1222_37173 (189)	REF7	Hb_001741_110 (189)	BDHL01043746_ 202,000 (177)
SRPP10	scaffold0624_516697 (148)	SRPP8	Hb_002016_040 (148)	BDHL01010403_150100 (148)

2015), *Hevea* rubber transferase is the CPT responsible for synthesizing high molecular weight natural rubber (Cornish 2014). Identification and functional characterisation of plant CPTs have lagged behind those in yeast and bacteria. This explains why CPT gene identification in plants has been reliant on sequence similarities with non-plant homologs (Oh et al. 2000; Cunillera et al. 2000; Schmidt et al. 2010; Liu et al. 2011; Akhtar et al. 2013).

Despite its key role in rubber polymerisation, CPT is not among the most highly expressed genes observed in *Hevea* latex transcriptome profiling (Ko et al. 2003; Chow et al. 2007). The earliest record of *Hevea* CPT sequences were a set of 12 ESTs deposited by Dupont, USA in the NCBI Nucleotide Database in 2002 (AY124464-74 and AY124934) without any characterization information. Asawatreratanakul et al. (2003) reported the cloning of two CPT transcripts, HRT1 and HRT2. HRT2 enabled synthesis of medium to long chain rubber in vitro. A more detailed picture of the *Hevea cis*-prenyltransfrase gene family comes from the recent genome assemblies. Initially, Rahman et al. (2013) identified eight CPTs (CPT 1–8) from a RRIM 600 draft genome, which were categorized into three different phylogenetic clades, namely CPT sequences with high similarity to HRT1 and HRT2, to undecaprenyl diphosphate synthase (UPPS) and to dehydrodolichyl diphosphate synthase (DHDDS). Subsequently, more than ten CPT genes were detected in the Reyan 7-33-97 genome assembly (Tang et al. 2016) and eight CPTs (designated CPT1, CPT2, CPT3, CPT 4, CPT5, CPT6, CPT7 and CPTL) were reported in an updated assembly of the RRIM 600 genome (Lau et al. 2016). The CPT2 isoform was identified to be the same as HRT2 (Asawatreratanakul et al. 2003). Other than that, neither of these subsequent genome-deduced CPT forms included classification into subfamilies. More recently, the multiple *Hevea* CPT sequences were resolved and updated by Uthup et al. (2019) through extensive analysis and curation of available genomic and transcriptomic resources. Based on sequence similarities to known CPT subfamilies, a distinction was made between the CPT sequences identified from the RRIM 600 genome (Lau et al. 2016), those which make high molecular weight rubber and those that synthesise shorter chain polyprenols. CPT2 and CPT3 were found to be the same and both CPT1 and CPT2/3 corresponded to the previously characterised *Hevea* HRT1 and HRT2 sequences. The remaining CPT members (CPT 4, CPT5, CPT6, CPT7 and CPTL) showed strong homology with *Hevea* DHDDS which is responsible for making dolichols. In conclusion, only two major isoforms of CPT that have been cloned so far appear to be associated with rubber biosynthesis. As with the REF/SRPP gene family, *Hevea* CPTs and DHDDS should be early targets in the development of a standardized nomenclature for rubber biosynthesis-related genes.

Without comprehensive functional characterization of cloned CPT sequences, determination of bona fide CPT(s) involved in rubber formation in *Hevea* cannot be ascertained. Moreover, a *Hevea* CPT enzyme has not been purified to date and no peptide sequences are available for validation of DNA sequences (Cornish and Xie 2012). IPP incorporation in vitro by recombinant *Hevea* CPTs has been reported independently previously (Asawatreratanakul et al. 2003; Ko et al. 2003) but under different rubber biosynthesis assay conditions. However, *Hevea* CPT expressed in yeast and *Arabidopsis* could not produce any rubber (Takahashi et al. 2012). Taking a cue from the discovery of CPT-binding proteins in plants (Brasher et al. 2015; Epping et al. 2015; Qu et al. 2015), it was proposed that other *Hevea*-specific components are required to promote the formation of *cis*-polyisoprene. Using a yeast 2-hybrid system, a Nogo B receptor homologue protein (HRBP) was isolated. HRBP acts as bridging protein forming a ternary complex with REF and HRT1 that is required for *cis*-polyisoprene synthesis (Yamashita et al. 2016). Surprisingly, blast searches of the Reyan 7-33-97 genome assembly and public *Hevea* sequence resources with a dandelion homolog (Epping et al. 2015) did not yield any Nogo B receptor hits (as noted by Tang et al. 2016). Given its significant role, the gene family and genome organization of HRBP should be explored further for application in elucidating topological requirements of membrane proteins in rubber polymerization.

Rubber Biosynthesis Stimulator and Inhibitor Proteins

Unlike REF, SRPP and CPT, the molecular genetics of a rubber biosynthesis stimulator and an inhibitor protein have been relatively less well characterised. RBSP (rubber biosynthesis stimulator protein) and RBIP (rubber biosynthesis inhibitor protein) were initially detected through in vitro rubber biosynthesis assays containing washed rubber particles combined with different fractions of re-constituted C-serum (equivalent to the latex cytosol) (Archer and Audley 1987). As their names suggest, these proteins either increased or decreased the incorporation of radioactively labelled IPP in the assays. Not long after, Yusof et al. (1998, 2000) undertook the biochemical purification and characterisation of these proteins and further confirmed their effects by similar assays. These studies indicated that RBIP blocks the incorporation of IPP (polymerization stage) into the elongating rubber molecule (Yusof et al. 1998) while RBSP stimulates the initiation stage (Yusof et al. 2000), as postulated by Archer and Audley (1987).

Mass spectroscopy showed purified RBIP and RBSP to have molecular weights of 43.7 kDa and 13.1 kDa respectively. Peptide sequencing revealed high similarity of RBIP to patatin, a potato protein. Beyond the cloning of the Hev b 7 allergen which was found to encode a patatin-like protein (Kostyal et al. 1998), the molecular genetics of RBIP have not been investigated.

However, latex patatin-like proteins described by other studies indicate there are likely to be several different isoforms in latex (Yusof et al. 1998). Peptide sequences of RBSP showed homology to the eukaryotic initiation factor 5A (eIF-5A), a factor involved in formation of the first peptide during protein synthesis (Park et al. 1997). Screening a cDNA library identified a family of nine isoforms encoding seven variant proteins (17.5–17.6 kDa in molecular weight), six of which were full-length (Chow et al. 2003). In addition to assays utilising C-serum as the source of RBSP, lysates of five recombinant proteins (insufficient expression of the sixth isoform) were demonstrated to enhance IPP incorporation when added in increasing amounts to in vitro rubber biosynthesis assays (Chow et al. 2006).

Without full protein information, it cannot be ascertained whether the Hev b 7 sequence or any of the eIF-5A cDNAs represent the purified native proteins. It is not clear how the RBIP and RBSP proteins enable inhibitory or stimulatory effects during rubber biosynthesis. Since both originate from the C-serum, we at least know that they are cytosolic proteins, and could somehow play an accessory role in rubber polymerisation. It was interesting to note the presence of an eIF-5A in a recent proteomic profiling of rubber particle proteins (Xiang et al. 2012). Suffice to add that the molecular genetics and functional analysis of RBSP and RBIP need to be revived after a long hiatus.

Unusual Protein Subunits

In an effort to identify other essential members of a membrane-bound complex facilitating *cis*-polyisoprene polymerization, a photoaffinity labeling approach was recently adopted (Cornish et al. 2018). Allylic diphosphates linked with benzophenone as photo-affinity labels were used as substrates of rubber transferase in assays containing washed rubber particles from three species, including *Hevea*. Following ultra-violet irradiation and analysis of labelled target proteins on SDS-gels, two small proteins (1.5 kDa and 36.5 kDa) and a much larger one (241 kDa) were identified in *Hevea*. The large protein is membrane-bound while the two small proteins are strongly associated with the large protein, thus suggesting this to be a three-dimensional structure of the rubber transferase complex (Cornish et al. 2018). The current availability of growing genomic and transcriptomic sequence resources in *Hevea* is expected to identify the genes before long. Given their extreme molecular weights and assuming the largest subunit of 241 kDa is not a product of protein aggregation, one challenge could be their absence from transcriptomic resources due to low transcript abundance.

7.3 Prospects for the Future

A growing range of genomic technologies are available for the investigation of the metabolic complexity underlying the productivity of crop plants. Natural rubber biosynthesis is no exception. However, there are many biological and technical hurdles posed by a non-model tree species like *Hevea* that need to be overcome. Preliminary investigations are often needed to address limitations such as access to samples and feasibility of sampling and experimental design. Integrated design of experiments is necessary to enable comparison of findings across the spectrum of multiple physiological conditions, growth conditions, developmental stage, genetic diversity and even seasonal variations where rubber is cultivated (in tropical and sub-tropical regions). The following sections discuss perspectives on three aspects that can advance rubber biosynthesis studies and the prospects of exploring novel investigations that arise from the accumulating *Hevea* genomic and biological resources.

7.3.1 Pathway Level Analysis

IPP is the key player in isoprenoid biosynthesis. Rubber biosynthesis pathways in *Hevea* may be viewed at the pre- and post-IPP steps of the isoprenoid network. While there is much interest in elucidating the gene families and genome organization of these steps, pathway level regulation of rubber biosynthesis has hardly been investigated in *Hevea* latex.

7.3.1.1 Pre-IPP: A Tale of Two Pathways

The MVA and MEP pathways in plants produce independent pools of IPP in the cytosol and plastid respectively. Cross-talk occur between these two pathways where IPP is one of the "exchange" intermediates between the cytosol and the plastidial compartment (Hemmerlin et al. 2012). The relative contribution of MVA and MEP pathways to fulfill cellular needs for IPP has been reported in many plants. For some isoprenoids, IPP is drawn from both MVA and MEP pathways while for others, either one may supply the substrate (Lipko and Swiezewska 2016). Pathway characterization studies have utilised techniques such as co-expression network analysis, metabolic labeling of pathway-specific compounds, effect of pathway inhibitors and analysis of mutants (Vranová et al. 2013, Lipko and Swiezewska 2016, Jozwiak et al. 2017). However, most of these approaches are challenging to implement in a tree species like *Hevea*.

Spatial separation of the MVA and MEP pathways is thought to be a means of efficient and optimal production of isoprenoids from a variety of carbon sources (Hemmerlin et al. 2012; Lipko and Swiezewska 2016). In *Hevea*, rubber must at least use IPP generated from the cytosolic MVA pathway. A *Hevea*-specific plastid, called the Frey-Wyssling particle, is the equivalent of a carotenoid-containing chromoplast (Gomez and Moir 1979). Latex of rubber tree clones containing high levels of carotenoids appears yellowish to the naked eye compared with the creamy white latex of clones having lower carotenoid content. An opportunity thus arose for assessing the relative activity of MVA and MEP pathways in two rubber clones with distinctly different degree of yellowness in latex colour (Chow et al. 2012). Initially, MEP pathway gene expression was expected to be higher in the yellow clone (PB 235). However, the reverse was observed. Compared with clone PB 235, the creamy white latex of clone (RRIM 600) showed higher expression of selected MEP pathway genes and at the same time, lower expression of several MVA pathway genes. The rubber content of latex from experimental trees of both clones were found to be almost the same. On this basis, it was deduced that in RRIM 600 latex, IPP made in the Frey-Wyssling particles is transported out to supplement the cytosolic pool of IPP for rubber formation. From here, two models were proposed to explain the subcellular partitioning of IPP made by the MVA and MEP pathways for rubber, either of which would apply depending on the carotenoid level in different clones (Chow et al. 2012). While high carotenoid content in latex is a clonal trait (Eaton and Fullerton 1929; Sakdapipanich and Rojruthai 2014), the models proposed could at the same time, be modulated by environmental factors as well. To test this model in future, a similar experiment comparing gene expression in rubber clones with extremes in latex carotenoid content (such as RRIM 600 and PB 235) should include transcript expression of all MVA and MEP steps and a larger number of tree replicates, together with a quantitative measure of latex carotenoids. Previously, Sando et al. (2008b) suggested that the MEP pathway supplies IPP for carotenoids rather than rubber based on feeding RRIM 600 seedling plants with a radioactively labelled intermediate of the pathway. This suggests that IPP partitioning could depend on the age of the rubber plant.

Because the investigation of pathway-driven regulation is very limited in *Hevea*, current hypotheses for the contribution of MVA and MEP pathways to rubber remain equivocal. In the meantime, much can be gleaned from developments in isoprenoid biosynthesis studies in other plants. Efforts to elucidate the regulation of genes encoding isoprenoid biosynthetic pathways in several species including the model plant, *Arabidopsis* have been extensively reviewed (Vranová et al. 2013; Lipko and Swiezewska 2016; Rodríguez-Concepción and Boronat 2015). According to Rodríguez-Concepción and Boronat (2015), transcriptional regulation of isoprenoid biosynthetic genes results in coarse level control over the MVA and MEP pathways, while their metabolic flux is fine-tuned by post-transcriptional and post-translational regulation of enzyme levels and activity. Both pathways respond to internal and environmental cues but there appears to be little overall transcriptional connectivity between the MVA and MEP pathways (Vranová et al. 2013; Rodríguez-Concepción and Boronat 2015).

7.3.1.2 Post-IPP: Lessons in Sharing

Biochemical studies in the past have identified sterols, tocotrienols, tocopherols, plastoquinone, carotenoids and dolichols in latex (Dunphy et al. 1965; Kekwick 1989; Tateyama et al. 1999; Phatthiya et al. 2007; Hasma and Subramaniam 1986). More recently, by adopting a sequencing approach, transcripts assembled from a combination of Sanger and RNA-seq reads were used to survey post-IPP branches (Chow et al. 2012). Based on the KEGG pathway 'terpenoid backbone biosynthesis' (Kanehisa et al. 2004), a large number of potential isoprenoid end products were detected based on the discovery of eight of eleven potential branches (including rubber) with matches to transcripts (Chow et al. 2012). The commercial and pharmaceutical significance of bioactive secondary metabolites such as carotenoids, artemisinin and Taxol has driven the characterisation of related genes and pathways towards genetically engineered production in plant or microbial hosts (Kirby and Keasling, 2009). A clearer understanding of IPP production and its relative utilization by isoprenoid compounds downstream can assist in developing biotechnological strategies for increasing flux in the rubber branch, thus enhancing tree productivity. Currently, there is no information on how rubber and non-rubber branches compete for the IPP substrate or how this is regulated. Integration of pathway transcript expression with data on the abundance of all branch products would be a useful approach to shed light on this question. The growing *Hevea* genomic and transcriptomic resources will allow the generation of a high quality reference transcript set. This will be crucial for identification and curation of sequences encoding the post-IPP steps which produce functionally and structurally diverse isoprenoid products. However, there is a paucity of methods optimized for extraction and quantitation of such metabolites and their intermediates in *Hevea* tissues. This situation is beginning to change. A method based on liquid chromatography tandem mass spectrometry (LC–MS/MS) was recently developed for quantitation of pathway intermediates, DMAPP and FPP, in *Hevea* latex (Zhang et al. 2018). In addition, high performance liquid chromatography (HPLC) was also developed for detection and characterization of latex carotenoids (Bahari et al. 2019 manuscript in preparation).

7.3.2 Particle Level Analysis

Rubber particles constitute 35–45% of latex by volume and >90% of dried latex content (Chrestin et al. 1997). Unlike many plant organelles, they are robust and easily recoverable by centrifugation of fresh latex, resulting in large rubber particles in the cream layer (Zone 1) and a denser fraction of small rubber particles (Zone 2) (Moir 1959). In *Hevea* latex, the mean diameters of small and large rubber particles are 0.2 and 1 μm, respectively (Yeang et al. 1995; Cornish 2001). According to Cornish (2001), *Hevea* rubber particle proteins are the most complicated among rubber-producing plant species examined. Previous profiling has recorded more than 80 different proteins with molecular weights of 5 to >200 kDa (Cornish 2001). Recent application of advanced proteomics techniques and bioinformatics has seen higher throughput of rubber particle protein discovery, functional classification into putative groups and new molecular handles for analyzing variant isoforms (Xiang et al. 2012; Dai et al. 2013a, 2017). GO functional classification of the total rubber particle proteome found that 8% of proteins identified did not have any annotations (Dai et al. 2013a). About 74% of unique transcripts generated from EST sequencing were similarly not identified (Chow et al. 2007). Although annotation databases may improve over time, these observations suggest that a proportion of latex transcripts of unknown function are among those associated with rubber particle membranes. These are worth exploring especially in the quest to identify proteins of the *Hevea* rubber transferase complex.

Almost all that is known about rubber biosynthesis was deduced from in vitro assays containing enzymatically active rubber particles (Cornish 2014) though clearly, preparation methods and assay conditions may limit what is possible to analyse. Regulation of rubber molecular weight in *Hevea* is dependent on factors such as rate of chain transfer of the rubber molecule, properties of the rubber transferase, and concentration and the identity of substrates and activators (Cornish et al. 2000; da Costa et al. 2005; Cornish and Xie 2012). Nonetheless, these may not be the only relevant parameters as such a situation would entail a very tight control of substrate and activator concentration given that *Hevea* rubber has low polydispersity (Cornish 2014). Thus, it has been suggested that other mechanisms of molecular weight control are likely (Cornish 2014). Elucidating this may require a deeper understanding of the genes and proteins regulating another latex trait, rubber molecular weight distribution (MWD). The *Hevea* rubber tree, like *Ficus elastica* and guayule, can synthesize both high and low molecular weight *cis*-polyisoprene (Cornish et al. 2000). The rubber MWD profile is generated by gel permeation chromatography (GPC) based on the analysis of molecular sizes of species of different chain lengths. Three types of MWD have been observed based on GPC analysis of latex of more than 40 rubber tree clones: a unimodal type consisting of a single peak and two bimodal types differentiated by even and uneven heights of the two peaks (Subramaniam 1975, 1980; Ong 2000; Eng et al. 2001). This trait has long been considered genetic, however published data does not contain sufficient detail for further evaluation. To investigate this, assessment of GPC as a phenotyping assay for clonal MWD has been initiated (Chow and Sajari 2018). Using fresh latex of 18 clones growing in the same rubber trial, replicated GPC measurements indicate that rubber MWD is more likely controlled by genetics rather than by environmental factors, assuming sampled clones are true-to-type. Having a reliable and facile MWD phenotyping assay is essential for linking the trait to gene families or SNPs associated with this trait.

MWD has also been analysed at the level of particle size. Large rubber particles of the rubber cream (Zone 1, Moir 1959) show a bimodal distribution while small rubber particles from Zone 2 (adjacent fraction under Zone 1, Moir 1959) display a unimodal type (Yeang et al. 1995). The clonal MWD would, of course, be a combination of rubber molecules from the whole particle population. Zone 2 rubber particles also have a higher average molecular weight as shown by GPC analysis, this concurring with Subramaniam (1980). Coupled with distinctive physical differences between the two particle populations, including a higher sensitivity of small rubber particles to destabilisation, the authors proposed the possibility of a two-enzyme system of regulation: one synthesizing higher and the other lower molecular weight rubber. Studies on in vitro assays containing rubber particles that had been fractionated into a series of sizes by differential centrifugation may provide some insight into the functioning of two (or perhaps more) enzyme systems. Previously, Ohya et al. (2000) prepared a series of six fractions of *Hevea* latex particles of varying sizes from the Zone 1 rubber cream for IPP incorporation assays. Fractions containing the smaller range of particle sizes clearly showed much higher biosynthesis activity than the larger ones. In a more recent study, Yamashita et al. (2018) similarly isolated five size subsets of *Hevea* latex particles from the rubber cream and observed the same result. Additionally, Yamashita et al. (2018) showed correlation between high levels of HRT1/2 forms of CPT protein expression with rubber transferase activity in two fractions of the smaller rubber particles based on a Western immunoblot. It was noted that three fractions of larger particles were distinctly different from the two fractions of small rubber particles in two aspects. Firstly, the three fractions of larger particles showed bimodal rubber distribution while the other two fractions were of unimodal distribution (consistent with Yeang et al. 1995). Secondly, from silver-stained SDS-PAGE, the predicted 14 kDa REF and 24 kDa SRPP proteins appeared in the two fractions of smaller rubber particles while variant protein sizes were observed in three fractions of larger rubber particles. Findings by these three groups (Yeang et al. 1995; Ohya et al. 2000; Yamashita et al. 2018) are consistent in several key aspects and by chance, rubber particles were prepared from the same clone, RRIM 600. Many more isoforms of REF, SRPP and CPT are now known (see Sect. 7.2.2.2). It would be worthwhile to supplement classification of rubber particle size fractions by immunodetection of REF, SRPP and CPT isoforms and correlating the protein profiles with in vitro assays and GPC analysis of each particle size subset. This might lead to the identification of specific REF, SRPP and CPT isoforms supporting the enzyme system in each particle size fraction. Stringent controls would be needed for such in vitro investigations.

The composition of rubber particle membrane lipids and proteins is highly species-specific, perhaps without a conserved structure (Cornish 2001). It was further suggested that particle membranes in different rubber-producing species possibly utilise readily available cellular lipids during their development. Lipid droplets or oil bodies are organelles involved in lipid and energy homeostasis in living cells. The diversity and significance of their functions in cell physiology were only realised in recent years after being considered a functionally inert organelle (Murphy 2012; Chapman et al. 2012; Huang 2018). Like rubber particles, lipid droplets possess a phospholipid monolayer embedded with membrane proteins, which encapsulates neutral lipids instead of rubber. As such, lipid droplets have been thought to be analogous with *Hevea* rubber particles (Cornish 2001; Berthelot et al. 2014). In plants, functions of lipid droplets encompass stress response, pathogen resistance, hormone metabolism, and anther and pollen coat development (Chapman et al. 2012). Understanding of these functions have been achieved via characterization of the major lipid droplet membrane proteins, oleosins, caleosins and steroleosins (Chapman et al. 2012). In contrast to rubber particles, lipid droplet membrane proteins are highly conserved, whereby this could be related to the need for a precise mechanism for this organelle to export products (Cornish 2001). In a surprising twist, a new class of lipid droplet structural proteins with high similarities to SRPP in rubber-producing plants were discovered by proteome analysis (Horn et al. 2013; Gidda et al. 2016). These proteins are unlike the conserved oleosins, caleosins and steroleosins, and thus named liquid droplet-associated proteins (LDAPs). Tang et al. (2016) also showed one of the REF/SRPP gene family members, SRPP2, to be clustered together with the *Arabidopsis* homolog of LDAP, At3g05500. Dai et al. (2017) suggested that REF and SRPP isoforms may play structural roles similar to oleosins; this is because many REF and SRPP protein isoforms are acidic and thus enable negatively charged particles to be uniformly dispersed in the cytosol. To date, there is no report or any comprehensive survey of lipid droplet proteins in *Hevea* rubber particle proteomes. A phylogenetic analysis of LDAPs inclusive of the all SRPP variants found in the rubber genome (Tang et al. 2016) will be useful. New aspects of unique evolutionary relationship between rubber particles and lipid droplets may emerge from comparative analysis of their membrane proteomes and functions.

7.3.3 Applications of SNP Haplotype Structuring

Ultimately, the goal of functional genomics investigations of rubber biosynthesis is to enable a more precise and rapid path to generating high-yielding tree clones, be it in terms of latex volume or rubber content. Whilst molecular elucidation of fundamental processes such as *cis*-polyisoprene formation is essential, most likely marker technology, primarily based on SNPs, is the approach that would bring the fastest returns. SNP markers have been deployed on a large scale for developing *Hevea* genetic maps (Pootakham et al. 2015; Conson et al. 2018; Xia et al. 2018; An et al. 2019). Development of SNP linkage maps associating phenotypes to trait QTLs, including genomic prediction models, are underway (Souza et al. 2013, 2019; Chanroj et al. 2017; Conson et al. 2018; de Souza et al. 2018; Rosa et al. 2018; Xia et al. 2018; An et al. 2019). However, the best publicly available rubber genomes today still only have a portion of the genomic scaffolds anchored to genetic maps. Due to the high level of heterozygosity (Gouvêa et al. 2010; de Souza et al. 2015) and large proportion of sequence repeats (Tang et al. 2016; Lau et al. 2016; Pootakham et al. 2017) in *Hevea*, the construction of one or more phased genomes would be ideal for haplotype analysis. A haplotype allele is a combination of SNPs that are located in close proximity in a region of genomic DNA which is inherited as a single block within a population (Kunihisa et al. 2016; Huang et al. 2017). Knowledge of linkage disequilibrium (LD) of the number of markers in a haplotype block can be useful for selection and heritability analysis of genes that are associated with important phenotypic traits. Haplotype alleles potentially have greater accuracy in trait selection, even at the level of individual isoforms of a gene family.

Existing *Hevea* genome and transcript annotations present opportunities to assemble haplotype blocks of genomic regions harbouring candidate genes related to rubber biosynthesis. The quality of SNPs in such regions may be readily supported by the abundance of *Hevea* transcriptome data, particularly RNA-seq reads of highly expressed genes. To date, haplotype analysis in *Hevea* is quite rudimentary and has only been reported in three genes involved in rubber biosynthesis. Haplotypes for farnesyl diphosphate synthase (FPPS), hydroxymethylglutaryl-CoA synthase (HMGS) and CPT genes were identified by amplification of genomic DNA of five diverse rubber clones based on their parentage and breeding origins (RRII 105, RRII 118, RRIM 600, RRIC 52 and GT 1) (Uthup et al. 2013, 2016, 2019). These studies reported seven FPPS haplotypes, eight HMGS haplotypes and eight CPT haplotypes based on the analysis of 20–25 SNPs in the respective genes. This shows that haplotype analysis of rubber biosynthesis related genes is promising given the high degree of haplotype diversity within a small collection of cultivated clones. In each case, LD analysis also indicated that allelic variations arose from recombination rather than point mutations.

Since the introduction of the Wickham rubber materials, rubber has undergone a century or more of genetic improvement in countries which received the early batches of germplasm material. For example, in Malaysia, systematic rubber breeding was established in the 1920s by the Rubber Research Institute and the Prang Besar Research Station, producing RRIM and PB clones respectively. Mass selection of progenies of controlled crosses was primarily based on the capacity to produce high quantity and quality of rubber. Haplotypes of rubber biosynthesis related genes may be utilised to trace the effect of phenotypic selection on changes at the genome level, including the inheritance of such alterations in the successive generations of crosses. An example of this application is in the 100-year old rice breeding program in Japan where a comprehensive survey of cultivars and ancestral landraces using genome-wide SNP haplotypes shed light on the population structure of modern varieties and indicated the decrease in genetic distances among them due to breeding practices (Yamamoto et al. 2010; Yonemaru et al. 2012; Shinada et al. 2014). In *Hevea*, a case in point is scaffold_1222 in the Reyan 7-33-97 genome assembly which harbours 12 REF and SRPP gene isoforms (Tang et al. 2016). Further sequencing of scaffold_1222 could generate combinations of haplotype blocks for this REF/SRPP cluster for clones within a pedigree map. The structural diversity of scaffold_1222 haplotypes in pedigree clones may uncover the nature of 'selection pressure' exerted by rubber breeders on this DNA region, thus potentially revealing features which underlie high rubber yields. Concurrently, this approach could provide a means of mining novel haplotype alleles from existing collections of wild Brazilian germplasm genotypes. Due to the significance of scaffold_1222 in rubber biosynthesis (see Sect. REF and SRPP), judicious linking of its haplotypes to breeders' historical field data can be beneficial for identifying specific SNP sets associated with high tree productivity, or even narrow down the contributing REF and SRPP isoforms. This information will help in the selection of the best parents for crossings and subsequent progeny evaluation. In reality, obtaining reliable phenotypic and pedigree performance data from rubber breeders is fraught with many problems. Breeders' data from multiple trials differing in aspects such as plant age, data processing, trial control clones, environmental and climatic conditions, need to be sufficiently validated for each pedigree clone. To overcome such issues, the parallel development of appropriate statistical approaches and trial designs to facilitate integration of data for haplotype analysis is crucial.

7.4 Concluding Remarks

The advent of genomic technologies is a watershed in the study of natural rubber biosynthesis. Recent improvements in genome scaffolding technologies such as Hi-C and optical mapping now have the potential to generate high quality phased pseudo-molecule quality reference sequences for one or more rubber tree clones. Having reliable chromosomal assignment of SNPs or gene-specific SNPs in phased genomes will strengthen the application of a range of SNP genotyping technologies to reveal the full range of diversity in wild and cultivated *Hevea* germplasm, involving both *Hevea brasiliensis* and related species. Today, *Hevea* transcriptome data on different tree clones, tissues and physiological backgrounds is more abundant than it has ever been before. Utilisation of new emerging approaches to generate reference transcriptomes and analyse RNA-seq transcript expression will boost the investigations into many aspects of rubber biosynthesis regulation. Advances in proteomics will add impact to such studies by enabling more sensitive resolution of protein components related to latex rubber biosynthesis. Taken together, these developments will continue to bring opportunities to yield new insights into rubber biosynthesis and to translate them into improvements in both the yield and quality of natural rubber from *Hevea*.

References

Akhtar TA, Matsuba Y, Schauvinhold I, Yu G, Lees HA, Klein SE, Pichersky E (2013) The tomato *cis*–prenyltransferase gene family. Plant J 73(4):640–652

Altschul SF, Madden TL, Schäffer AA, Zhang J, Zhang Z, Miller W (1997) Gapped BLAST and PSI-BLAST: a new generation of protein database search programs. Nucleic Acids Res 25(17):3389–3402

An Z, Wang Q, Hu Y, Zhao Y, Li Y, Cheng H, Huang H (2012) Co-extraction of high-quality RNA and DNA from rubber tree (*Hevea brasiliensis*). Afr J Biotechnol 11(39):9308–9314

An Z, Li Y, Xie L, Zhai Q, Huang H (2013) A rapid and economical method for low molecular weight RNA isolation from a wide variety of plant species. Biosci Biotechnol Biochem 77(7):1599–1601

An Z, Zhao Y, Zhang X, Huang X, Hu Y, Cheng H, Li X, Huang H (2019) A high-density genetic map and QTL mapping on growth and latex yield-related traits in *Hevea brasiliensis* Müll Arg. Ind Crop Prod 132:440–448

Aoki Y, Takahashi S, Takayama D, Ogata Y, Sakurai N, Suzuki H, Asawatreratanakul K, Wititsuwannakul D, Wititsuwannakul R, Shibata D, Koyama T, Nakayama T (2014) Identification of laticifer-specific genes and their promoter regions from a natural rubber producing plant *Hevea brasiliensis*. Plant Sci 225:1–8

Archer BL, Audley BG (1987) New aspects of rubber biosynthesis. Bot J Linn Soc 94:181–196

Asawatreratanakul K, Zhang YW, Wititsuwannakul D, Wititsuwannakul R, Takahashi S, Rattanapittayaporn A, Koyama T (2003) Molecular cloning, expression and characterisation of cDNA encoding *cis*-prenyltransferases from *Hevea brasiliensis*. Eur J Biochem 270:4671–4680

Attanyaka DPSTG, Kekwick RGO, Franklin FCH (1991) Molecular cloning and nucleotide sequencing of the rubber elongation factor gene from *Hevea brasiliensis*. Plant Mol Biol 16:1079–1081

Bahari A (2019) Approaches to understanding diversity in rubber and carotenoid synthesis in *Hevea brasiliensis* latex. PhD Thesis, University of Dundee, United Kingdom

Bahari A, Campbell R, Chow K-S, Marshall D, Taylor M (2019) Determination of major carotenoid and expression profiling of key genes underpinning carotenoid formation in *Hevea brasiliensis* latex. Manuscript in preparation.

Berthelot K, Lecomte S, Estevez Y, Peruch F (2014) *Hevea brasiliensis* REF (Hev b 1) and SRPP (Hev b 3): An overview on rubber particles proteins. Biochimie 106:1–9

Brasher MI, Surmacz L, Leong B, Pitcher J, Swiezewska E, Pichersky E, Akhtar TA (2015) A two-component enzyme complex is required for dolichol biosynthesis in tomato. Plant J 82(6):903–914

Brown D, Feeney M, Ahmadi M, Lonoce C, Sajari R, Di Cola A, Frigerio L (2017) Subcellular localization and interactions among rubber particle proteins from *Hevea brasiliensis*. J Exp Bot 68(18):5045–5055

Chanroj V, Rattanawong R, Phumichai T, Tangphatsornruang S, Ukoskit K (2017) Genome-wide association mapping of latex yield and girth in Amazonian accessions of *Hevea brasiliensis* grown in a suboptimal climate zone. Genomics 109(5):475–484

Chao J, Chen Y, Wu S, Tian W-M (2015) Comparative transcriptome analysis of latex from rubber tree clone CATAS8-79 and PR107 reveals new cues for the regulation of latex regeneration and duration of latex flow. BMC Plant Biol 15(1):104

Chapman KD, Dyer JM, Mullen RT (2012) Biogenesis and functions of lipid droplets in plants: thematic review series: lipid droplet synthesis and metabolism: from yeast to man. J Lipid Res 53(2):215–226

Cheong KF, Koh CL (1993) Nucleotide sequence of the chloroplast gene for the large sub-unit of ribulose-l,5-bisphosphate carboxylase/oxygenase from *Hevea brasiliensis*. J nat Rubb Res 8(2):146–153

Cherian S, Ryu SB, Cornish K (2019) Natural rubber biosynthesis in plants, the rubber transferase complex, and metabolic engineering progress and prospects. Plant Biotechnol J 17(11):2041–2061

Chow K-S, Sajari R (2018) Studies on the genetic control of clonal variation in rubber molecular weight distribution. In: Proceedings of the IRRDB International Rubber Conference, Ivory Coast, 22–24 October 2018

Chow K-S, Mohd-Lim S, Han K-H (2003) Characterisation of a family eIF-5A cDNAs from *Hevea brasiliensis*. J Rubber Res 6:107–120

Chow K-S, Yusof F, Mohd-Lim S, Abdullah L (2006) An assessment of stimulation of *Hevea brasiliensis* rubber biosynthesis by eIF-5A protein-enriched bacterial lysates. J Rubber Res 9:251–259

Chow K-S, Wan K-L, Mat-Isa M-N, Bahari A, Tan S-H, Harikrishna K, Yeang HY (2007) Insights into rubber biosynthesis from transcriptome analysis of *Hevea brasiliensis* latex. J Exp Bot 58:2429–2440

Chow K-S, Mohd-Noor M-I, Bahari A, Ahmad-Kamal G, Alias H, Zainorlina M-Z, Hoh C-C, Wan K-L (2012) Metabolic routes affecting rubber biosynthesis in *Hevea brasiliensis* latex. J Exp Bot 63:1863–1871

Chow K-S, Ahmad-Kamal G, Hoh C-C, Zainorlina MZ (2014) RNA sequencing read depth requirement for optimal transcriptome coverage in *Hevea brasiliensis*. BMC Res Notes 7:69

Chow K-S, Khoo J-S, Zainuddin ZM, Ng S-M, Hoh C-C (2019) Utility of PacBio Iso-Seq for transcript and gene discovery in *Hevea* latex. J Rubber Res 22 (4):169–186

Chrestin H, Gidrol X, Kush A (1997) Towards a latex molecular diagnostic of yield potential and the genetic engineering of the rubber tree. Euphytica 96(1):77–82

Conson ARO, Taniguti CH, Amadeu RR, Andreotti IAA, de Souza LM, dos Santos LHB, Rosa JRBF, Mantello CC, da Silva CC, Scaloppi Junior EJ, Ribeiro RV, Le Guen V, Garcia AAF, Gonçalves PdS, de Souza AP (2018) High-resolution genetic map and QTL analysis of growth-related traits of *Hevea brasiliensis* cultivated under suboptimal temperature and humidity conditions. Front Plant Sci 9:1255

Cornish K (2001) Similarities and differences in rubber biochemistry among plant species. Phytochemistry 57:1123–1134

Cornish K (2014) Biosynthesis of natural rubber (NR) in different rubber-producing species. In: Kohjiya S, Ikeda Y (eds) Chemistry, manufacture and applications of natural rubber. Woodhead Publishing, pp 3–29

Cornish K, Xie W (2012) Natural rubber biosynthesis in plants: Rubber transferase. In: Hopwood DA (ed) Methods in Enzymology, vol 515. Academic Press, pp 63–82

Cornish K, Castillón J, Scott DJ (2000) Rubber molecular weight regulation, in vitro, in plant species that produce high and low molecular weights *in vivo*. Biomacromol 1(4):632–641

Cornish K, Scott DJ, Xie W, Mau CJD, Zheng YF, Liu X-h, Prestwich GD (2018) Unusual subunits are directly involved in binding substrates for natural rubber biosynthesis in multiple plant species. Phytochemistry 156:55–72

Cunillera N, Arró M, Forés O, Manzano D, Ferrer A (2000) Characterization of dehydrodolichyl diphosphate synthase of *Arabidopsis thaliana*, a key enzyme in dolichol biosynthesis. FEBS Lett 477(3):170–174

Czuppon AB, Chen Z, Rennert S, Engelke T, Meyer HE, Heber M, Baur X (1993) The rubber elongation factor of rubber trees (*Hevea brasiliensis*) is the major allergen in latex. J Allergy Clin Immunol 92(5):690–697

da Costa BM, Keasling JD, Cornish K (2005) Regulation of rubber biosynthetic rate and molecular weight in *Hevea brasiliensis* by metal cofactor. Biomacromol 6 (1):279–289

Dai L, Kang G, Li Y, Nie Z, Duan C, Zeng R (2013a) In-depth proteome analysis of the rubber particle of *Hevea brasiliensis* (para rubber tree). Plant Mol Biol 82(1):155–168

Dai L, Li Y, Nie ZY, Kang GJ, Duan CF, Zeng RZ (2013b) Effects of different solutions on extraction and separation of rubber particle proteins of *Hevea brasiliensis* Muell Arg. Zhiwu Shengli Xuebao 49:1089–1098

Dai L, Kang G, Nie Z, Li Y, Zeng R (2016) Comparative proteomic analysis of latex from *Hevea brasiliensis* treated with Ethrel and methyl jasmonate using iTRAQ-coupled two-dimensional LC–MS/MS. J Proteomics 132:167–175

Dai L, Nie Z, Kang G, Li Y, Zeng R (2017) Identification and subcellular localization analysis of two rubber elongation factor isoforms on *Hevea brasiliensis* rubber particles. Plant Physiol Biochem 111:97–106

de Faÿ E, Hébant CH, Jacob JL (1989) Cytology and cytochemistry of the laticiferous system. In: Physiology of rubber tree latex. CRC Press Inc., Boca Raton, USA

de Souza LM, Le Guen V, Cerqueira-Silva CBM, Silva CC, Mantello CC, Conson ARO, Vianna JPG, Zucchi MI, Scaloppi Junior EJ, Fialho JdF, de Moraes MLT, Gonçalves PdS, de Souza AP (2015) Genetic diversity strategy for the management and use of rubber genetic resources: more than 1,000 wild and cultivated accessions in a 100-genotype core collection. PLoS ONE 10(7):e0134607

de Souza LM, dos Santos LHB, Rosa JRBF, da Silva CC, Mantello CC, Conson ARO, Scaloppi Junior EJ, Fialho JdF, de Moraes MLT, Gonçalves PdS, Margarido GRA, Garcia AAF, Le Guen V, de Souza AP (2018) Linkage disequilibrium and population structure in wild and cultivated populations of rubber tree (*Hevea brasiliensis*). bioRxiv:360610

Dennis MS, Henzel WJ, Bell J, Kohr W, Light DR (1989) Amino acid sequence of rubber elongation factor protein associated with rubber particles in *Hevea* latex. J Biol Chem 264:18618–18626

Dunphy PJ, Whittle KJ, Morton RA (1965) Identification and estimation of tocotrienols in *Hevea* latex. Nature 207:521–522

Eaton BJ, Fullerton G (1929) The identification and quantitative estimation of the yellow pigment of raw rubber. J Rubber Res Inst Malaya 1(3):135–148

Eng AH, Othman H, Hasma H, Ramli O, Masahuling B, Muniandy V, Kawahara S (2001) Some properties of natural rubber from latex-timber clones. J Rubber Res 4:164–176

Epping J, van Deenen N, Niephaus E, Stolze A, Fricke J, Huber C, Eisenreich W, Twyman RM, Prüfer D, Schulze Gronover C (2015) A rubber transferase activator is necessary for natural rubber biosynthesis in dandelion. Nat Plants 1:15048

Gidda SK, Park S, Pyc M, Yurchenko O, Cai Y, Wu P, Andrews DW, Chapman KD, Dyer JM, Mullen RT (2016) Lipid droplet-associated proteins (LDAPs) are required for the dynamic regulation of neutral lipid compartmentation in plant cells. Plant Physiol 170 (4):2052–2071

Gomez JB, Moir GFJ (1979) Ultracytology of latex vessels in *Hevea brasiliensis*. MRRDB Monograph No. 4. Malaysian Rubber Research and Development Board, Kuala Lumpur, Malaysia

Gouvêa LRL, Rubiano LB, Chioratto AF, Zucchi MI, Goncalves PdS (2010) Genetic divergence of rubber tree estimated by multivariate techniques and microsatellite markers. Genet Mol Biol 33(2):308–318

Goyvaerts E, Dennis M, Light D, Chua N-H (1991) Cloning and sequencing of the cdna encoding the rubber elongation factor of *Hevea brasiliensis*. Plant Physiol 97(1):317–321

Grabińska KA, Park EJ, Sessa WC (2016) *cis*-Prenyltransferase: new insights into protein glycosylation, rubber synthesis, and human diseases. J Biol Chem 291(35):18582–18590

Han K-H, Shin DH, Yang J, Kim IJ, Oh SK, Chow K-S (2000) Genes expressed in the latex of *Hevea brasiliensis*. Tree Physiol 20(8):503–510

Hasma H, Subramaniam A (1986) Composition of lipids in latex of *Hevea brasiliensis* clone RRIM 501. J nat Rubb Res 1:30–40

Hemmerlin A, Harwood JL, Bach TJ (2012) A raison d'être for two distinct pathways in the early steps of plant isoprenoid biosynthesis. Prog Lipid Res 51 (2):95–148

Horn PJ, James CN, Gidda S, Kilaru A, Dyer JM, Mullen RT, Ohlrogge JB, Chapman K (2013) Identification of a new class of lipid droplet-associated proteins in plants. Plant Physiol 162:1926–1936

Huang AHC (2018) Plant lipid droplets and their associated proteins: potential for rapid advances. Plant Physiol 176(3):1894–1918

Huang M, Tu J, Lu Z (2017) Recent advances in experimental whole genome haplotyping methods. Int J Mol Sci 18:1944

Jozwiak A, Lipko A, Kania M, Danikiewicz W, Surmacz L, Witek A, Wojcik J, Zdanowski K, Pączkowski C, Chojnacki T, Poznanski J, Swiezewska E (2017) Modeling of dolichol mass spectra isotopic envelopes as a tool to monitor isoprenoid biosynthesis. Plant Physiol 174(2):857–874

Kanehisa M, Goto S, Kawashima S, Okuno Y, Hattori M (2004) The KEGG resource for deciphering the genome. Nucleic Acids Res 32(suppl_1):D277-D280

Kekwick RGO (1989) The formation of isoprenoids in *Hevea* latex. In: Physiology of rubber tree. CRC Press Ltd., Boca Raton, Florida, USA

Kim EY, Park KY, Seo YS, Kim WT (2016) *Arabidopsis* small rubber particle protein homolog srps play dual roles as positive factors for tissue growth and development and in drought stress responses. Plant Physiol 170(4):2494–2510

Kirby J, Keasling JD (2009) Biosynthesis of plant isoprenoids: perspectives for microbial engineering. Annu Rev Plant Biol 60:335–355

Ko JH, Chow K-S, Han KH (2003) Transcriptome analysis reveals novel features of the molecular events occurring in the laticifers of *Hevea brasiliensis* (para rubber tree). Plant Mol Biol 53:479–492

Kostyal DA, Hickey VL, Noti JD, Sussman GL, Beezhold DH (1998) Cloning and characterization of a latex allergen (Hev b 7): homology to patatin, a plant PLA2. Clin Exp Immunol 112(3):355–362

Kunihisa M, Moriya S, Abe K, Okada K, Haji T, Hayashi T, Kawahara Y, Itoh R, Itoh T, Katayose Y, Kanamori H, Matsumoto T, Mori S, Sasaki H, Matsumoto T, Nishitani C, Terakami S, Yamamoto T (2016) Genomic dissection of a 'Fuji' apple cultivar: re-sequencing, SNP marker development, definition of haplotypes, and QTL detection. Breed Sci 66(4):499–515

Kush A, Goyvaerts E, Chye ML, Chua NH (1990) Laticifer-specific gene expression in *Hevea brasiliensis* (rubber tree). Proc Natl Acad Sci USA 87(5):1787–1790

Lau N-S, Makita Y, Kawashima M, Taylor TD, Kondo S, Othman AS, Shu-Chien AC, Matsui M (2016) The rubber tree genome shows expansion of gene family associated with rubber biosynthesis. Sci Rep 6:28594

Li D, Deng Z, Qin B, Liu X, Men Z (2012) *De novo* assembly and characterization of bark transcriptome using Illumina sequencing and development of EST-SSR markers in rubber tree (*Hevea brasiliensis* Muell. Arg.). BMC Genom 13(1):192

Li D, Hao L, Liu H, Zhao M, Deng Z, Li Y, Zeng R, Tian W (2015) Next-generation sequencing, assembly, and comparative analyses of the latex transcriptomes from two elite *Hevea brasiliensis* varieties. Tree Genet Genomes 11(5):98

Li D, Zeng R, Li Y, Zhao M, Chao J, Li Y, Wang K, Zhu L, Tian W-M, Liang C (2016a) Gene expression analysis and SNP/InDel discovery to investigate yield heterosis of two rubber tree F1 hybrids. Sci Rep 6:24984

Li H-L, Guo D, Zhu J-H, Wang Y, Chen X-T, Peng S-Q (2016b) Comparative transcriptome analysis of latex reveals molecular mechanisms underlying increased rubber yield in *Hevea brasiliensis* self-rooting juvenile clones. Front Plant Sci 7:1204

Light DR, Dennis MS (1989) Purification of a prenyltransferase that elongates cis-polyisoprene rubber from the latex of *Hevea brasiliensis*. J Biol Chem 264:18589–18597

Lipko A, Swiezewska E (2016) Isoprenoid generating systems in plants—a handy toolbox how to assess contribution of the mevalonate and methylerythritol phosphate pathways to the biosynthetic process. Prog Lipid Res 63:70–92

Liu M-C, Wang B-J, Huang J-K, Wang C-S (2011) Expression, localization and function of a *cis*-prenyltransferase in the tapetum and microspores of lily anthers. Plant Cell Physiol 52(9):1487–1500

Makita Y, Ng KK, Veera Singham G, Kawashima M, Hirakawa H, Sato S, Othman AS, Matsui M (2017) Large-scale collection of full-length cDNA and transcriptome analysis in *Hevea brasiliensis*. DNA Res 24 (2):159–167

Mantello CC, Cardoso-Silva CB, da Silva CC, de Souza LM, Scaloppi Junior EJ, Goncalves PdS, Vicentini R, de Souza AP (2014) De novo assembly and trannscriptome analysis of the rubber tree (*Hevea brasiliensis*) and SNP markers development for rubber biosythesis. PLoS ONE 9:e102665

Moir GFJ (1959) Ultracentrifugation and staining of *Hevea* latex. Nature 84:1626

Murphy DJ (2007) People, plants and genes: the story of crops and humanity. Oxford University Press

Murphy DJ (2012) The dynamic roles of intracellular lipid droplets: from archaea to mammals. Protoplasma 249 (3):541–585

Newman T, de Bruijn FJ, Green P, Keegstra K, Kende H, McIntosh L, Ohlrogge J, Raikhel N, Somerville S, Thomashow M, Retzel E, Somerville C (1994) Genes Galore: a summary of methods for accessing results from large-scale partial sequencing of anonymous *Arabidopsis* cDNA clones. Plant Physiol 106(4):1241

Oh SK, Kang H, Shin DH, Yang J, Chow K-S, Yeang HY, Wagner B, Breiteneder H, Han K-H (1999) Isolation, characterization, and functional analysis of a novel cDNA clone encoding a small rubber particle protein from *Hevea brasiliensis*. J Biol Chem 274:17132–17138

Oh SK, Han KH, Ryu SB, Kang H (2000) Molecular cloning, expression, and functional analysis of a *cis*-prenyltransferase from *Arabidopsis thaliana*: implications in rubber biosynthesis. J Biol Chem 275 (24):18482–18488

Ohya N, Tanaka Y, Wititsuwaannakul D, Koyama T (2000) Activity of rubber transferase and rubber particle size in *Hevea* latex. J Rubber Res 3(4):214–221

Ong EL (2000) Characterisation of new latex-timber clones of natural rubber. J Appl Polym 78:1517–1521

Park MH, Lee YB, Joe YA (1997) Hypusine is essential for eukaryotic cell proliferation. Biol Signals 6 (3):115–123

Phatthiya A, Takahashi S, Chareonthiphakorn N, Koyama T, Wititsuwannakul D, Wititsuwannakul R (2007) Cloning and expression of the gene encoding solanesyl diphosphate synthase from *Hevea brasiliensis*. Plant Sci 172:824–831

Pootakham W, Chanprasert J, Jomchai N, Sangsrakru D, Yoocha T, Therawattanasuk K, Tangphatsornruang S (2011) Single nucleotide polymorphism marker development in the rubber tree, *Hevea brasiliensis* (Euphorbiaceae). Am J Bot 98(11):e337–e338

Pootakham W, Ruang-Areerate P, Jomchai N, Sonthirod C, Sangsrakru D, Yoocha T, Theerawattanasuk K, Nirapathpongporn K, Romruensukharom P, Tragoonrung S, Tangphatsornruang S (2015) Construction of a high-density integrated genetic linkage map of rubber tree (*Hevea brasiliensis*) using genotyping-by-sequencing (GBS). Front Plant Sci 6:367

Pootakham W, Sonthirod C, Naktang C, Ruang-Areerate P, Yoocha T, Sangsrakru D, Theerawattanasuk K, Rattanawong R, Lekawipat N, Tangphatsornruang S (2017) De novo hybrid assembly of the rubber tree genome reveals evidence of paleotetraploidy in *Hevea* species. Sci Rep 7:41457

Qu Y, Chakrabarty R, Tran HT, Kwon EJ, Kwon M, Nguyen TD, Ro DK (2015) A lettuce (*Lactuca sativa*)

homolog of human Nogo-B receptor interacts with *cis*-prenyltransferase and is necessary for natural rubber biosynthesis. J Biol Chem 290(4):1898–1914

Rahman AYA, Usharaj AO, Misra BB, Thottathil GP, Jayasekarana K, Feng Y, Hou S, Ong SY, Ng FL, Lee LS, Tan HS, Sakaff MK, Teh BS, Khoo BF, Badai SS, Ab-Aziz N, Yuryev A, Knudsen B, Dionne-Laporte A, Mchunu NP, Yu Q, Langston BJ, Freitas TA, Young AG, Chen R, Wang L, Najimudin N, Saito JA, Alam M (2013) Draft genome sequence of the rubber tree *Hevea brasiliensis*. BMC Genom 14:75

Rodríguez-Concepción M, Boronat A (2015) Breaking new ground in the regulation of the early steps of plant isoprenoid biosynthesis. Curr Opin Plant Biol 25:17–22

Rosa JRBF, Mantello CC, Garcia D, de Souza LM, da Silva CC, Gazaffi R, da Silva CC, Toledo-Silva G, Cubry P, Garcia AAF, de Souza AP, Le Guen V (2018) QTL detection for growth and latex production in a full-sib rubber tree population cultivated under suboptimal climate conditions. BMC Plant Biol 18 (1):223

Sakdapipanich JT, Rojruthai P (2014) Natural rubber: biosynthesis, structure, properties and application. In: Natural rubber materials: volume 1: blends and IPNs, vol 1. The Royal Society of Chemistry, pp 28–52

Sando T, Takaoka C, Mukai Y, Yamashita A, Hattori M, Ogasawara N, Fukusaki E, Kobayashi A (2008a) Cloning and characterization of mevalonate pathway genes in a natural rubber producing plant, *Hevea brasiliensis*. Biosci Biotechnol Biochem 72:2049–2060

Sando T, Takeno S, Watanabe N, Okumoto H, Kuzuyama T, Yamashita A, Hattori M, Ogasawara N, Fukusaki E, Kobayashi A (2008b) Cloning and characterization of the 2-C-methyl-D-erythritol 4-phosphate (MEP) pathway genes of a natural-rubber producing plant, *Hevea brasiliensis*. Biosci Biotechnol Biochem 72:2903–2917

Schmidt T, Hillebrand A, Wurbs D, Wahler D, Lenders M, Gronover CS, Prüfer D (2010) Molecular cloning and characterization of rubber biosynthetic genes from *Taraxacum koksaghyz*. Plant Mol Biol 28:277–284

Shinada H, Yamamoto T, Yamamoto E, Hori K, Yonemaru J, Matsuba S, Fujino K (2014) Historical changes in population structure during rice breeding programs in the northern limits of rice cultivation. Theor Appl Genet 127(4):995–1004

Souza LM, Gazaffi R, Mantello CC, Silva CC, Garcia D, Le Guen V, Cardoso SEA, Garcia AAF, Souza AP (2013) QTL Mapping of growth-related traits in a full-sib family of rubber tree (*Hevea brasiliensis*) evaluated in a sub-tropical climate. PLoS ONE 8(4):e61238

Souza LM, Francisco FR, Gonçalves PS, Scaloppi Junior EJ, Le Guen V, Fritsche-Neto R, Souza AP (2019) Genomic selection in rubber tree breeding: a comparison of models and methods for managing G×E interactions. Front Plant Sci 10:1353

Subramaniam A (1975) Molecular weight and other properties of natural rubber: a study of clonal variations. In: Proceedings of the International Rubber Conference, Kuala Lumpur, Malaysia

Subramaniam A (1980) Molecular weight and molecular weight distribution of natural rubber. In: Rubber Research Institute of Malaysia Technology Bulletin, vol 4. Rubber Research Institute of Malaysia, Malaysia

Takahashi S, Lee H-J, Yamashita S, Koyama T (2012) Characterization of *cis*-prenyltransferases from the rubber producing plant *Hevea brasiliensis* heterologously expressed in yeast and plant cells. Plant Biotechnol 29(4):411–417

Tan D, Hu X, Fu L, Kumpeangkeaw A, Ding Z, Sun X, Zhang J (2017) Comparative morphology and transcriptome analysis reveals distinct functions of the primary and secondary laticifer cells in the rubber tree. Sci Rep 7(1):3126

Tang C, Qi J, Li H, Zhang C, Wang Y (2007) A convenient and efficient protocol for isolating high-quality RNA from latex of *Hevea brasiliensis* (para rubber tree). J Biochem Bioph Methods 70(5):749–754

Tang C, Xiao X, Li H, Fan Y, Yang J, Qi J, Li H (2013) Comparative analysis of latex transcriptome reveals putative molecular mechanisms underlying super productivity of *Hevea brasiliensis*. PLoS ONE 8: e75307

Tang C, Yang M, Fang Y, Luo Y, Gao S, Xiao X, An Z, Zhou B, Zhang B, Tan X, Yeang H-Y, Qin Y, Yang J, Lin Q, Mei H, Montoro P, Long X, Qi J, Hua Y, He Z, Sun M, Li W, Zeng X, Cheng H, Liu Y, Yang J, Tian W, Zhuang N, Zeng R, Li D, He P, Li Z, Zou Z, Li S, Li C, Wang J, Wei D, Lai C-Q, Luo W, Yu J, Hu S, Huang H (2016) The rubber tree genome reveals new insights into rubber production and species adaptation. Nat Plants 2:16073

Tateyama S, Wititsuwannakul R, Wititsuwannakul D, Sagami H, Ogura K (1999) Dolichols of rubber plant, ginkgo and pine. Phytochemistry 51:11–15

Thulaseedharan A, Sajeevan RS, Saleena A (2012) Detection and isolation of good quality DNA from the latex of *Hevea brasiliensis*. Nat Rubber Res 25:77–85

Tong Z, Wang D, Sun Y, Yang Q, Meng X, Wang L, Feng W, Li L, Wurtele E, Wang X (2017) Comparative proteomics of rubber latex revealed multiple protein species of ref/srpp family respond diversely to ethylene stimulation among different rubber tree clones. Int J Mol Sci 18(5):958

Triwitayakorn K, Chatkulkawin P, Kanjanawattanawong S, Sraphet S, Yoocha T, Sangsrakru D, Chanprasert J, Ngamphiw C, Jomchai N, Therawattanasuk K, Tangphatsornruang S (2011) Transcriptome sequencing of *Hevea brasiliensis* for development of microsatellite markers and construction of a genetic linkage map. DNA Res 18(6):471–482

Uthup TK, Saha T, Ravindran M, Bini K (2013) Impact of an intragenic retrotransposon on the structural integrity and evolution of a major isoprenoid biosynthesis pathway gene in *Hevea brasiliensis*. Plant Physiol Biochem 73:176–188

Uthup TK, Rajamani A, Ravindran M, Saha T (2016) Molecular evolution and functional characterisation of haplotypes of an important rubber biosynthesis gene in *Hevea brasiliensis*. Plant Biol 18(4):720–728

Uthup TK, Rajamani A, Ravindran M, Saha T (2019) Distinguishing CPT gene family members and vetting the sequence structure of a putative rubber synthesizing variant in *Hevea brasiliensis*. Gene 689:183–193

Vranová E, Coman D, Gruissem W (2013) Network analysis of the MVA and MEP pathways for isoprenoid synthesis. Annu Rev Plant Biol 64(1):665–700

Wang X, Shi M, Lu X, Ma R, Wu C, Guo A, Peng M, Tian W (2010) A method for protein extraction from different subcellular fractions of laticifer latex in *Hevea brasiliensis* compatible with 2-DE and MS. Proteome Sci 8:35–35

Wang D, Sun Y, Chang L, Tong Z, Xie Q, Jin X, Zhu L, He P, Li H, Wang X (2018) Subcellular proteome profiles of different latex fractions revealed washed solutions from rubber particles contain crucial enzymes for natural rubber biosynthesis. J Proteomics 182:53–64

Xia Z, Xu H, Zhai J, Li D, Luo H, He C, Huang X (2011) RNA-Seq analysis and *de novo* transcriptome assembly of *Hevea brasiliensis*. Plant Mol Biol 77(3)::299–308

Xia Z, Liu K, Zhang S, Yu W, Zou M, He L, Wang W (2018) An ultra-high density map allowed for mapping QTL and candidate genes controlling dry latex yield in rubber tree. Ind Crops Prod 120:351–356

Xiang Q, Xia K, Dai L, Kang G, Li Y, Nie Z, Duan C, Zheng R (2012) Proteome analysis of the large and the small rubber particles of *Hevea brasiliensis* using 2D-DIGE. Plant Physiol Biochem 60:207–213

Yamamoto T, Nagasaki H, Yonemaru J-i, Ebana K, Nakajima M, Shibaya T, Yano M (2010) Fine definition of the pedigree haplotypes of closely related rice cultivars by means of genome-wide discovery of single-nucleotide polymorphisms. BMC Genom 11 (1):267

Yamashita S, Yamaguchi H, Waki T, Aoki Y, Mizuno M, Yanbe F, Ishii T, Funaki A, Tozawa Y, Miyagi-Inoue Y, Fushihara K, Nakayama T, Takahashi S (2016) Identification and reconstitution of the rubber biosynthetic machinery on rubber particles from *Hevea brasiliensis*. elife 5:e19022

Yamashita S, Mizuno M, Hayashi H, Yamaguchi H, Miyagi-Inoue Y, Fushihara K, Koyama T, Nakayama T, Takahashi S (2018) Purification and characterization of small and large rubber particles from *Hevea brasiliensis*. Biosci Biotechnol Biochem 82(6):1011–1020

Yeang HY, Yip E, Samsidar H (1995) Characterisation of Zone 1 and Zone 2 rubber particles in *Hevea brasiliensis* latex. J Nat Rubber Res 10:108–123

Yeang HY, Cheong KF, Sunderasan E, Hamilton RG, Cardosa MJ (1996) The 14.6 kd rubber elongation factor (Hev b 1) and 24 kd (Hev b 3) rubber particle proteins are recognised by IgE from patients with spina bifida and latex allergy. J Allergy Clin Immunol 98:628–639

Yeang HY, Ward MA, Zamri ASM, Dennis MS, Light DR (1998) Amino acid sequence similarity of Hev b 3 to two previously reported 27- and 23-kDa latex proteins allergenic to spina bifida patients. Allergy 53(5):513–517

Yonemaru J-i, Yamamoto T, Ebana K, Yamamoto E, Nagasaki H, Shibaya T, Yano M (2012) Genome-wide haplotype changes produced by artificial selection during modern rice breeding in Japan. PLoS ONE 7 (3):e32982

Yusof F, Ward MA, Walker JM (1998) Purification and characterisation of an inhibitor of a rubber biosynthesis from C-serum of *Hevea brasiliensis* latex. J Rubber Res 1:95–110

Yusof F, Chow K-S, Ward MA, Walker JM (2000) A stimulator protein of rubber biosynthesis from *Hevea brasiliensis* latex. J Rubber Res 3(4):232–247

Zhang X, Guo T, Xiang T, Dong Y, Zhang J, Zhang L (2018) Quantitation of isoprenoids for natural rubber biosynthesis in natural rubber latex by liquid chromatography with tandem mass spectrometry. J Chromatogr A 1558:115–119

8 Current Progress in Transcriptomics and Proteomics of Latex Physiology and Metabolism in the *Hevea brasiliensis* Rubber Tree

Dejun Li, Shaohua Wu, and Longjun Dai

Abstract

Enhancing the productivity of the rubber tree (*Hevea brasiliensis*) through breeding is crucial for meeting the increasing global demand for natural rubber (NR). In spite of existing fundamental knowledge of laticifer physiology and metabolism, further elucidation of aspects which have direct effects on latex yield is crucial. These include investigations of rubber biosynthesis and latex flow, laticifer cell development, tapping panel dryness (TPD) and tree tolerance to biotic and abiotic stresses. In recent years, transcriptomic and proteomic approaches have provided insights into genes, proteins and molecular mechanisms regulating these aspects.

8.1 Introduction

Latex is the cytoplasm of the laticifer cell, in which *cis*-polyisoprene is synthesized. More than 2000 plant species produce natural rubber (NR), but the rubber tree (*Hevea brasiliensis*) is the only available source for large-scale commercial production of NR. NR from rubber tree latex is an excellent industrial raw material with physical and chemical properties that cannot be fully replaced by synthetic rubber (Van Beilen and Poirier 2007). With rapid development of the world economy, global demand for NR will continue to increase till 2020 and beyond (http://www.dailymirror.lk/31402/meeting-increasing-global-demand-for-natural-rubber).

Latex physiology and metabolism are closely associated with complex biochemical pathways and their genes and proteins, which eventually determine the latex-yielding capacity of different rubber tree clones. In this chapter, we first describe molecular mechanisms underlying high yield, followed by tolerance to biotic and abiotic stresses (cold tolerance and South American Leaf Blight (SALB) resistance). In the section on laticifer cell development, we focus on the morphologies and functions of laticifer cells, including genes and other factors affecting laticifer differentiation. Subsequently, genes and proteins related to rubber biosynthesis and latex flow and the effects of exogenous ethylene (ET) and jasmonate (JA) on these processes are described. The tapping panel dryness (TPD) section discusses the physiology of latex in affected trees, TPD-related gene expression profiling and the role of ET stimulation and reactive oxygen species (ROS) in TPD.

D. Li (✉) · S. Wu · L. Dai
Key Laboratory of Biology and Genetic Resources of Rubber Tree, Ministry of Agriculture and Rural Affairs, Rubber Research Institute, Chinese Academy of Tropical Agricultural Sciences, Danzhou 571737, People's Republic of China
e-mail: djli@catas.cn

M. Matsui and K.-S. Chow (eds.), *The Rubber Tree Genome*, Compendium of Plant Genomes,
https://doi.org/10.1007/978-3-030-42258-5_8

8.2 Molecular Mechanisms Underlying High Yield in Rubber Trees

Breeding high-yielding rubber tree clones (or varieties) is one of the primary approaches to meet increasing global demand for NR. Great progress made in breeding rubber tree clones resulted in increased NR yields from 650 to 1600 kg/ha (Priyadarshan et al. 2009). However, latex production of the highest-yielding clone available is still lower than the theoretical yield of rubber trees (Paardekooper 1989). Conventional rubber tree breeding is confronted by several challenges, including the long breeding period ($\sim$30 years), the low fertility of female flowers and the difficulties in efficient utilization of wild germplasms. Genetic transformation, gene editing and marker-assisted selection (MAS) are promising techniques to overcome these barriers, but their applications rely on a detailed elucidation of molecular mechanisms underlying high latex yield in rubber tree.

8.2.1 Latex High-Yielding Mechanisms Elucidated by Analyzing Self-rooting JCs and DCs

Rubber trees are usually propagated by grafting scions from elite rubber tree clones onto unselected rootstocks, thus producing bud-grafted clones (Clément-Demange et al. 2007; Hua et al. 2010). Bud-grafted clones usually manifest significant intra-clonal variations in growth and latex yield due to genetic differences and interactions between stock and scion (Chandrashekar et al. 1997; Clément-Demange et al. 2007; Hua et al. 2010). An alternative propagation material is self-rooting juvenile clones (JCs), which are derived from internal integuments of immature fruits, or somatic embryos of anthers obtained through primary somatic embryogenesis (Wang et al. 1980; Carron and Enjalric 1982; Chen et al. 2002; Clément-Demange et al. 2007; Hua et al. 2010). In contrast to donor clones (DCs), self-rooting JCs possess better phenotypes in terms of higher rubber yield and number of laticifers, and larger trunk girth (Liu et al. 1985; Yang and Mo 1994; Chen et al. 1998, 2002; Yuan et al. 1998).

Using latex from self-rooting JCs and DCs as experiment materials, Li et al. (2011) first identified and analyzed differentially expressed proteins (DEPs) between self-rooting JCs and DCs. Compared with DCs, 13 and 11 proteins were up- and down-regulated in self-rooting JCs. The DEPs mentioned above were classified into eight categories including carbohydrate and energy metabolism, transport, signal translocation, transcriptional regulation-related, secondary metabolism, protein synthesis and degradation, nucleoside acid process and lipid metabolism, suggesting that these DEPs may be closely associated with molecular mechanisms involved in the difference between self-rooting JCs and DCs. Three years later, Li et al. (2014) identified 176 differentially expressed latex genes (DEGs) between DCs and self-rooting JCs based on the suppression subtractive hybridization (SSH) method. A total of 81 and 95 DEGs were down- and up-regulated, respectively, in self-rooting JCs, in contrast to DCs. These DEGs were associated with stress/defense response, metabolism and energy, rubber biosynthesis, protein metabolism, signal transduction, transcription and post-transcription. They suggested that the DEGs in rubber biosynthesis and ROS metabolism might play important roles in the high-yielding property of self-rooting JCs. To further reveal the molecular mechanism of high yield in self-rooting JCs, Li et al. (2016c) carried out comparative analysis of the latex transcriptomes of self-rooting JCs and DCs. A total of 1,716 genes were identified as DEGs between self-rooting JCs and DCs. The expression of several genes in ROS scavenging, hormone metabolism and carbohydrate metabolism were increased in self-rooting JCs, thus providing a molecular basis for the high yield of self-rooting JCs. In addition, they found that some genes encoding epigenetic modification enzymes were differentially expressed, and this may lead to the variations in gene expression profiles between self-rooting JCs and DCs.

8.2.2 Latex High-Yielding Mechanisms Elucidated by Analyzing Super-Productivity

In a rubber plantation of Mengla County in Yunnan Province of China, a rubber tree of clone PR 107 planted in 1963 produced 126 kg of dry rubber in 2002, over 20-fold higher than the average rubber trees (Huang 2003); the super high-yielding PR 107 was named as SY 107. By comparing SY 107 and the control (PR 107), Tang et al. (2013) found that the mean latex yield of SY 107 was 4.12 kg per tapping, whereas the mean latex yields of two controls were 0.32 and 0.83 kg, respectively. The values of total solid and inorganic phosphorus content were not significantly different between SY 107 and the controls. In contrast to the control trees, sucrose and thiol content in SY 107 latex were significantly lower. To elucidate molecular mechanisms underlying super productivity, a comprehensive cDNA-AFLP transcript profiling was used to analyze latex from SY 107 and the controls. Of 453 differentially expressed (DE) transcript-derived fragments, 215 with known or partially known functions were grouped into 10 functional categories. The largest category was transcription and protein synthesis, followed by defense and stress, primary and secondary metabolism, and others. Tang et al. (2013) suggested that the notable characteristics of the super high-yielding phenotype may be attributed to higher sucrose-loading capability, improved general metabolism, rubber biosynthesis-preferred sugar utilization and high efficiency of stress alleviation; however, there was little correlation with the genes in the rubber biosynthesis pathway.

8.2.3 Latex High-Yielding Mechanisms Elucidated by Analyzing Different Rubber Clones

Li et al. (2015) compared the latex transcriptomes from rubber tree clones RRIM 600 and Reyan 7-20-59 to reveal molecular mechanisms regulating high latex yield. In contrast to the lower-yielding clone (RRIM 600), 2,513 and 1,391 genes were up- and down-regulated, respectively, in Reyan 7-20-59. The DEGs were significantly enriched in plant–pathogen interactions, phenylpropanoid biosynthesis, phenylalanine metabolism, biosynthesis of secondary metabolites, ubiquinone and other terpenoid-quinone biosynthesis, and photosynthesis pathways. In contrast with the results reported by Tang et al. (2013), the genes related to the rubber biosynthesis pathway, such as *cis*-prenyltransferase (*CPT*), geranyl diphosphate synthase (*GPPS*), 3-hydroxy-3-methylglutaryl-CoA reductase (*HMGR*), 3-hydroxy-3-methylglutaryl-CoA synthase (*HMGS*), farnesyl diphosphate synthase (*FPPS*) and 1-deoxy-D-xylulose-5-phosphate synthase (*DXS*), were differentially expressed between the two rubber tree clones. Compared to RRIM 600, the DEGs encoding *FPPS*, *GPPS*, *HMGS2*, *DXS* and *HMGR1* were up-regulated in Reyan 7-20-59. Based on the aforementioned results, Li et al. (2015) suggested that high yield in rubber trees is likely to be attributed to greater supply of isopentenyl pyrophosphate (IPP), higher utilization of IPP for making rubber and increased expression of genes related to the rubber biosynthesis pathway.

Rubber tree clones that are extensively cultivated today have been bred using Wickham germplasm material for approximately one century (Priyadarshan et al. 2009). High-yielding genotypes of the rubber tree are often selected from hybrids; therefore, analysis of the genetic makeup, variation and gene expression patterns between the parental clones and their elite hybrids is very helpful for rubber tree breeding.

By comparing the parents and their elite hybrids, Li et al. (2016b) found that the offspring clones possessed a much higher heterozygosity ratio compared with their parents. Latex yield in the selected offspring hybrids is positively correlated with genome heterozygosity, suggesting that high genome heterozygosity is closely associated with latex yield increase in rubber tree breeding. Reyan 7-33-97 was found to be genetically closer to RRIM 600 while Reyan 7-20-59 was closer to PR 107, this being in agreement with the corresponding phenotypic similarities and gene expression patterns of the clones.

8.3 Rubber Tree Responses to Biotic and Abiotic Stresses

Biotic and abiotic stresses have serious effects on rubber tree growth and development as well as in the sustainable development of the NR industry, especially in non-traditional planting regions. Biotic and abiotic stresses are key limiting factors on latex yield. This section focuses on two main stresses challenging the productivity of rubber trees: cold and SALB disease.

8.3.1 Response to Cold

In 2017, cold-responsive microRNAs from the leaf were identified and validated in the rubber tree clone, RRIM 600. Among 42 novel miRNAs related to cold response, miR169 was closely associated with cold tolerance, while the miRNAs, miR159 and miR482, also indicated a modest correlation with cold tolerance (Kuruvilla et al. 2017). With the cold-resistant rubber tree clone CATAS 93-114 and the cold-sensitive clone Reken 501 as experimental materials, Cheng et al. (2018) found that 7,870 genes were identified as DEGs between the leaves of the two clones. Compared with Reken 501, 614, 694, 829 and 799 DEGs were identified at 0, 2, 8 and 24 h of chilling treatment, respectively, in CATAS 93-114. Moreover, a large number of genes were up- or down-regulated between 2 and 8 h of chilling in CATAS 93-114, indicating a greater response toward cold by CATAS 93-114 compared to Reken 501. Gene Ontology (GO) analyses of the DEGs demonstrated that genes related to defense functions were more active during the late stage of chilling and more classes of genes were activated in CATAS 93-114 than in Reken 501 after 8 h of chilling treatment. Interestingly, the expression patterns of genes involved in abscisic acid (ABA) metabolism and signaling, the ABA-independent pathway and early signal perception were distinctly different between the two clones.

Using leaves of the same two clones, Deng et al. (2018a) reported that 1,919 and 2,929 genes were up- and down-regulated, respectively, in CATAS 93-114 without cold treatment when compared to Reken 501. With 4 °C cold stress, the up-regulated genes were 1,501 after 1 h treatment and 1,285 after 24 h treatment; the corresponding down-regulated genes were 7,567 and 5,482 after 1 and 24 h treatments, respectively. In contrast to Reken 501, 9,158 genes were identified in CATAS 93-114 after 1 h cold treatment. Moreover, functional annotation of the DEGs suggests that the rubber tree may cope with cold stress by down-regulating the genes related to auxin and ET signaling and by activating ROS scavengers and heat shock modules.

Based on a similar research strategy, Mantello et al. (2019) sequenced leaf tissues at 0, 1.5, 12 and 24 h after cold treatment in clones RRIM 600 and GT 1, and then compared and analyzed their transcriptomes. Comparison between RRIM 600 and GT 1 showed that 1,356, 1,368, 1,179 and 1,909 genes were differently expressed at 0, 1.5, 12 and 24 h after cold treatment, respectively. The up-regulated genes in RRIM 600 and GT 1 were classified into 83 and 183 GO terms, respectively. The DEGs were closely associated with signal transduction, ROS scavenging, the mitogen-activated protein kinase (MAPK) signaling pathway, photosynthesis activity and stomata closure.

The rubber tree clone, Reyan 7-33-97, which is high-yielding and cold-resistant, was studied in its response to low-temperature treatment (Gong et al. 2018). Three cDNA libraries from leaves, namely CT (control), LT2 (cold treatment at 4 °C for 2 h) and LT24 (cold treatment at 4 °C for 24 h) were generated for RNA-seq and comparative transcriptome analysis. In contrast to CT, 1,457 and 2,328 genes were differentially expressed in LT2 and LT24, respectively. The DEGs were significantly enriched in flavonoid biosynthesis, plant hormone signal transduction, phenylpropanoid biosynthesis, cutin, suberine and wax biosynthesis, phenylalanine metabolism, pentose and glucuronate interconversions, and starch and sucrose metabolism pathways. The functions of some cold-response transcription factor families that were differentially expressed included ARR-B, BES1, B3, bHLH, C2H, Dof, ERF, CO-like, FAR1, G2-like, SRS, GRAS, GRF, HD-ZIP, MIKC-MADS, HSF, LBD, M-type MADS, NAC, MYB, MYB-related, RAV, TALE, TCP, WRKY, Trihelix, WOX, YABBY and ZF-HD. A total of 1070 transcripts were identified to be related to cold response; and 109 and 961 transcripts were down- and up-regulated by cold treatment, respectively. In another study, the ethylene-responsive transcription factor (ERF) was reported to be closely associated with cold tolerance in the leaf samples of rubber tree clone, RRIM 600 (Sathik et al. 2018). Also, rubber trees treated with exogenous methyl jasmonate can enhance its cold tolerance through the involvement of the *HbICE2* gene in the leaves of Reyan 7-33-97 (Chen et al. 2019).

8.3.2 Response to South American Leaf Blight (SALB) Disease

SALB is a disease caused by *Microcyclus ulei* which can decrease latex yield by delaying growth and development of the rubber tree. Using the SSH method, Garcia et al. (2011) assessed the biological progression of SALB infection by comparing leaf transcriptomes of a partially resistant clone (MDF 180) and a susceptible one (PB 314) which had been inoculated with *M. ulei*. Leaf tissues corresponding to 6 h post-infection (hpi) to 58 days post-infection (dpi) were collected, and 5 cDNA libraries were prepared for enriching the genes induced by *M. ulei* at different stages of the infection. The number of non-redundant sequences (NRSs) that were exclusive to MDF 180 and PB 314 were 767 and 709, respectively, while 147 NRSs were common to both MDF 180 and PB 314. At 6-72 hpi, about 66.4% of the NRSs were exclusive to PB 314, while only 28.9% were specific to MDF 180. At 4-28 dpi, a total of 83 NRSs (11.8%) were specific to PB 314 while 594 (84.4%) were specific to MDF 180. Interestingly, the number of NRSs clearly decreased in the later stages of the infection in PB 314. The number of NRSs in PB 314 decreased from 727 to 69 and finally to 18 at 6-72 hpi, 4-28 dpi and 34-58 dpi, respectively. In comparison, the number of NRSs specific to each library of MDF 180 increased throughout the infection process. In MDF 180, the DEGs involved in the defense response to *M. ulei* were found to include pathogenesis-related (PR) proteins, R genes and the proteins related to ROS detoxification and phenol metabolism.

Hurtado Páez et al. (2015) selected a secondary clone (FX 3864) that is resistant to the *M. ulei* isolate GCL012, as experimental material. After infection with the isolate GCL012, they compared and analyzed the leaf transcriptomes of the FX 3864 clone at 0 and 48 hpi. In total, 86 DEGs were identified to be related to the defense response of FX 3864 toward GCL012. Seven putative members from the AP2/ERF ET-dependent superfamily were down-regulated. A rise in salicylic acid (SA) was attributed to the up-regulation of three genes related to cell wall synthesis and remodeling, as well as down-regulation of the putative gene, CPR5. These results suggested that the defense response of FX 3864 against the GCL012 isolate was likely to be associated with the antagonistic effect of SA, ET and JA pathways.

8.4 Laticifer Cell Development

Based on the histology of rubber tree tissues, there are two types of laticifers: primary laticifers and secondary laticifers (Gomez 1975, 1982; Tian et al. 2015a). In the course of primary growth, the young epicormic shoots only produced primary laticifers derived from the apical meristem. The primary laticifer cells were almost randomly distributed in the phloem, thereby making them hard to count. Owing to the disappearance of the primary laticifer cells in mature rubber trees, they are not relevant to latex yield. The differentiation of secondary laticifers started from the vascular cambia, and was distributed in the secondary phloem. In mature rubber trees, there is a significant positive correlation between the number of secondary laticifers and latex yield (Gomez 1975, 1982; Hao and Wu 2000).

8.4.1 Morphologies and Comparative Transcriptomics of Primary and Secondary Laticifer Cells

The developmental characteristics of the primary and secondary laticifer networks were investigated using optical and electron microscope (Wu et al. 2000; Tan et al. 2017). The initial laticifer tubes present a series of features, such as slender, straight and un-branched, and are often non-articulated. Initially, the primary laticifer tubes began intrusive growth at the lateral wall and expanded toward the neighboring parenchyma cells in multiple directions. Necklace-like structures gradually formed with the development of the primary laticifer tubes. Next, these primary laticifers connected to each other to form a three-dimensional laticiferous network (Tan et al. 2017). With bark growth and stem thickening, the primary laticifer cells are eventually no longer observed in bark cross-sections of the rubber tree trunk (Tian et al. 2015a).

Secondary laticifers emerge as rows and multiple rings around the trunk of the rubber tree. The secondary laticifer network has thick, straight and smooth cell walls. Some short bridges connected the neighboring laticifer tubes, while some transversal laticifer cells connected more distant and parallel laticifer tubes in the bark. In addition, neighboring cells were connected at different sites of the laticifer tubes (Tan et al. 2017). The number of secondary laticifers is a specific feature of different rubber tree clones, and is also affected by mechanical wounding and latex drainage (Hao and Wu 1982; Hao et al. 1984; Tian et al. 2003).

Latex from primary and secondary laticifers was collected from virgin rubber trees of clone Reyan 7-33-97 for transcriptome sequencing (Tan et al. 2017). In total, 131,078 transcripts and 103,704 unigenes were assembled using the Trinity software. Of 844 DEGs, 597 and 247 were up- and down-regulated, respectively, in the primary laticifers compared to the secondary laticifers. The up-regulated unigenes in the primary laticifers were associated with cell wall modification, defense against biotic stresses and rubber biosynthesis, while the up-regulated unigenes in the secondary laticifers were associated with abiotic stresses and dormancy. These results suggested that the primary laticifers were involved in response to biotic stresses while the secondary laticifers play a key role in defense against abiotic stresses.

8.4.2 Factors Inducing Secondary Laticifer Differentiation

Owing to the importance of the secondary laticifers in rubber biosynthesis and storage, laticifer studies focus mainly on secondary laticifer differentiation. In the secondary phloem of the rubber tree trunk, the developmental course of the secondary laticifers consists of differentiation, development, maturity and senium from the inside out (Sun 1997; Sando et al. 2009; Wu et al. 2000). Secondary laticifer differentiation is influenced by environmental factors and induced by some chemicals.

Rubber tree exploitation can accelerate secondary laticifer differentiation and lead to an increase of 1–3 times in the number of laticifer rings in the area of the tapped bark. No secondary laticifers were observed in cross-sections of bark that was subjected to mechanical wounding with little latex flow. These results indicate that latex flow has a major effect on the formation of secondary laticifers in the course of tree exploitation (Hao and Wu 1982, 2000; Hao et al. 1984).

In epicormic shoots, the first extension unit (EU) of the young stems contains primary laticifers, but no secondary laticifers (Wu et al. 2002; Tian et al. 2003). Mechanical wounding can induce the formation of secondary laticifers in the stem of epicormic shoots. This is because of the production of H_2O_2 and JA in the process of dehydration (Tian et al. 2015b). However, the wound-induced effect on laticifer differentiation was partial, and the secondary laticifers formed only at the wounded site.

JA plays an important role in plant defense and development. JA also has an obvious effect on secondary laticifer differentiation whereby its effect is dependent on the concentration of JA applied to the rubber tree. Compared to the lowest concentration (0.005%), a higher JA concentration (0.1%) can induce the formation of more secondary laticifers (Hao and Wu 2000). JA treatment does not change the structure of secondary laticifers other than produce additional laticifers in the treated sites of bark (Tian et al. 2003). Linolenic acid (LA) is another compound which also has apparent effects on the formation of secondary laticifers (Hao and Wu 2000). Besides JA and LA, laticifer differentiation can also be induced by the the histone deacetylation inhibitor, trichostatin A (TSA) (Zhang et al. 2016). As with JA, the effect of TSA on secondary laticifer differentiation was also dependent on its concentration. Laticifer cells were present in sites treated with 10 nmol/L TSA, and more secondary laticifer cells were detected when TSA concentration reached 1 μmol/L. The effects of three other chemicals, namely ABA, ethephon and SA, on laticifer differentiation were analyzed previously. The experimental results indicated that these chemicals could not induce laticifer differentiation (Hao and Wu 2000).

8.4.3 Expression Profiling of Genes Related to Secondary Laticifer Differentiation

In 2000, Hao and Wu reported that secondary laticifers appeared in JA- and LA-treated epicormic shoots of the rubber tree. Coronatine (COR), a toxin isolated from *Pseudomonas syringae* pv. Atropurpurea, can structurally and functionally mimic the active form of JA (Ichihara et al. 1977; Benedetti et al. 1995; Fonseca et al. 2009; Wasternack and Xie 2010). In 2015, the effect of COR on secondary laticifer differentiation was tested in the young stem of epicormic shoots (Zhang et al. 2015). The experimental procedures were as follows: Firstly, the stem surface of EU1 was scraped with a sharp razor to remove the epidermis cuticle and part of the cortex. Then, the surface was immediately applied with 20 μM COR and wrapped with a piece of polyethylene membrane. Finally, the morphology of the secondary laticifer cells in cross-sections was observed with an optical microscope. As far as the effect on inducing secondary laticifer differentiation was concerned, they found that COR was much more effective than JA (Zhang et al. 2015).

Based on the experimental procedure above, the authors also identified the genes and regulatory networks related to secondary laticifer differentiation using the SSH method (Zhang et al. 2015). In total, 147 and 109 unigenes were annotated in the forward and reverse SSH libraries, respectively. These unigenes were mainly involved in stress/defense response, plant hormone signal transduction and structure development. The results from functional annotation and expression analyses suggested that Ca^{2+} signal transduction and redox reactions might play a role in secondary laticifer differentiation, while polygalacturonase and translation initiation factors were likely involved in the division of cambium initials during secondary laticifer differentiation induced with COR in the rubber tree (Zhang et al. 2015).

One year later, Wu et al. (2016) further identified and analyzed genes related to secondary laticifer differentiation at the transcriptome level. Of 50,548 annotated unigenes, 15,780 and 19,824 were found to be differentially expressed at early and late stages of COR treatment, respectively. At the early stage, there were 8,646 up-regulated unigenes and 7,134 down-regulated unigenes, while 7,711 up-regulated unigenes and 12,113 down-regulated unigenes were identified at the late stage. The results from annotation and expression profiling of the DEGs suggested that JA-mediated signaling and the CLAVATA-MAPK-WOX signaling pathway may play a pivotal role in the process of secondary laticifer formation in the rubber tree (Wu et al. 2016).

By comparing bark samples with differentiated and non-differentiated secondary laticifers, Loh et al. (2016) identified DEGs using NimbleGen microarrays. Compared with ET-treated samples, 491 filtered differentially expressed (FDE) transcripts were identified in both JA- and LA-treated bark samples. Based on the Eukaryotic Orthologous Group analysis, these FDE transcripts were shown to be related to multiple functional categories, thus suggesting that laticifer differentiation is regulated by diverse pathways. Gene families such as cytochrome P450, ATP-binding cassette transporter, short-chain dehydrogenase/reductase (SDR) and cinnamyl alcohol dehydrogenase were associated with secondary metabolism. In addition, the FDE transcripts also encoded genes related to cell division, cell wall synthesis and cell differentiation. Based on the analyses of FDE transcripts, Loh et al. (2016) speculated that the most abundant transcript, SDR65C, may be associated with laticifer differentiation.

Taken together, secondary laticifer differentiation is a highly complex process which involves multiple pathways such as JA signaling, the CLV-MAPK-WOX signal pathway and Ca^{2+} signal transduction.

8.4.4 Analysis of Selected Genes Involved in Laticifer Development

In 2003, transgenic rubber trees overexpressing a superoxide dismutase (SOD) gene was first reported (Jayashree et al. 2003). Subsequently, in 2012, the function of a *Hevea* gene, copper zinc SOD, was analyzed using transgenic technology (Leclercq et al. 2012). More recently, transgenic rubber trees overexpressing *HbERF-IXc5* were analyzed (Lestari et al. 2018). In contrast to the wild type (WT), transgenic lines overexpressing *HbERF-IXc5* showed clearly enhanced plant growth and tolerance to abiotic stresses. Interestingly, more primary latex cells were observed in the leaf midrib of *HbERF-IXc5*-overexpressing plants. In green and lignified stems, the line TS19A46 overexpressing *HbERF-IXc5* had much more primary and secondary latex cells compared to WT. Moreover, the line TS19A46 possessed more secondary laticifers in young green stems. These results indicated that *HbERF-IXc5* may be associated with laticifer differentiation (Lestari et al. 2018).

Jasmonate ZIM-domain (JAZ) proteins are the repressors of jasmonate signaling which play a key role in the process of secondary laticifer differentiation. In rubber trees, the expression of *HbJAZ5.0* and *HbJAZ10.0b* were clearly increased after COR treatment (Chao et al. 2019). In addition, the promoters of the two genes have a *cis*-regulatory element (CAT-box) which is likely to be connected with their expression in the meristem. Based on these findings, Chao et al. (2019) speculated that the two genes, *HbJAZ5.0* and *HbJAZ10.0b*, might be involved in laticifer differentiation.

8.5 Proteomic and Transcriptomic Studies of Rubber Biosynthesis and Latex Flow

Rubber biosynthesis and latex flow are closely related to latex yield in rubber tree. The former and the latter determine the rubber-producing capacity in latex and the volume of latex exuded, respectively. Both are also intrinsic inherited characteristics of rubber tree clones.

8.5.1 Genes and Proteins Involved in Rubber Biosynthesis and Latex Flow

NR, or *cis*-1,4-polyisoprene polymer is derived from IPP, and is synthesized by rubber transferase (RT) in the laticifers. *cis*-1,4-polyisoprene synthesis involves three steps: initiation, elongation and termination. A total of 84 genes from 20 gene families involved in rubber biosynthesis were identified at the genome level: 18 mevalonate (MVA) pathway genes, 22 2-C-methyl-D-erythritol-4-phosphate (MEP) pathway genes, 15 initiator synthesis genes and 29 rubber elongation-related genes (Tang et al. 2016). An ubiquitin-proteasome protein degradation mechanism was also proposed to be involved in the termination step (Wang et al. 2018).

In the rubber tree, IPP supply for synthesizing *cis*-1,4-polyisoprene may come from one or both the MVA and the MEP pathways (Chow et al. 2012) based on gene expression profiles (Chow et al. 2012; Tang et al. 2016). *HMGR1* in the MVA pathway was more highly expressed in a rubber tree clone with higher latex yield compared to a lower yielding one (Chao et al. 2015). Interestingly, latex yield showed a positive correlation with HMGR enzyme activity and the accumulation of *Hbhmgr1* transcripts in rubber trees overexpressing the *Hbhmgr1* gene (Jayashree et al. 2018).

It was reported that dimethylallyl pyrophosphate, geranyl pyrophosphate, farnesyl pyrophosphate (FPP), geranylgeranyl pyrophosphate or *cis*-allylic pyrophosphate could act as oligomeric allylic pyrophosphate initiators in rubber chain initiation in vitro (Cornish 2001; Chiang et al. 2011). FPP showed a lower binding constant compared with other initiators, and the cytosolic location of FPP suggested that FPP is the major initiator for rubber biosynthesis (Cornish 2001; da Costa et al. 2005; Espy et al. 2006; Xie et al. 2008).

The rubber elongation factor (REF) and the small rubber particle protein (SRPP) were reported to be required for rubber biosynthesis (Dennis and Light 1989; Oh et al. 1999; Kim et al. 2004). Genome studies indicated that the REF/SRPP family contained more than 10 members (Rahman et al. 2013; Lau et al. 2016; Tang et al. 2016). Results from 16-BAC/SDS PAGE-based proteomic analyses also showed the presence of these REF/SRPP homologous proteins (Dai et al. 2013, 2017). Some members of the REF/SRPP family are highly expressed in latex at both gene and protein levels (Chow et al. 2007; Dai et al. 2016; Tang et al. 2016). In contrast to SRPPs which are abundantly localized on small rubber particles (SRPs), different REF isoforms were either located mainly on large rubber particles (LRPs) or expressed equally on both SRPs and LRPs (Berthelot et al. 2012; Xiang et al. 2012; Dai et al. 2017). HbREF was bound to the rubber particle lipid monolayer more tightly than HbSRPP (Berthelot et al. 2014b; Brown et al. 2017). In addition, HbSRPP and HbREF have aggregative properties and probably play important roles in rubber particle stabilization and coagulation (Berthelot et al. 2012, 2014a). Priya et al. (2007) demonstrated that in latex, the abundance of REF transcripts was 3–5 times higher in high-yielding rubber tree clones than in low-yielding ones, implying a positive correlation between REF gene expression level and latex yield. Dai et al. (2017) also proposed that the REF/SRPP family proteins act as structural proteins and are likely to play a role in packaging/storing NR and stabilizing rubber particles in rubber tree latex. Some serine residues in REF/SRPP isoforms were found to be phosphorylated and N-terminal acetylated (Wang et al. 2015; Habib et al. 2018), suggesting that phosphorylation and acetylation of REF/SRPP proteins might also be involved in rubber biosynthesis.

Asawatreratanakul et al. (2003) cloned two CPT genes from rubber tree latex and named them as *HRT1* and *HRT2*. Although *HRT1* and *HRT2* are expressed mainly in latex (Asawatreratanakul et al. 2003; Brown et al. 2017), enzyme activity assays showed that only HRT2 could catalyze the synthesis of *cis*-1,4-polyisoprene with a size range of 2×10^3–1×10^4 Da without addition of any washed bottom fraction components (Asawatreratanakul et al. 2003). Interestingly, comparative transcriptomic analysis showed that *HRT2* transcripts were overexpressed in a rubber tree clone with higher latex yield (Chao et al. 2015). Genome analyses of rubber tree showed that the CPT family contained several members (Rahman et al. 2013; Lau et al. 2016; Tang et al. 2016). Several CPT isoforms were identified by the 1D-LC MS method (Dai et al. 2013; Yamashita et al. 2016). Also, two CPT proteins were separated by 16-BAC/SDS PAGE as distinct protein spots and identified in each round of the two-step extractions (Dai et al. 2017).

CPT-like/CPT-binding proteins (CPTL/CPTBP) are recently isolated proteins that are involved in rubber biosynthesis. These are homologs of the human Nogo-B receptor (NgBR) and are localized mainly on ER and rubber particles in rubber-producing plants (Brown et al. 2017; Qu et al. 2015). The NgBR homologs usually contain one or more N-terminal transmembrane domains and a C-terminal CPT-like domain (Yamashita et al. 2016; Liu et al. 2018). They do not contain catalytic residues conserved among CPTs, but the RxG motif in their C-terminal is critical for prenyltransferase activity (Grabińska et al. 2017). By reconstituting the rubber biosynthetic machinery on detergent-washed rubber particles (WRPs) in vitro, Yamashita et al. (2016) identified new components of the RT complex. The HRT1-REF bridging protein (HRBP), a member of the CPTL/CPTBP clade, played important roles in introducing recombinant CPTs to WRPs. By interacting with HRT1/2 and REF, CPTL/CPTBP may act as a bridge between HRT1 and REF to form a complex, suggesting that the rubber biosynthetic machinery consists of CPT, HRBP and REF (Yamashita et al. 2016; Brown et al. 2017).

Besides rubber biosynthesis, latex flow is another important limiting factor on latex yield. Wei et al. (2015) found that REF and HMGS transcripts were down-regulated in rubber trees with long duration of latex flow. The rate and duration of latex flow are determined mainly by latex coagulation. Hevamine (or rubber tree chitinase), β-1,3-glucanase and prohevein/hevein are the most abundant proteins in lutoids which are associated with latex coagulation (including rubber particle aggregation) (Gidrol et al. 1994; Koningsveld et al. 1996; Shi et al. 2010). A basic chitinase was suggested to remove the N-acetyl-D-glucosamine group from the hevein receptor, and inhibit hevein-meditated rubber particle aggregation (Chrestin et al. 1997). A basic chitinase and several β-1,3-glucanase isoforms (with basic isoelectric points) were identified from washed rubber particles (Dai et al. 2012, 2017); these basic proteins were difficult to be removed from rubber particles by washing. Contrary to the results of Chrestin et al. (1997), Wang et al. (2013) reported that basic chitinase and β-1,3-glucanase played a positive role in accumulating rubber particle aggregation in in vitro experiments. The expression of a gene encoding a 44 kDa protein which is antagonistic to rubber particle aggregation was higher in the latex of CATAS 8-79 than that of CATAS 7-33-97 at all stages of latex flow (Chao et al. 2017).

8.5.2 Roles of ET and JA in Rubber Biosynthesis and Latex Flow

ET or ethephon (an ET releaser), and JA are widely used to increase NR production. Proteomic results from rubber tree latex suggested that exogenous application of ethephon could effectively increase latex production not only by enhancing carbohydrate catabolism and energy production instead of rubber biosynthesis itself, but also by inhibiting the expression of some proteins related to rubber particle aggregation to prolong latex flow (Wang et al. 2015). Most proteins in the MVA pathway including HMGS, HMGR, mevalonate diphosphate decarboxylase and mevalonate kinase were inhibited by ethephon, whereas an isoform of acetyl-CoA acetyltransferase was induced by ethephon (Wang et al. 2015). In addition, a CPT isoform catalyzing IPP polymerization into NR was found to be induced by ethylene (Wang et al. 2015). The most abundant proteins, REF and SRPP, were not significantly changed upon ethylene treatment at either gene or protein level (Wang et al. 2015). The results from transcriptomic analyses in rubber tree bark are similar to the findings by Wang et al. (2015); the rate-limiting enzymes of glycolysis were up-regulated and DEGs were not enriched significantly for the MVA pathway (Liu et al. 2016). When ethephon was exogenously applied to rubber tree bark, a chitinase isoform which is likely to facilitate latex flow (Chrestin et al. 1997) was up-regulated in latex, while a sieve element occlusion protein which probably facilitates latex coagulation was down-regulated (Dai et al. 2016). Two REFs with molecular weights 19.6 and 27.3 kDa were found to be ethylene-responsive only in a high-yielding clone, but not in a low-yielding one (Tong et al. 2017).

In addition, transcriptomic analyses of exogenous application of methyl jasmonic acid (MeJA) on bark were performed in rubber trees (Liu et al. 2018). The authors found that a *HMGS* gene was down-regulated by exogenous MeJA, whereas the genes encoding FPS and CPT were up-regulated, possibly leading to increased rubber production in the latex. Deng et al. (2018b) found that an HbCOI1-HbJAZ3-HbMYC2 module existed in laticifer cells. This module up-regulated genes encoding HbFPS1 (a farnesyl pyrophosphate synthase) and HbSRPP1 (the most abundant SRPP isoform), resulting in increased rubber biosynthesis.

8.6 Progress on Rubber Tree TPD

The typical feature of TPD is the appearance of partial or complete dry zones (no latex flow) on the tapping panel. Additionally, other symptoms such as bark browning, thickening, and even flaking usually occur in TPD-affected rubber trees (Sookmark et al. 2002). TPD has become a serious threat to latex yields. The losses from TPD-affected rubber trees can amount to 12–20% of latex yield in different rubber planting countries (Jiang and Zhou 1997; Okoma et al. 2011).

8.6.1 Utilization of Latex Physiological Parameters in Early Diagnosis of TPD

Eschbach et al. (1984) reported that latex physiological parameters reflect the metabolism level of the laticiferous system including the status of rubber biosynthesis and latex flow in rubber trees. Zeng et al. (1997) found that RNA, sucrose, Pi and R-SH contents in healthy rubber trees are higher than those in TPD-affected ones, whereas RNase activity is higher in TPD-affected trees than in healthy trees. A study of TPD susceptibility of the rubber tree clone, PB 260, by Putranto et al. (2015) reported that the sucrose, Pi and R-SH contents were all higher in healthy rubber trees. By measuring and analyzing latex physiological parameters of rubber tree clones with different degrees of TPD, Guo et al. (2016) found that dry rubber content, sucrose content, Mg^{2+} content and lutoid bursting index rise with increasing TPD severity. Latex thiol, inorganic phosphorus and pH decreased when the severity of TPD of rubber tree increased. Moreover, latex thiol and inorganic phosphorus contents were significantly lower in TPD trees than in healthy ones. There is no obvious content change in crude enzyme liquid protein between healthy and TPD-affected rubber trees. More recently, Tristama et al. (2019) analyzed the physiological parameters of rubber trees with high and low metabolism and suggested that the decline of peroxidase activity and Pi could be utilized to detect TPD incidence in rubber trees.

8.6.2 ET and TPD

Since ethephon was found to enhance rubber production, it has been utilized in rubber tree plantations worldwide. It was reported that ethylene stimulation usually increases latex yield by 1.5–2 folds (Coupé and Chrestin 1989). However, it is widely accepted that TPD can be induced by overstimulation with ET (Fan and Yang 1984; Chrestin 1989; Faridah et al. 1996). The genes involved in ET biosynthesis, perception and signaling showed differential expression in latex and bark, and some of these genes were strictly regulated by TPD occurrence and development, suggesting that ET is mainly synthesized in bark and perceived in both tissues during TPD (Putranto et al. 2015). The expression of *HbERF-I* was lower in latex than in bark, while *HbERF-VIII* indicated a similar expression level in both tissues. Since ERF-VIII was shown to be related to programmed cell death (PCD) (Ogata et al. 2013), accumulation of *HbERF-VIII* transcripts was suggested to play potential roles in early PCD of TPD-affected rubber trees (Putranto et al. 2015). Changes in hormone signalling pathways and consequently TPD during ethephon stimulation were revealed by latex transcriptome sequencing (Montoro et al. 2018). In total, 8,111 DEGs were related to ethephon response in slightly-affected TPD trees, while 728 DEGs were associated with ethephon-induced severe TPD trees. The authors suggested that a biological network of responses to ET and TPD highlighted the major influence of metabolic processes and the response to stimulus, especially JA depression and wounding, in TPD trees induced by ET (Montoro et al. 2018). ET also plays important roles in regulating ROS homeostasis. Compared with untreated rubber trees, stimulation with high concentrations of ethephon results in lower SOD activity and higher levels of free radicals (Das et al. 1998). After ethephon stimulation, the contents of glutathione S-transferase and SOD decreased, whereas the contents of monodehydroascorbate peroxidase and peroxidase increased in latex (Wang et al. 2015). These results suggested that ET might be closely connected to rubber tree TPD by regulating PCD, metabolic processes, ROS homeostasis and stress, as well as hormone responses.

8.6.3 ROS and TPD

ROS homeostasis is vital for plant growth and development. Huang et al. (2018) reported that metallothioneins (*HbMTs*) played important roles in latex regeneration and species adaptation by regulating ROS homeostasis. In rubber trees, environmental stresses, tapping and metabolic activities during latex regeneration can result in ROS production. ROS accumulation and burst is an important cause of TPD occurrence. Lutoids are organelles which constitute 10–20% of fresh latex volume. It was proposed that ROS are mainly generated from lutoids in latex (Chrestin 1989). NAD(P)H oxidase was reported to be the main source of ROS, especially when the laticifers were under abiotic stress (Cretin and Bangratz 1983; Chrestin et al. 1984). Lutoidic NAD(P)H oxidase can generate a variety of toxic oxygen species, which can degrade the unsaturated lipids of cellular membranes (Chrestin et al. 1984). Three main antioxidants, namely ascorbate, thiol and tocotrienol exist in latex. In contrast to healthy rubber trees, NAD(P)H oxidase and peroxidase activities were abnormally high in TPD-affected rubber trees, whereas antioxidant substances such as thiols, variable peroxidase, ascorbic acids, catalase and SOD were down-regulated (Eschbach et al. 1984; Zeng et al. 1997; Coupé and Chrestin 1989). The aforementioned results suggested that the balance between ROS production and scavenging was broken, thus leading to ROS accumulation and burst in the rubber tree. In laticifer cells, excessive oxidative stress can destroy cellular membranes and induce the lutoid burst, finally leading to latex coagulation in situ (Chrestin 1989). Among 407 genes from 30 redox-related gene families, Zhang et al. (2019) found that 13 genes were targeted by 11 microRNAs. In addition, an important mutation in the miR398 binding site targeting the cytosolic copper zinc SOD was first reported in their study.

8.6.4 Expression Profiling of TPD-Related Genes or Proteins

Using latex from healthy and TPD-affected rubber trees without ethylene stimulation, Venkatachalam et al. (2007) identified 134 TPD-related genes using the SSH method. Compared with healthy rubber trees, 47 and 87 unigenes were down- and up-regulated, respectively, in TPD-affected rubber trees. Among eight functional categories, stress/defense-related genes constituted the largest one, followed by protein synthesis, cell division and growth, and cell metabolism. Interestingly, one gene related to rubber biosynthesis, *HbHMGS-CoA*, was up-regulated in TPD-affected rubber trees. They suggested that many TPD-related genes up-regulated in TPD trees may lead to PCD during the onset of the syndrome. In addition, a link between PCD and TPD syndrome was built by the genes identified in their study (Venkatachalam et al. 2007). Three years later, Li et al. (2010) identified 237 TPD-related unigenes from the latex of TPD and healthy rubber trees using the SSH method. The 237 unique genes were involved in 10 functional categories such as stress/defense response, protein metabolism, rubber biosynthesis, and transcription and post-transcription. The preferentially expressed genes were connected to stress/defense response, and metabolism and energy in TPD and healthy rubber trees, respectively. They suggested that ROS production and scavenging, PCD, ubiquitin proteasome pathway (UPP) and rubber biosynthesis might be important pathways in TPD occurrence and development (Li et al. 2010). Subsequently, Liu et al. (2015) also identified 22,577 DEGs between the bark of healthy and TPD rubber trees by comparative transcriptome analysis. A majority of genes involved in rubber biosynthesis and JA synthesis were found to be differentially expressed. They suggested that the expression of most genes related to rubber biosynthesis and JA synthesis were obviously decreased, which probably resulted in TPD occurrence. One year later, Li et al. (2016a) found that 8,137 unigenes were differentially expressed between the bark of TPD-affected (initial stage) and healthy rubber trees stimulated with ethylene. Compared with healthy rubber trees, 2,485 and 5,652 unigenes were down- and up-regulated, respectively, in TPD-affected rubber trees. The TPD-related genes were significantly enriched in five KEGG pathways and eight GO terms that were closely connected to PCD, ROS metabolism and rubber biosynthesis, whereby this is consistent with previous results from rubber tree latex. They suggested that rubber tree TPD is a complicated biological process that is related to many genes. The lower latex yield of TPD-affected trees might be attributed to lower IPP availability for rubber biosynthesis and down-regulation of the genes related to post-IPP steps of the rubber biosynthesis pathway (Li et al. 2016a). In a more recent study, Liu et al. (2019) found that proteins which interacted with metacaspase (*HbMC1*) are closely associated with PCD, ROS metabolism and rubber biosynthesis, thus suggesting that *HbMC1* was probably involved in TPD by regulating the PCD process induced by ROS in the rubber tree (Liu et al. 2019).

Unlike TPD-related genes, there are relatively less reports on identifying and analyzing TPD-related proteins. Dian et al. (1995) found that five proteins in the cytosolic compartment of latex were associated with TPD. Compared with healthy rubber trees, two proteins (26 and 14.5 kDa) were dramatically increased, while three other proteins (55, 34 and 21 kDa) were not significantly decreased in TPD-affected rubber trees. Interestingly, the 26 kDa protein is involved in latex coagulation, and its accumulation is specifically associated with TPD development and inhibited by ET treatment. The 14.5 kDa protein is preferentially accumulated during the severe stages of TPD tree. Seven years later, Sookmark et al. (2002) also identified three TPD-related polypeptides in latex. Two polypeptides (P15 and P22) accumulated in the latex of TPD-affected rubber trees, and a polypeptide (P29) was only expressed in the latex of TPD-affected rubber trees, but not in healthy ones. P22 and P15 were identified as SRPP (Hev b3) and REF (Hev b1), respectively. P29 was identified as a new member of the patatin-like protein family. The results from Northern blot analyses indicated that REF and SRPP were highly up-regulated by tapping-induced metabolic activation, but not by wounding per se, or by ABA or ET treatments. However, the expression of REF and SRPP was not significantly different between healthy and TPD-affected rubber trees, indicating that their accumulation was not transcriptionally regulated in the latex of these trees. The TPD-related proteins reported to date are likely to be associated with partial or complete stoppage of latex flow in TPD-affected rubber trees.

8.7 Future Prospects

Increasing latex yield through breeding high-yielding rubber tree clones is one of the key approaches to producing sufficient rubber for world consumption. Other non-conventional approaches such as genetic transformation, marker-assisted trait selection and gene editing are also very helpful for generating productive rubber tree clones. However, a deeper understanding of the molecular mechanisms underlying high latex yield and its relationship with major latex physiological and metabolic processes are important pre-requisites for successful application of these technologies. In spite of active application of transcriptomic and proteomic analyses of important processes such as rubber biosynthesis and latex flow, stress tolerance, laticifer cell development and TPD, further studies need to be carried out before the findings can be gainfully utilized by breeders. Insights on molecular mechanisms still need to be elucidated in greater detail together with the biological functions of key genes related to these processes. Integration of results from the studies to date will provide a solid foundation of fundamental information for designing strategies of breeding high-yielding rubber trees using approaches such as gene editing, genetic transformation, marker-assisted breeding and genome-wide association studies.

References

Asawatreratanakul K, Zhang YW, Wititsuwannakul D, Wititsuwannakul R, Takahashi S, Rattanapittayaporn A, Koyama T (2003) Molecular cloning, expression and characterization of cDNA encoding cis-prenyltransferases from *Hevea brasiliensis*. A key factor participating in natural rubber biosynthesis. Eur J Biochem 270(23):4671–4680

Benedetti CE, Xie D, Turner JG (1995) Coi1-dependent expression of an *Arabidopsis* vegetative storage protein in flowers and siliques and in response to coronatine or methyl jasmonate. Plant Physiol 109 (2):567–572

Berthelot K, Lecomte S, Estevez Y, Coulary-Salin B, Bentaleb A, Cullin C, Deffieux A, Peruch F (2012) Rubber elongation factor (REF), a major allergen component in *Hevea brasiliensis* latex has amyloid properties. PLoS One 7(10):e48065

Berthelot K, Lecomte S, Estevez Y, Peruch F (2014a) *Hevea brasiliensis* REF (Hev b 1) and SRPP (Hev b 3): an overview on rubber particle proteins. Biochimie 106:1–9

Berthelot K, Lecomte S, Estevez Y, Coulary-Salin B, Peruch F (2014b) Homologous *Hevea brasiliensis* REF (Hevb1) and SRPP (Hevb3) present different auto-assembling. BBA Proteins Proteom 1844 (2):473–485

Brown D, Feeney M, Ahmadi M, Lonoce C, Sajari R, Di Cola A, Frigerio L (2017) Subcellular localization and interactions among rubber particle proteins from *Hevea brasiliensis*. J Exp Bot 68(18):5045–5055

Carron MP, Enjalric F (1982) Studies on vegetative micropropagation of *Hevea brasiliensis* by somatic embryogenesis and *in vitro* microcutting. In: Fujiwara A (ed) Plant tissue culture. Maruzen Co., Tokyo, pp 751–752

Chandrashekar TR, Mydin KK, Alice J, Varghese YA, Saraswathyamma CK (1997) Intraclonal variability for yield in rubber (*Hevea brasiliensis*). Ind J Nat Rubb Res 10:43–47

Chao J, Chen Y, Wu S, Tian W (2015) Comparative transcriptome analysis of latex from rubber tree clone CATAS8-79 and PR107 reveals new cues for the regulation of latex regeneration and duration of latex flow. BMC Plant Biol 15(1):104

Chao J, Yang S, Chen Y, Tian WM (2017) Transcript profiling of *Hevea brasiliensis* during latex flow. Front Plant Sci 8:1904

Chao J, Zhao Y, Jin J, Wu S, Deng X, Chen Y, Tian WM (2019) Genome-wide identification and characterization of the *JAZ* gene family in rubber tree (*Hevea brasiliensis*). Front Genet 10:372

Chen WJ, Wang X, Yan S, Huang X, Yuan HM (2019) The ICE-like transcription factor *HbICE2* is involved in jasmonate-regulated cold tolerance in the rubber tree (*Hevea brasiliensis*). Plant Cell Rep 38(6):699–714

Chen XT, Wang ZY, Wu HD, Xie YP, Zheng WR (1998) *In vitro* micropropagation of juvenile self-root clones of *Hevea brasiliensis*. Acta Agron Sin 24(2):225–230

Chen XT, Wang ZY, Wu HD, Zhang XJ (2002) A new planting material of *Hevea brasiliensis*-self-rooting juvenile-type clone. Chin J Trop Crops 23(1):19–23

Cheng H, Chen X, Fang J, An Z, Hu Y, Huang H (2018) Comparative transcriptome analysis reveals an early gene expression profile that contributes to cold resistance in *Hevea brasiliensis* (the Para rubber tree). Tree Physiol 38(9):1409–1423

Chiang CK, Xie W, McMahan C, Puskas JE (2011) Unraveling the mystery of natural rubber biosynthesis. Part I: investigation of the composition and growth of in vitro natural rubber using high resolution size exclusion choromatography. Rubber Chem Technol 84:166–177

Chow KS, Mat-Isa MN, Bahari A, Ghazali AK, Alias H, Mohd-Zainuddin Z, Hoh CC, Wan KL (2012) Metabolic routes affecting rubber biosynthesis in *Hevea brasiliensis* latex. J Exp Bot 63(5): 1863–1871

Chow KS, Wan KL, Isa MN, Bahari A, Tan SH, Harikrishna K, Yeang HY (2007) Insights into rubber biosynthesis from transcriptome analysis of *Hevea brasiliensis* latex. J Exp Bot 58(10):2429–2440

Chrestin H (1989) Biochemical aspects of bark dryness induced by overstimulation of rubber trees with ethrel. In: d'Auzac J, Jacob JL, Chrestin H (eds) Physiology of rubber tree latex. CRC Press Inc., Boca Raton, pp 432–439

Chrestin H, Bangratz J, d'Auzac J, Jacob J (1984) Role of the lutoidic tonoplast in the senescence and degeneration of the laticifers of *Hevea brasiliensis*. Zeitschrift fiir Pflanzenphysiologie 114(3):261–268

Chrestin H, Gidrol X, Kush A (1997) Towards a latex molecular diagnostic of yield potential and the genetic engineering of the rubber tree. Euphytica 96(1):77–82

Clément-Demange A, Priyadarshan PM, Hoa TTT, Venkatachalam P (2007) *Hevea* rubber breeding and genetics. In: Janick Jules (ed) Plant breeding reviews. John Wiley and Sons Inc., Hoboken, NJ, pp 177–281

Cornish K (2001) Similarities and differences in rubber biochemistry among plant species. Phytochemistry 57:1123–1134

Coupé M, Chrestin H (1989) Physico-chemical and biochemical mechanisms of the hormonal (ethylene) stimulation: early biochemical events induced, in *Hevea* latex, by hormonal bark stimulation. In: d'Auzac J, Jacob JL, Chrestin H (eds) Physiology of rubber tree latex. CRC Press Inc., Boca Raton, pp 295–319

Cretin H, Bangratz J (1983) Une activite enzymatique endogene NAD(P)H dependante, responsable de la degradation peroxydative des organites membranaires et de la coagulation precoce, ou *in situ*, du latex d'*Hevea brasiliensis*. Comptes Rendus Hebdomadaires des Seances de l'Academie des Sciences Serie 3:101–106

da Costa BM, Keasling JD, Cornish K (2005) Regulation of rubber biosynthetic rate and molecular weight in *Hevea brasiliensis* by metal cofactor. Biomacromolecules 6(1):279–289

Dai L, Kang G, Li Y, Nie Z, Duan C, Zeng R (2013) In-depth proteome analysis of the rubber particle of *Hevea brasiliensis* (para rubber tree). Plant Mol Biol 82(1–2):155–168

Dai L, Kang G, Nie Z, Li Y, Zeng R (2016) Comparative proteomic analysis of latex from *Hevea brasiliensis* treated with Ethrel and methyl jasmonate using iTRAQ-coupled two-dimensional LC–MS/MS. J Proteomics 132:167–175

Dai L, Nie Z, Kang G, Li Y, Zeng R (2017) Identification and subcellular localization analysis of two rubber elongation factor isoforms on *Hevea brasiliensis* rubber particles. Plant Physiol Biochem 111:97–106

Dai L, Xiang Q, Li Y, Nie Z, Kang G, Duan C, Zeng R (2012) Rubber particle protein analysis of *Hevea brasiliensis* by two dimensional 16-BAC/SDS-PAGE and mass spectrometry. Sci Agric Sin 45(11):2328–2338

Das G, Raj S, Pothen J, Sethuraj MR, Sen-Mandi S (1998) Status of free radical and its scavenging system with stimulation in *Hevea brasiliensis*. Plant Physiol Biochem 25:47–50

Deng X, Wang J, Li Y, Wu S, Yang S, Chao J, Chen Y, Zhang S, Shi M, Tian W (2018a) Comparative transcriptome analysis reveals phytohormone signalings, heat shock module and ROS scavenger mediate the cold-tolerance of rubber tree. Sci Rep 8(1):4931

Deng X, Guo D, Yang S, Shi M, Chao J, Li H, Peng S, Tian W (2018b) Jasmonate signalling in the regulation of rubber biosynthesis in laticifer cells of rubber tree *Hevea brasiliensis*. J Exp Bot 69(15):3559–3571

Dennis MS, Light DR (1989) Rubber elongation factor from *Hevea brasiliensis*: Identification, characterization, and role in rubber biosynthesis. J Biol Chem 264 (31):18608–18617

Dian K, Sangare A, Diopoh JK (1995) Evidence for specific variation of protein pattern during tapping panel dryness condition development in *Hevea brasiliensis*. Plant Sci 105:207–216

Eschbach JM, Roussel D, Van de Sype H, Jacob JL, D'Auzac J (1984) Relationship between yield and clonal physiological characteristics of latex from *Hevea brasiliensis*. Physiologie Vegetale 22(3):295–304

Espy SC, Keasling JD, Castillón J, Cornish K (2006) Initiator-independent and initiator-dependent rubber biosynthesis in *Ficus elastica*. Arch Biochem Biophys 448(1–2):13–22

Fan SW, Yang SQ (1984) Cause of disease and hypothesis on tapping panel dryness of *Hevea brasiliensis*. Chin J Trop Crops Res 18:43–48

Faridah Y, Siti Arija MA, Ghandimathi H (1996) Changes in some physiological latex parameters in relation to over exploitation and onset of induced tapping panel dryness. J Nat Rubber Res 10:182–186

Fonseca S, Chini A, Hamberg M, Adie B, Porzel A, Kramell R, Miersch O, Wasternack C, Solano R (2009) (+)-7-iso-Jasmonoyl-L-isoleucine is the endogenous bioactive jasmonate. Nat Chem Biol 5 (5):344–350

Garcia D, Carels N, Koop DM, de Sousa LA, Andrade Junior SJd, Pujade-Renaud V, Reis Mattos CR, de Mattos Cascardo JC (2011) EST profiling of resistant and susceptible *Hevea* infected by *Microcyclus ulei*. Physiol Mol Plant Pathol 76(2):126–136

Gidrol X, Chrestin H, Tan HL, Kush A (1994) Hevein, a lectin-like protein from *Hevea brasiliensis* (rubber tree) is involved in the coagulation of latex. J Biol Chem 269(12):9278–9283

Gomez JB (1975) Comparative ultracytology of young and mature latex vessels in *Hevea brasiliensis*. In: Proceedings of International Rubber Conference 1975, Rubber Research Institute Malaysia, Kuala Lumpur, pp 143–164

Gomez JB (1982) Anatomy of *Hevea* and its influence on latex production. Malaysian Rubber Research and Development Board, Monograph No. 7, Kuala Lumpur

Gong XX, Yan BY, Hu J, Yang CP, Li YJ, Liu JP, Liao WB (2018) Transcriptome profiling of rubber tree (*Hevea brasiliensis*) discovers candidate regulators of the cold stress response. Genes Genom 40(11):1181–1197

Guo XL, Sun L, Hu YY, Liu JP, Wang ZH, Yuan K (2016) Analysis on latex physiological parameters of *Hevea brasiliensis* plants at different tapping panel dryness extents. J South Agr 47(9):1553–1557

Grabińska KA, Edani BH, Park EJ, Kraehling JR, Sessa WC (2017) A conserved carboxy-terminal RxG motif in the NgBR subunit of *cis*-prenyltransferase is critical for prenyltransferase activity. J Biol Chem 292(42):17351–17361

Habib MAH, Gan CY, Abdul Latiff A, Ismail MN (2018) Unrestrictive identification of post-translational modifications in *Hevea brasiliensis* latex. Biochem Cell Biol 30:1–7

Hao BZ, Wu JL, Yun CY (1984) Acceleration of laticifer differentiation in *Hevea brasiliensis* by latex drainage. Chin J Trop Crops 5(2):19–23

Hao BZ, Wu JL (1982) Effects of wound (tapping) on laticifer differentiation in *Hevea brasiliensis*. Acta Botanica Sinica 24(4):388–391

Hao BZ, Wu JL (2000) Laticifer differentiation in *Hevea brasiliensis*: induction by exogenous jasmonic acid and linolenic acid. Ann Bot 85:37–43

Hua YW, Huang TD, Huang HS (2010) Micropropagation of self-rooting juvenile clones by secondary somatic embryogenesis in *Hevea brasiliensis*. Plant Breeding 129(2):202–207

Huang ZM (2003) A rubber-yielding king tree in a Mengla rubber plantation of Yunnan province. Trop Agr Sci Technol 26(S):51

Huang Y, Fang Y, Long X, Liu L, Wang J, Zhu J, Ma Y, Qin Y, Qi J, Hu X, Tang C (2018) Characterization of the rubber tree metallothionein family reveals a role in mitigating the effects of reactive oxygen species associated with physiological stress. Tree Physiol 38 (6):911–924

Hurtado Páez UA, García Romero IA, Restrepo Restrepo S, Aristizábal Gutiérrez FA, Montoya Castaño D (2015) Assembly and analysis of differential transcriptome responses of *Hevea brasiliensis* on interaction with *Microcyclus ulei*. PLoS One 10(8): e0134837

Ichihara A, Shiraishi K, Sato H, Sakamura S, Nishiyama K, Sakai R, Furusaki A, Matsumoto T (1977) The structure of coronatine. J Am Chem Soc 99(2):636–637

Jayashree R, Nazeem PA, Rekha K, Sreelatha S, Thulaseedharan A, Krishnakumar R, Kala RG, Vineetha M, Leda P, Jinu U, Venkatachalam P (2018) Over-expression of 3-hydroxy-3-methylglutaryl-coenzyme A reductase 1 (*hmgr1*) gene under super-promoter for enhanced latex biosynthesis in rubber tree (*Hevea brasiliensis* Muell. Arg.). Plant Physiol Biochem 127:414–424

Jayashree R, Rekha K, Venkatachalam P, Uratsu SL, Dandekar AM, Kumari Jayasree P, Kala RG, Priya P, Sushma Kumari S, Sobha S, Ashokan MP, Sethuraj MR, Thulaseedharan (2003) Genetic transformation and regeneration of rubber tree (*Hevea brasiliensis* Muell. Arg) transgenic plants with a constitutive version of an anti-oxidative stress superoxide dismutase gene. Plant Cell Rep 22(3):201–209

Jiang JS, Zhou ZY (1997) Prospects of science and technology for natural rubber in China. Chin J Trop Crops 21:1–7

Kim IJ, Ryu SB, Kwak YS, Kang H (2004) A novel cDNA from *Parthenium argentatum* Gray enhances the rubber biosynthetic activity in vitro. J Exp Bot 55 (396):377–385

Koningsveld GAV, Schreuder HA, Soedjanaatmadja UMS, Beintema JJ (1996) Chitinase and beta-1,3-glucanase in the lutoid-body fraction of *Hevea* latex. Phytochemistry 43(1):29–37

Kuruvilla L, Sathik MBM, Thomas M, Luke LP, Sumesh KV (2017) Identification and validation of cold responsive microRNAs of *Hevea brasiliensis* using high throughput sequencing. J Crop Sci Biotech 20(5):369–377

Lau NS, Makita Y, Kawashima M, Taylor TD, Kondo S, Othman AS, Shu-Chien AC, Matsui M (2016) The rubber tree genome shows expansion of gene family associated with rubber biosynthesis. Sci Rep 6:28594

Leclercq J, Martin F, Sanier C, Clement-Vidal A, Fabre D, Oliver G, Lardet L, Ayar A, Peyramard Montoro P (2012) Over-expression of a cytosolic isoform of the *HbCuZnSOD* gene in *Hevea brasiliensis* changes its response to a water deficit. Plant Mol Biol 80(3):255–272

Lestari R, Rio M, Martin F, Leclercq J, Woraathasin N, Roques S, Dessailly F, Clément-Vidal A, Sanier C, Fabre D, Melliti S, Suharsono S, Montoro P (2018) Overexpression of *Hevea brasiliensis* ethylene response factor *HbERF-IXc5* enhances growth and tolerance to abiotic stress and affects laticifer differentiation. Plant Biotechnol J 16(1):322–336

Li D, Deng Z, Chen C, Xia Z, Wu M, He P, Chen S (2010) Identification and characterization of genes associated with tapping panel dryness from *Hevea brasiliensis* latex using suppression subtractive hybridization. BMC Plant Biol 10:140

Li D, Hao L, Liu H, Zhao M, Deng Z, Li Y, Zeng R, Tian W (2015) Next-generation sequencing, assembly and comparative analyses of the latex transcriptomes from two elite *Hevea brasiliensis* varieties. Tree Genet Genomes 11(5):98

Li D, Wang X, Deng Z, Liu H, Yang H, He G (2016a) Transcriptome analyses reveal molecular mechanism underlying tapping panel dryness of rubber tree (*Hevea brasiliensis*). Sci Rep 6:23540

Li D, Zeng R, Li Y, Zhao M, Chao J, Li Y, Wang K, Zhu L, Tian WM, Liang C (2016b) Gene expression analysis and SNP/InDel discovery to investigate yield heterosis of two rubber tree F1 hybrids. Sci Rep 6:24984

Li HL, Guo D, Lan FY, Tian WM, Peng SQ (2011) Protein differential expression in the latex from *Hevea brasiliensis* between self-rooting juvenile clones and donor clones. Acta Physiol Plant 33:1853–1859

Li HL, Guo D, Peng SQ (2014) Differential gene expression profiles in latex from *Hevea brasiliensis* between self-rooting juvenile and donor clones. Plant Growth Regul 74:65–71

Li HL, Guo D, Zhu JH, Wang Y, Chen XT, Peng SQ (2016c) Comparative transcriptome analysis of latex reveals molecular mechanisms underlying increased rubber yield in *Hevea brasiliensis* self-rooting juvenile clones. Front Plant Sci 7:1204

Liu H, Wei Y, Deng Z, Yang H, Dai L, Li D (2019) Involvement of HbMC1-mediated cell death in tapping panel dryness of rubber tree (*Hevea brasiliensis*). Tree Physiol 39(3):391–403

Liu JP, Hu J, Liu YH, Yang CP, Zhuang YF, Guo XL, Li YJ, Zhang L (2018) Transcriptome analysis of *Hevea brasiliensis* in response to exogenous methyl jasmonate provides novel insights into regulation of jasmonate-elicited rubber biosynthesis. Physiol Mol Biol Plants 24(3):349–358

Liu JP, Xia ZQ, Tian XY, Li YJ (2015) Transcriptome sequencing and analysis of rubber tree (*Hevea brasiliensis* Muell.) to discover putative genes associated with tapping panel dryness (TPD). BMC Genom 16:398

Liu JP, Zhuang YF, Guo XL, Li YJ (2016) Molecular mechanism of ethylene stimulation of latex yield in rubber tree (*Hevea brasiliensis*) revealed by *de novo* sequencing and transcriptome analysis. BMC Genom 17(1):257

Liu SQ, Yuan XH, Huang X, Xu LY (1985) Comparative studies on yield and properties of juvenile type- and its mature type clones. Trop Crop Res 3:1–5

Loh SC, Thottathil GP, Othman AS (2016) Identification of differentially expressed genes and signalling pathways in bark of *Hevea brasiliensis* seedlings associated with secondary laticifer differentiation using gene expression microarray. Plant Physiol Biochem 107:45–55

Mantello CC, Boatwright L, da Silva CC, Scaloppi EJ Jr, de Souza Goncalves P, Barbazuk WB, Pereira de Souza A (2019) Deep expression analysis reveals distinct cold-response strategies in rubber tree (*Hevea brasiliensis*). BMC Genom 20(1):455

Montoro P, Wu S, Favreau B, Herlinawati E, Labrune C, Martin-Magniette ML, Pointet S, Rio M, Leclercq J, Ismawanto S, Kuswanhadi (2018) Transcriptome analysis in *Hevea brasiliensis* latex revealed changes in hormone signalling pathways during ethephon stimulation and consequent tapping panel dryness. Sci Rep 8(1):8483

Ogata T, Kida Y, Tochigi M, Matsushita Y (2013) Analysis of the cell death-inducing ability of the ethylene response factors in group VIII of the AP2/ERF family. Plant Sci 209:12–23

Oh SK, Kang H, Shin DH, Yang J, Chow KS, Yeang HY, Wagner B, Breiteneder H, Han KH (1999) Isolation, characterization, and functional analysis of a novel cDNA clone encoding a small rubber particle protein from *Hevea brasiliensis*. J Biol Chem 274(24):17132–17138

Okoma KM, Dian K, Obouayeba S, Elabo AAE, N'guetta ASP (2011) Seasonal variation of tapping panel dryness expression in rubber tree *Hevea brasiliensis* Muell. arg. in Cote d'Ivoire. Agr Biol J N Am 2:559–569

Paardekooper EC (1989) Exploitation of the rubber tree. In: Webster CC, Baulkwill WJ (eds) Rubber. Longman Scientific and Technical, Essex, UK, pp 349–414

Priya P, Venkatachalam P, Thulaseedharan A (2007) Differential expression pattern of rubber elongation factor (REF) mRNA transcripts from high and low yielding clones of rubber tree (*Hevea brasiliensis* Muell. Arg.). Plant Cell Rep 26(10):1833–1838

Priyadarshan PM, Goncalves PS, Omokhafe KO (2009) Breeding *Hevea* rubber. In: Mohan JS, Priyadarshan PM (eds) Breeding plantation tree crops: tropical species. Springer, New York, pp 469–522

Putranto RA, Herlinawati E, Rio M, Leclercq J, Piyatrakul P, Gohet E, Sanier C, Oktavia F, Pirrello J, Kuswanhadi Montoro P (2015) Involvement of ethylene in the latex metabolism and tapping panel dryness of *Hevea brasiliensis*. Int J Mol Sci 16 (8):17885–17908

Qu Y, Chakrabarty R, Tran HT, Kwon EJ, Kwon M, Nguyen TD, Ro DK (2015) A lettuce (*Lactuca sativa*) homolog of human Nogo-B receptor interacts with *cis*-prenyltransferase and is necessary for natural rubber biosynthesis. J Biol Chem 290(4):1898–1914

Rahman AY, Usharraj AO, Misra BB, Thottathil GP, Jayasekaran K, Feng Y, Hou S, Ong SY, Ng FL, Lee LS, Tan HS, Sakaff MK, Teh BS, Khoo BF, Badai SS, Aziz NA, Yuryev A, Knudsen B, Dionne-Laporte A, Mchunu NP, Yu Q, Langston BJ, Freitas TA, Young AG, Chen R, Wang L, Najimudin N, Saito JA, Alam M (2013) Draft genome sequence of the rubber tree *Hevea brasiliensis*. BMC Genom 14:75

Sando T, Hayashi T, Takeda T, Akiyama Y, Nakazawa Y, Fukusaki E, Kobayashi A (2009) Histochemical study of detailed laticifer structure and rubber biosynthesis-related protein localization in *Hevea brasiliensis* using spectral confocal laser scanning microscopy. Planta 230(1):215–225

Sathik MBM, Luke LP, Rajamani A, Kuruvilla L, Sumesh KV, Thomas M (2018) De novo transcriptome analysis of abiotic stress-responsive transcripts of *Hevea brasiliensis*. Mol Breeding 38:32

Shi MJ, Hao BZ, Wu JL, Tian WM (2010) Progress in mechanism for the regulation of the laticifer wound plug formation in rubber tree (*Hevea brasiliensis* MueU.Arg.). Chin J Trop Crops 11:2043–2050

Sookmark U, Pujade-Renaud V, Chrestin H, Lacote R, Naiyanetr C, Seguin M, Romruensukharom P, Narangajavana J (2002) Characterization of polypeptides accumulated in the latex cytosol of rubber trees affected by the tapping panel dryness syndrome. Plant Cell Physiol 43(11):1323–1333

Sun Q (1997) Study on development of articulated laticiferous tissue of *Hevea brasiliensis*. J Xiamen Univ (Nat Sci) 36(6):935–941

Tan D, Hu X, Fu L, Kumpeangkeaw A, Ding Z, Sun X, Zhang J (2017) Comparative morphology and transcriptome analysis reveals distinct functions of the primary and secondary laticifer cells in the rubber tree. Sci Rep 7(1):3126

Tang C, Xiao X, Li H, Fan Y, Yang J, Qi J, Li H (2013) Comparative analysis of latex transcriptome reveals putative molecular mechanisms underlying super productivity of *Hevea brasiliensis*. PLoS One 8(9): e75307

Tang C, Yang M, Fang Y, Luo Y, Gao S, Xiao X, An Z, Zhou B, Zhang B, Tan X, Yeang HY, Qin Y, Yang J, Lin Q, Mei H, Montoro P, Long X, Qi J, Hua Y, He Z, Sun M, Li W, Zeng X, Cheng H, Liu Y, Yang J, Tian W, Zhuang N, Zeng R, Li D, He P, Li Z, Zou Z, Li S, Li C, Wang J, Wei D, Lai CQ, Luo W, Yu J, Hu S, Huang H (2016) The rubber tree genome reveals new insights into rubber production and species adaptation. Nat Plants 2(6):16073

Tian WM, Shi MJ, Yu FY, Wu JL, Hao BZ, Cui KM (2003) Localized effects of mechanical wounding and exogenous jasmoic acid on the induction of secondary laticifer differentiation in relation to the distribution of jasmonic acid in *Hevea brasiliensis*. Acta Bot Sin 45 (11):1366–1372

Tian WM, Shi MJ, Tan HY, Wu JL, Hao BZ (eds) (2015a) Structure and development of the bark in *Hevea brasiliensis*. In: China Science Publishers, Beijing, pp 15–81

Tian WM, Yang SG, Shi MJ, Zhang SX Wu JL (2015b) Mechanical wounding-induced laticifer differentiation in rubber tree: an indicative role of dehydration, hydrogen peroxide, and jasmonates. J Plant Physiol 182:95–103

Tong Z, Wang D, Sun Y, Yang Q, Meng X, Wang L, Feng W, Li L, Wurtele ES, Wang X (2017) Comparative proteomics of rubber latex revealed multiple protein species of REF/SRPP family respond diversely to ethylene stimulation among different rubber tree clones. Int J Mol Sci 18(5):958

Tristama R, Mawaddah PAS, Ade-fipriani L, Junaidi (2019) Physiological status of high and low metabolism *Hevea* clones in the difference stage of tapping panel dryness. Biodiversitas 20(1):267–273

Van Beilen JB, Poirier Y (2007) Establishment of new crops for the production of natural rubber. Trends Biotechnol 25(11):522–529

Venkatachalam P, Thulaseedharan A, Raghothama K (2007) Identification of expression profiles of tapping panel dryness (TPD) associated genes from the latex of rubber tree (*Hevea brasiliensis* Muell. Arg). Planta 226(2):499–515

Wang D, Sun Y, Chang L, Tong Z, Xie Q, Jin X, Zhu L, He P, Li H, Wang X (2018) Subcellular proteome profiles of different latex fractions revealed washed solutions from rubber particles contain crucial enzymes for natural rubber biosynthesis. J Proteomics 182:53–64

Wang X, Shi M, Wang D, Chen Y, Cai F, Zhang S, Wang L, Tong Z, Tian WM (2013) Comparative proteomics of primary and secondary lutoids reveals that chitinase and glucanase play a crucial combined role in rubber particle aggregation in *Hevea brasiliensis*. J Proteome Res 12(11):5146–5159

Wang X, Wang D, Sun Y, Yang Q, Chang L, Wang L, Meng X, Huang Q, Jin X, Tong Z (2015) Comprehensive proteomics analysis of laticifer latex reveals new insights into ethylene stimulation of natural rubber production. Sci Rep 5:13778

Wang ZY, Zeng XS, Chen CQ, Wu HY, Li QY, Fan GJ, Lu WJ (1980) Induction of rubber plantlets from anther of *Hevea brasiliensis in vitro*. Chin J Trop Crops 1(1):16–27

Wasternack C, Xie D (2010) The genuine ligand of a jasmonic acid receptor: improved analysis of jasmonates is now required. Plant Signal Behav 5 (4):337–340

Wei F, Luo S, Zheng Q, Qiu J, Yang W, Wu M, Xiao X (2015) Transcriptome sequencing and comparative analysis reveal long-term flowing mechanisms in *Hevea brasiliensis* latex. Gene 556(2):153–162

Wu JL, Tan HY, Zeng RZ, Hao BZ (2000) Primary laticifer differentiation of *Hevea brasiliensis* in relation to shoot growth. Chin J Trop Crops 21(4):1–6

Wu JL, Hao BZ, Tan HY (2002) Wound-induced laticifer differentiation in *Hevea brasiliensis* shoots mediated by jasmonic acid. J Rubb Res 5(1):53–63

Wu S, Zhang S, Chao J, Deng X, Chen Y, Shi M, Tian WM (2016) Transcriptome analysis of the signalling networks in coronatine-induced secondary laticifer differentiation from vascular cambia in rubber trees. Sci Rep 6:36384

Xiang Q, Xia K, Dai L, Kang G, Li Y, Nie Z, Duan C, Zeng R (2012) Proteome analysis of the large and the small rubber particles of *Hevea brasiliensis* using 2D-DIGE. Plant Physiol Biochem 60:207–213

Xie W, McMahan CM, DeGraw Distefano MD, Cornish K, Whalen MC, Shintani DK (2008) Initiation of rubber biosynthesis: in vitro comparisons of benzophenone-modified diphosphate analogues in three rubber-producing species. Phytochemistry 69 (14):2539–2545

Yamashita S, Yamaguchi H, Waki T, Aoki Y, Mizuno M, Yanbe F, Ishii T, Funaki A, Tozawa Y, Miyagi-Inoue Y, Fushihara K, Nakayama T, Takahashi S (2016) Identification and reconstitution of the rubber biosynthetic machinery on rubber particles from *Hevea brasiliensis*. eLife 5: e19022

Yang SQ, Mo YY (1994) Some physiological properties of latex from somatic plants of *Hevea brasiliensis*. Chin J Trop Crops 15(2):13–20

Yuan XH, Yang SQ, Xu LY, Wu JL, Hao BZ (1998) Characteristics related to higher rubber yield of *Hevea brasiliensis* juvenile-type clone GT1. J Rubb Res 1:125–132

Zeng RZ, Li Y, Yang SQ (1997) The relation between contents of nucleic acid and tapping panel dryness in latex from *Hevea brasiliensis*. Chin J Trop Crops 18 (1):10–15

Zhang SX, Wu SH, Chen YY, Tian WM (2015) Analysis of differentially expressed genes associated with coronatine-induced laticifer differentiation in the rubber tree by subtractive hybridization suppression. PLoS One 10(7):e0132070

Zhang SX, Wu SH, Tian WM (2016) The secondary laticifer differentiation in rubber tree is induced by trichostatin A, an inhibitor of histone acetylation. Front Agr Sci Eng 3(4):357–362

Zhang Y, Leclercq J, Wu S, Ortega-Abboud E, Pointet S, Tang C, Hu S, Montoro P (2019) Genome-wide analysis in *Hevea brasiliensis* laticifers revealed species-specific post-transcriptional regulations of several redox-related genes. Sci Rep 9(1):5701

9 HeveaDB: A Hub for Rubber Tree Genetic and Genomic Resources

Han Cheng

Abstract

With rapid research advances in the genome of the *Hevea brasiliensis* rubber tree, more genomic and transcriptomic data have been generated. However, these data are still deposited in raw format in the National Centre for Biotechnology Information (NCBI) and other public databases, and therefore, data utilization is cumbersome for the scientists in the rubber research community. To overcome such issues, a comprehensive rubber tree data hub, HeveaDB, was constructed and maintained under the organization, the International Rubber Research and Development Board (IRRDB). The first version of the database, HeveaDB 1.0 (http://hevea.catas.cn), contains four versions of *H. brasiliensis* draft genomes, 99 next-generation sequencing (NGS) transcriptomes, 21,619 annotated EST sequences, 12 curated gene families, 30,200 gene annotations, phenotype data of 5,049 wild Brazilian germplasm, and information on 19,141 rubber clones of Wickham origin. A selection of bioinformatic tools are integrated into the database to facilitate the utilization of the data by scientists who are not skilled in bioinformatics. With periodic updating and improvement of content and function, HeveaDB will serve as a useful platform for the molecular biological and genomic studies of the rubber tree.

9.1 Introduction

Major progress in rubber breeding programs has produced *Hevea brasiliensis* clones with substantial increases in yield. However, due to the long lifecycle of the tree, the breeding program takes more than 30 years. Recent advances in genomics may accelerate crop breeding through the promising genome selection technique, in which genome-wide SNP markers are used to predict individual breeding values within a population. The adoption of next-generation sequencing (NGS) in *Hevea* genetics and genomics has greatly promoted the molecular investigations of rubber tree biology (Cheng et al. 2018b). Currently, four versions of draft genomes have been compiled (Rahman et al. 2013; Tang et al. 2016; Lau et al. 2016; Pootakham et al. 2017), and 142 transcriptomes were found in the NCBI SRA database by Jan 2018 (Xia et al. 2011; Triwitayakorn et al. 2011; Li et al. 2012; Chao et al. 2015; Liu et al. 2015; Fang et al. 2016; Cheng et al. 2018a). However, these data are still deposited in raw format in the National Centre for Biotechnology Information (NCBI) and other public databases, thus hindering utilization of the NGS

H. Cheng (✉)
Key Laboratory of Rubber Biology, Ministry of Agriculture, Rubber Research Institute, Chinese Academy of Tropical Agricultural Sciences, Haikou, Hainan, People's Republic of China
e-mail: forcheng@gmail.com

M. Matsui and K.-S. Chow (eds.), *The Rubber Tree Genome*, Compendium of Plant Genomes,
https://doi.org/10.1007/978-3-030-42258-5_9

data by scientists in the rubber research community. Moreover, huge amounts of phenotypic data on rubber tree clones and wild Brazilian germplasm materials need to be stored and conveniently mined using a comprehensive data hub. Being a non-model plant species, the rubber tree research community also has limited access to bioinformatic resources. In response to the situation, the International Rubber Research and Development Board (IRRDB) decided to initiate construction of an integrated genome data hub during the 2017 IRRDB International Rubber Conference in Jakarta, Indonesia. Such a hub would be used to store *Hevea* data such as draft genomes, germplasm re-sequencing data, clonal variation descriptions, transcriptomic and proteomic data and phenotypic information, and to provide a platform for data archiving, distribution and utilization. The first version of the database, HeveaDB 1.0 (http://hevea.catas.cn), was therefore constructed to provide online services to the rubber tree community. Before the construction of HeveaDB, another rubber genome database had been established to provide similar access to *Hevea* genome and transcriptome data (Makita et al. 2018). However, this database only compiled data generated from one research group, and did not integrate other public NGS data.

As a non-model plant, genome studies in the rubber tree lag behind those in other model plants such as *Arabidopsis* and rice. Currently, only draft genomes of the rubber tree have been published, all of which are presented as thousands of scaffolds. Among the four published rubber tree genomes, the Reyan 7-33-97 genome assembly (Tang et al. 2016) is highest in quality (Table 9.1). This version of the genome has 7,453 scaffolds with an N50 length of 1.28 Mb. Nonetheless, mapping these scaffolds onto 18 chromosomes of the rubber tree is a big challenge in *Hevea* genome research.

HeveaDB 1.0 consists of four key features: (1) A substantial collection of data including four versions of draft genomes of *H. brasiliensis*, 99 NGS transcriptomes, 21,619 annotated EST sequences, 12 curated gene families, 30,200 gene annotations, phenotype data of 5,049 wild Brazilian germplasm and information on 19,141 rubber clones of Wickham origin (Table 9.2). (2) A user-friendly interface to facilitate scientists to utilize data. Several tools are integrated in the database, including genome browsers, blast and blat tools, text searching, sequence retrieval and downloading. (3) An easy-to-use interface for users to retrieve and analyze gene expression (presented as heatmaps and as digital values) and correlations (displayed as co-expression

Table 9.1 Published draft genomes of the rubber tree

Institute	Clone	Genome size (Gb)	Scaffold N50 (kb)	Number of Scaffolds	Contig N50 (kb)	Year	References	Pubmed ID
USM[a], Malaysia	RRIM 600	2.15	2.97	608,017	1.26	2013	Rahman et al. (2013)	23375136
CATAS[b], China	Reyan 7-33-97	1.46	1,280	7,453	30.6	2016	Tang et al. (2016)	27255837
RIKEN[c], Japan/USM, Malaysia	RRIM 600	2.15	67.24	189,316	20.75	2016	Lau et al. (2016)	27339202
BIOTEC[d], Thailand	BPM 24	2.0	96.8	592,579	1.3	2017	Pootakham et al. (2017)	28150702

[a]USM, Universiti Sains Malaysia
[b]CATAS, Chinese Academy of Tropical Agricultural Sciences
[c]RIKEN, RIkagaku KENkyusho
[d]BIOTEC, National Center for Genetic Engineering and Biotechnology

Table 9.2 Data content and sources of each database in HeveaDB version 1.0

Category	Contents	Number	Data source	References
Genome	Draft genomes	4	NCBI	Rahman et al. (2013), Tang et al. (2016), Lau et al. (2016), Pootakham et al. (2017)
Transcriptome	NGS transcriptomes	99	NCBI	Xia et al. (2011), Triwitayakorn et al. (2011), Li et al. (2012), Chao et al. (2015), Liu et al. (2015), Fang et al. (2016), Cheng et al. (2018a)
Gene annotation	NR	30,200	Tang et al. (2016)	Tang et al. (2016)
	GO	43,792	Tang et al. (2016)	Tang et al. (2016)
	Pfam	43,877	Tang et al. (2016)	Tang et al. (2016)
Gene family	Curated gene family	12	This study	
EST	EST sequences	41,903	This study	Chow et al. (2007), Cheng et al. (2016)
	EST annotations	21,619	This study	
Germplasm	Rubber clones of Wickham origin	19,141	Collected from internally circulated publications	
	Wild Brazilian germplasm	5,049	Zhen and Hu (1994)	
Gene expression dataset	Gene expression	99	From analysis of 99 transcriptomes	
Co-expression network	WGCNA co-expression network constructed based on expression data	1	This study	

networks). (4) A mechanism to receive and share data among members of the rubber research community under the framework of the IRRDB.

9.2 Database Structure, Content, and Organization

The database has five major functional modules: Data, Search, Tools, Download, and Submit (Fig. 9.1). Basically, two types of rubber tree data are deposited in HeveaDB: sequence data (genome, transcriptome and markers) and germplasm phenotypic data. So far, four versions of genome drafts were published in rubber tree, among which the Reyan 7-33-97 genome has the least number of scaffolds and the longest N50 (Tang et al. 2016). HeveaDB 1.0 uses the Reyan 7-33-97 genome as the reference. In total, 30,200 functional genes were annotated from 43,792 predicted genes (Tang et al. 2016). The 43,792 protein-coding genes predicted from the reference genome can be accessed from the database pages organized by gene name. The annotated genes can be searched and viewed from a compiled gene page. The gene page contains general gene information, gene structure, gene expression, co-expression network, gene sequence (including CDS and deduced peptide), and annotation. Gene structural information may be extracted from the gff3 genome annotation file and viewed in the Gbrowse and Jbrowse plugins. Expression data were obtained from the published transcriptomes. In total, there were 142 transcriptomes in the NCBI SRA database (up to Jan 2018), among which 99 (>1 Gb data, >10 M

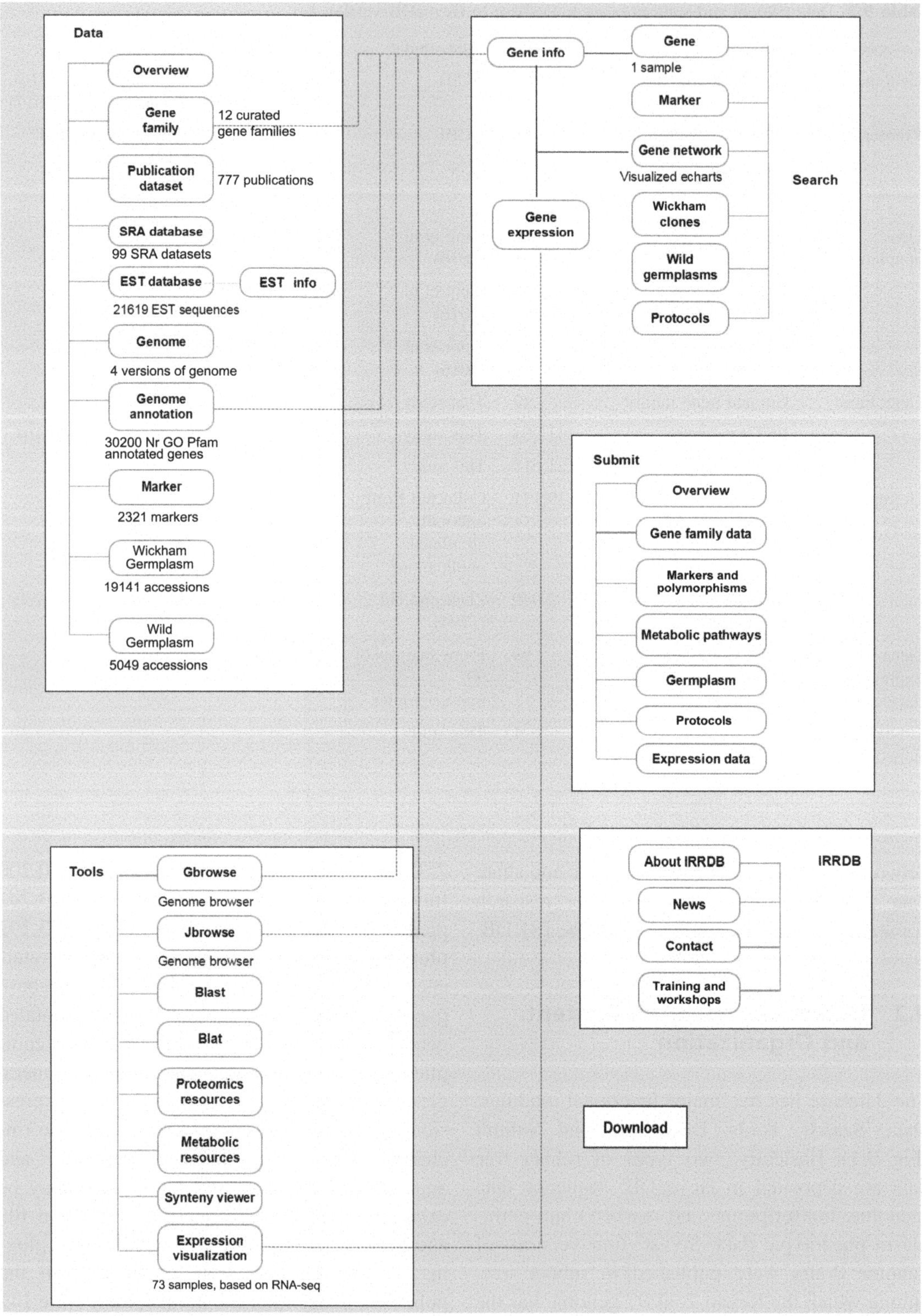

Fig. 9.1 The structure of HeveaDB version 1.0

reads) are of high quality and were therefore collected for calculating gene expression. In total, 41,903 EST sequences were collected among which 15,870 were sequenced from a transgene-ready full-length cDNA library (Cheng et al. 2016) and 2851 obtained from NCBI (Chow et al. 2007). The ESTs were annotated against the NCBI Nr, KEGG, Trembl, Swissprot, InterPro, and GO databases. Besides the sequence data, HeveaDB 1.0 also stores data on rubber tree germplasm genotypes. Phenotypic data of 5,049 wild germplasm (originating from the IRRDB 1981 Brazilian expedition) which were obtained from the Chinese rubber germplasm resources repository (Zhen and Hu 1994), and information on 19,141 rubber clones with origins in the Wickham germplasm have been deposited.

9.3 Database Utility

9.3.1 Genome Browser

The Reyan 7-33-97 reference genome (Tang et al. 2016) is visualized using the Gbrowse and Jbrowse plug-ins. Each gene is linked to its detailed information page by clicking the gene name in the browser. The Gbrowse gene page includes the following basic information: name, scaffold position, length, CDS and sequence. The Jbrowse page is displayed in a pop-up window. The primary gene data and attributes are displayed, including the name, position, length, and sequences of adjacent regions and their sub-features. The default scaffold in the genome browsers is scaffold0001; the user can shift to view other scaffolds by entering their names in the search form (Fig. 9.2).

9.3.2 Expression Visualization

From a total of 144 transcriptomes in the NCBI SRA database (up to Jan. 2018), 99 were selected (>1 Gb, Table 9.3). These transcriptomes belong to 74 samples with biological replicates. The gene expression values (in FPKM or fragments per kilobase of transcript per million fragments mapped) were retrieved from the 99 transcriptomes using the Reyan 7-33-97 genome (Tang et al. 2016) as reference. Gene expression may be further visualized as a heatmap in the Echart plug-in, or as downloadable

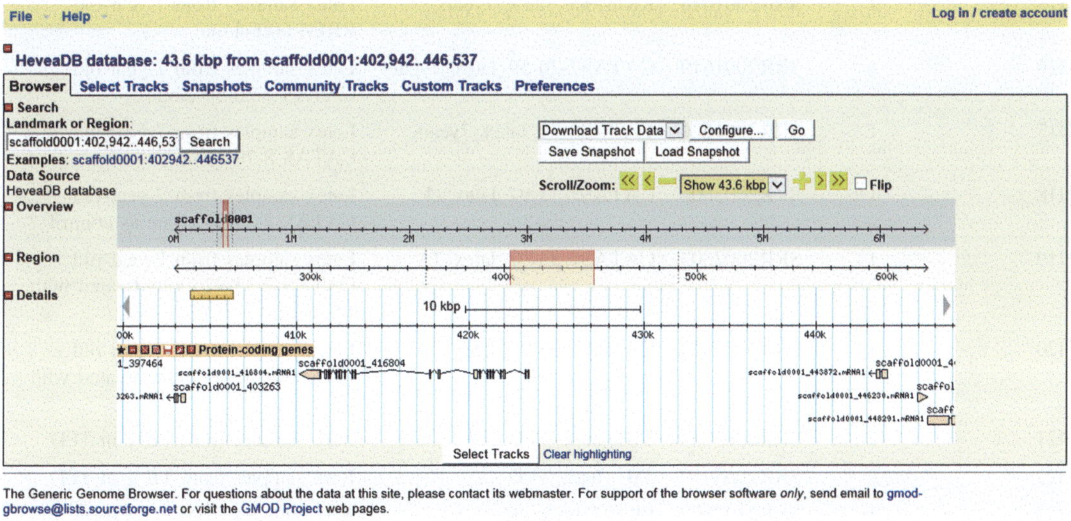

Fig. 9.2 Gbrowse page indicating gene annotations of scaffold0001

Table 9.3 Ninety-nine selected transcriptomes used for gene expression analysis

Transcriptome	Duplicates	SRA accession no.	Sample name	Description
H1	2	SRR1508164	RRIM600_latex_-	Latex samples from the RRIM 600 clone
H2	2	SRR1508165	RRIM600_latex_-	Latex samples from the RRIM 600 clone
H3	2	SRR1508166	CATAS7-20-59_latex_-	Latex samples from the CATAS 7-20-59 clone
H4	2	SRR1508167	CATAS7-20-59_latex_-	Latex samples from the CATAS 7-20-59 clone
H5	1	SRR1533844	PR 107_latex_3years	Latex samples from 3-year-old PR 107 clone
H6	1	SRR1533845	CATAS 8-79_latex_3years	Latex samples from 3-year-old CATAS 8-79 clone
H7	1	SRR1544255	PR 255_bark_-	Bark samples from PR 255 clone
H8	1	SRR1544256	GT1_bark_-	Bark samples from GT 1 clone
H9	2	SRR1588170	PR107_bark_healthy	Bark samples from 14-year-old healthy PR 107 clone
H10	2	SRR1611790	PR107_bark_healthy	Bark samples from 14-year-old healthy PR 107 clone
H11	1	SRR1611791	PR107_bark_BrownBlast	Bark samples from 14-year-old PR 107 clone with brown bast disease
H12	1	SRR1611792	PR107_bark_TPD	Bark samples from 14-year-old PR 107 clone with TPD disease
H13	1	SRR1648124	CATAS7-33-97_latex_-	Latex samples from the CATAS 7-33-97 clone
H14	1	SRR167696	CATAS7-33-97_bark-leaf	Bark and leaf mixed samples from the CATAS 7-33-97 clone
H15	1	SRR2001618	RRIM600_latex_7years	Latex samples from 7-year-old RRIM 600 clone
H16	1	SRR2001619	CATAS7-20-59_latex_7years	Latex samples from 7-year-old CATAS 7-20-59 clone
H17	1	SRR2001620	CATAS8-79_latex_7years	Latex samples from 7-year-old CATAS 8-79 clone
H18	1	SRR2002800	CATAS7-33-97_latex_ck	Latex samples from 7-year-old CATAS 7-33-97 clone as control
H19	1	SRR2002803	CATAS7-33-97_latex_JA	Latex samples from 7-year-old CATAS 7-33-97 clone treated with JA-2
H20	1	SRR2002804	CATAS7-33-97_latex_ET	Latex samples from 7-year-old CATAS 7-33-97 clone treated with ET-2
H21	2	SRR2063953	TB1_bark_TPD	Bark samples from TB 1 in TPD
H22	2	SRR2063959	TB1_bark_TPD	Bark samples from TB 1 in TPD
H23	1	SRR2147072	PR107_bark_ck	Bark samples from 15-year-old PR 107 clone as control

(continued)

Table 9.3 (continued)

Transcriptome	Duplicates	SRA accession no.	Sample name	Description
H24	1	SRR2147073	PR107_bark_ET8	Bark samples from 15-year-old PR 107 clone treated with 1.5% Ethephon for 8h
H25	1	SRR2147074	PR107_bark_ET24	Bark samples from 15-year-old PR 107 clone treated with 1.5% Ethephon for 24h
H26	1	SRR2156988	FX3864_leaf_GCL12C1s0h	Leaf samples from FX 3864 clone treated with GCL12C1s for 0h
H27	1	SRR2156992	FX3864_leaf_GCL12C1s48h	Leaf samples from FX 3864 clone treated with GCL12C1s for 48h
H28	1	SRR2157179	FX3864_leaf_GCL12C2s0h	Leaf samples from FX 3864 clone treated with GCL12Cs for 0h
H29	1	SRR3136158	CATAS7-33-97_bark_10years	Bark samples from 10-year-old CATAS 7-33-97 clone
H30	1	SRR3136159	CATAS7-33-97_leaf_10years	Leaf samples from 10-year-old CATAS 7-33-97 clone
H31	1	SRR3136162	CATAS7-33-97_latex_10years	Latex samples from 10-year-old CATAS 7-33-97 clone
H32	1	SRR3136165	CATAS7-33-97_flower_female	Female flower samples from 10-year-old CATAS 7-33-97 clone
H33	1	SRR3136166	CATAS7-33-97_flower_male	Male flower samples from 10-year-old CATAS 7-33-97 clone
H34	1	SRR3136168	CATAS7-33-97_seed_10years	Seeds samples from 10-year-old CATAS 7-33-97 clone
H35	1	SRR3136173	CATAS7-33-97_latex_ET0	Latex samples from 10-year-old CATAS 7-33-97 clone treated with Ethrel for 0h as control
H36	1	SRR3136176	CATAS7-33-97_latex_ET3	Latex samples from 10-year-old CATAS 7-33-97 clone treated with Ethrel for 3h
H37	1	SRR3136177	CATAS7-33-97_latex_ET12	Latex samples from 10-year-old CATAS 7-33-97 clone treated with Ethrel for 12h
H38	1	SRR3136178	CATAS7-33-97_latex_ET24	Latex samples from 10-year-old CATAS 7-33-97 clone treated with Ethrel for 24h
H39	1	SRR3136185	CATAS7-33-97_leaf_stageB	Leaf samples in stage B from 10-year-old CATAS 7-33-97 clone
H40	1	SRR3136188	CATAS7-33-97_leaf_stageBC	Leaf samples in stage B-C from 10-year-old CATAS 7-33-97 clone
H41	1	SRR3136190	CATAS7-33-97_leaf_stageC	Leaf samples in stage C from 10-year-old CATAS 7-33-97 clone
H42	1	SRR3136192	CATAS7-33-97_leaf_stageD	Leaf samples in stage D from 10-year-old CATAS 7-33-97 clone
H43	1	SRR3240371	PR107_leaf_1years	Leaf samples from 1-year-old PR 107 clone

(continued)

Table 9.3 (continued)

Transcriptome	Duplicates	SRA accession no.	Sample name	Description
H44	1	SRR3240372	RRIM600_leaf_1years	Leaf samples from 1-year-old RRIM 600 clone
H45	1	SRR3240373	BT3410_leaf_1years	Leaf samples from 1-year-old BT 3410 clone
H46	1	SRR3240374	Wenchang11_leaf_1years	Leaf samples from 1-year-old Wenchang11 clone
H47	1	SRR3423347	CATAS7-33-97_bark_earlyCK	Bark samples from CATAS 7-33-97 clone as early control
H48	1	SRR3423348	CATAS7-33-97_bark_lateCK	Bark samples from CATAS 7-33-97 clone as late control
H49	1	SRR3423349	CATAS7-3397_bark_earlyCoronatine	Bark samples from CATAS 7-33-97 clone treated with coronatine at early stage
H50	1	SRR3423350	CATAS7-33-97_bark_lateCoronatine	Bark samples from CATAS 7-33-97 clone treated with coronatine at late stage
H51	1	SRR5051416	CATAS88-13_bark_JA24-2mg	Bark samples from CATAS 88-13 clone treated with 2mg JA for 24h
H52	1	SRR5051417	CATAS88-13_bark_JA24-ck	Bark samples from CATAS 88-13 clone as control for 24h
H53	1	SRR5051418	CATAS88-13_bark_JA24-4mg	Bark samples from CATAS 88-13 clone treated with 4mg JA for 24h
H54	3	SRR5118395	CATAS7-33-97_secondaryLaticifer_7years	Secondary laticifer samples from 7-year-old CATAS 7-33-97 clone
H55	3	SRR5118396	CATAS7-33-97_primaryLaticifer_7years	Primary laticifer samples from 7-year-old CATAS 7-33-97 clone
H56	3	SRR5118397	CATAS7-33-97_secondaryLaticifer_7years	Secondary laticifer samples from 7-year-old CATAS 7-33-97 clone
H57	3	SRR5118398	CATAS7-33-97_primaryLaticifer_7years	Primary laticifer samples from 7-year-old CATAS 7-33-97 clone
H58	3	SRR5118399	CATAS7-33 97_secondaryLaticifer_7years	Secondary laticifer samples from 7-year-old CATAS 7-33-97 clone
H59	3	SRR5118400	CATAS7-33-97_primaryLaticifer_7years	Primary laticifer samples from 7-year-old CATAS 7-33-97 clone
H60	2	SRR5868395	CATAS7-33-97_latex_JCDC	Latex samples from 3-year-old CATAS 7-33-97 clone
H61	2	SRR5868396	CATAS7-33-97_latex_JCDC	Latex samples from 3-year-old CATAS 7-33-97 clone
H62	1	SRR611643	GT1_leaf_C2	Leaf samples from GT1 clone with *Corynespora cassiicola* tolerance as control
H63	1	SRR611644	RRII105_leaf_T1	Leaf samples from RRII 105 clone with *Corynespora cassiicola* tolerance as T1 treatment

(continued)

Table 9.3 (continued)

Transcriptome	Duplicates	SRA accession no.	Sample name	Description
H64	1	SRR611645	GT1_leaf_T2	Leaf samples from GT1 clone with *Corynespora cassiicola* tolerance as T2 treatment
H65	1	SRR6127583	CATAS7-33-97_stem_cold24	Stem samples from 1-year-old CATAS 7-33-97 clone treated under 4 °C for 24h
H66	1	SRR6127584	CATAS7-33-97_stem_cold2	Stem samples from 1-year-old CATAS 7-33-97 clone treated under 4 °C for 2h
H67	1	SRR6127585	CATAS7-33-97_stem_cold0	Stem samples from 1-year-old CATAS 7-33-97 clone treated under 4 °C for 0h
H68	1	SRR620233	RRIM600_leaf_ck	Leaf samples from 6 months old RRIM 600 clone as control
H69	1	SRR620234	RRIM600_leaf_drought	Leaf samples from 6 months old RRIM 600 clone treated with drought
H70	1	SRR620235	RRIM600_leaf_cold	Leaf samples from 6 months old RRIM 600 clone treated with cold stress
H71	1	SRR620236	RRII105_latex_ET	Latex samples from 20-year-old RRII 105 clones treated with ethylene
H72	1	SRR620237	RRII105_latex_ck	Latex samples from 20-year-old RRII 105 clones as control
H73	1	SRR854520	RRIM928_bark_-	Bark samples from the RRIM 928 clone
H74	1	SRR854521	RRIM928_latex_-	Latex samples from the RRIM 928 clone
H75	1	SRR854522	RRIM928_leaf_-	Leaf samples from the RRIM 928 clone
H76	3	SRR6668812	REKEN501_leaf_cold0h	Leaf samples from 2-year-old Reken 501 clone treated with cold for 0h
H77	3	SRR6668813	REKEN501_leaf_cold0h	Leaf samples from 2-year-old Reken 501 clone treated with cold for 0h
H78	3	SRR6668814	REKEN501_leaf_cold0h	Leaf samples from 2-year-old Reken 501 clone treated with cold for 0h
H79	3	SRR6668811	REKEN501_leaf_cold24h	Leaf samples from 2-year-old Reken 501 clone treated with cold for 24h
H80	3	SRR6668806	REKEN501_leaf_cold24h	Leaf samples from 2-year-old Reken 501 clone treated with cold for 24h
H81	3	SRR6668807	REKEN501_leaf_cold24h	Leaf samples from 2-year-old Reken 501 clone treated with cold for 24h
H82	3	SRR6668815	REKEN501_leaf_cold2h	Leaf samples from 2-year-old Reken 501 clone treated with cold for 2h

(continued)

Table 9.3 (continued)

Transcriptome	Duplicates	SRA accession no.	Sample name	Description
H83	3	SRR6668816	REKEN501_leaf_cold2h	Leaf samples from 2-year-old Reken 501 clone treated with cold for 2h
H84	3	SRR6668817	REKEN501_leaf_cold2h	Leaf samples from 2-year-old Reken 501 clone treated with cold for 2h
H85	3	SRR6668818	REKEN501_leaf_cold8h	Leaf samples from 2-year-old Reken 501 clone treated with cold for 8h
H86	3	SRR6668819	REKEN501_leaf_cold8h	Leaf samples from 2-year-old Reken 501 clone treated with cold for 8h
H87	3	SRR6668810	REKEN501_leaf_cold8h	Leaf samples from 2-year-old Reken 501 clone treated with cold for 8h
H88	3	SRR6668820	CATAS93-114_leaf_cold24h	Leaf samples from 2-year-old CATAS 93-114 clone treated with cold for 24h
H89	3	SRR6668823	CATAS93-114_leaf_cold24h	Leaf samples from 2-year-old CATAS 93-114 clone treated with cold for 24h
H90	3	SRR6668822	CATAS93-114_leaf_cold24h	Leaf samples from 2-year-old CATAS 93-114 clone treated with cold for 24h
H91	3	SRR6668808	CATAS93-114_leaf_cold0h	Leaf samples from 2-year-old CATAS 93-114 clone treated with cold for 0h
H92	3	SRR6668809	CATAS93-114_leaf_cold0h	Leaf samples from 2-year-old CATAS93-114 clone treated with cold for 0h
H93	3	SRR6668802	CATAS93-114_leaf_cold0h	Leaf samples from 2-year-old CATAS 93-114 clone treated with cold for 0h
H94	3	SRR6668803	CATAS93-114_leaf_cold2h	Leaf samples from 2-year-old CATAS 93-114 clone treated with cold for 2h
H95	3	SRR6668804	CATAS93-114_leaf_cold2h	Leaf samples from 2-year-old CATAS 93-114 clone treated with cold for 2h
H96	3	SRR6668805	CATAS93-114_leaf_cold2h	Leaf samples from 2-year-old CATAS 93-114 clone treated with cold for 2h
H97	3	SRR6668800	CATAS93-114_leaf_cold8h	Leaf samples from 2-year-old CATAS 93-114 clone treated with cold for 8h
H98	3	SRR6668801	CATAS93-114_leaf_cold8h	Leaf samples from 2-year-old CATAS 93-114 clone treated with cold for 8h
H99	3	SRR6668821	CATAS93-114_leaf_cold8h	Leaf samples from 2-year-old CATAS 93-114 clone treated with cold for 8h

Note In this table, CATAS 7-33-97 is the alternative name of the clone Reyan 7-33-97

digital FPKM values. A gene expression search system is also provided for users to investigate up to 20 genes at once.

9.3.3 Co-expression Network

A WGCNA (weighted gene co-expression network analysis) co-expression network was constructed with the gene expression data from 74 samples and used in the expression visualization module. The Pearson's coefficient value threshold was set as 0.3 to build the network. Some genes have too many adjacent nodes, making the network too complicated to be recognized. Only the top 20 nodes are therefore allowed to be displayed. The annotation information of adjacent node genes is also displayed below the network.

9.3.4 Data Search Systems

There are six search systems in HeveaDB 1.0: gene search, marker search, network search, gene expression search, sequence homology search (blast and blat), and germplasm search (Table 9.4). The gene search tool utilizes ID or annotation as keywords to search the annotated genes (Fig. 9.3a). If the keywords match a single gene, the page will be re-directed to the gene page. If there are multiple genes matching the keywords, a gene list will be returned (Fig. 9.3b). The returned gene list can also be changed by re-selecting the annotation types, or re-entering a new keyword.

The markers in HeveaDB are searchable and browsable. The high-density integrated genetic linkage maps presented in this database were constructed by using genotyping-by-sequencing (GBS) (Pootakham et al. 2015). On the linkage group page (http://hevea.catas.cn/marker/v1/toMarker), a user can click each linkage group map to view detailed marker information (Fig. 9.4) or alternatively, perform a search using the name of the marker.

The gene expression data in HeveaDB 1.0 consist of digital expression based on 99 published transcriptomes. These data were categorized according to clone, tissue, age, and treatment. In the gene expression search system, up to 20 genes can be searched at once (http://hevea.catas.cn/tool/v1/toExpression). Users can also select for the expression in specific samples by using the check boxes provided for clone, tissue, age, and treatment. The expression results can be displayed as a heatmap or an FPKM table depending on the users' needs (Fig. 9.5).

Table 9.4 Search engines embedded in HeveaDB version 1.0

Search engine	Database	Keyword type	Address
Gene Search	Gene annotation	Gene ID Protein ID Annotation	http://hevea.catas.cn/search/v1/toGeneSearch
Marker Search	Genetic map	Marker name	http://hevea.catas.cn/search/v1/toMarker
Network Search	Co-expression network	Gene ID	http://hevea.catas.cn/coexpression/v1/toCoexpression
Gene Expression Search	Gene expression database	Gene ID Sample names	http://hevea.catas.cn/tool/v1/toExpression
Sequence Homology Search	Genome database (blast, blat)	Query sequence	http://hevea.catas.cn/tool/v1/toBlat http://hevea.catas.cn/tool/v1/toBlast
Germplasm Search	Wickham germplasm Wild Brazilian germplasm	Clone accession	http://hevea.catas.cn/search/v1/toGermplasms http://hevea.catas.cn/search/v1/toPhenotype

Fig. 9.3 Gene search engine showing multiple genes returned from annotation keyword search. **a** Gene search page; **b** search results page

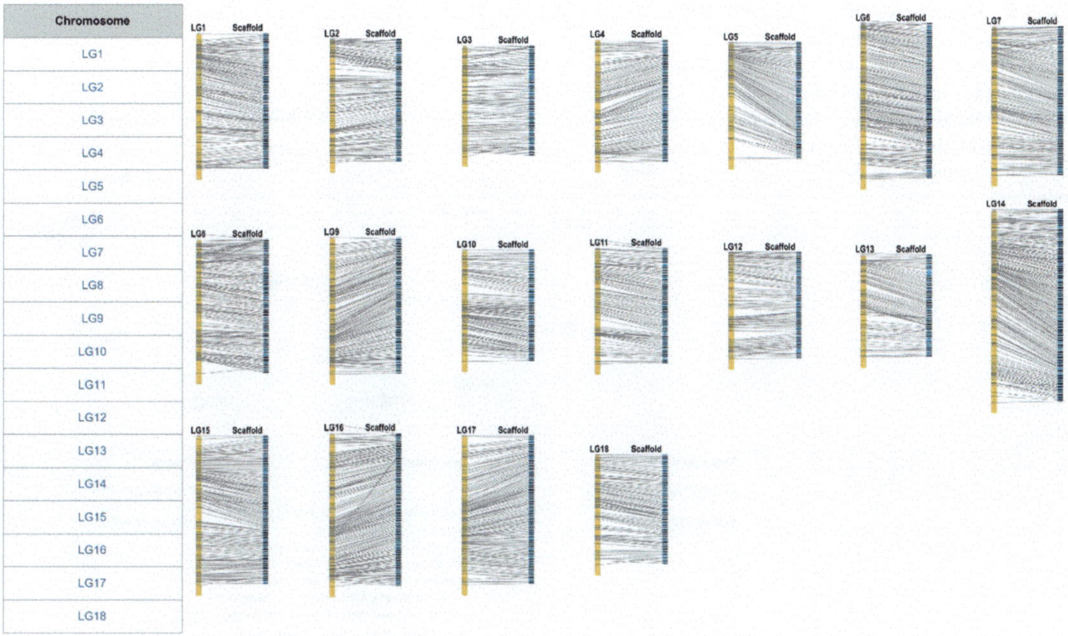

Fig. 9.4 A high-density integrated genetic linkage map constructed using genotyping-by-sequencing (GBS)

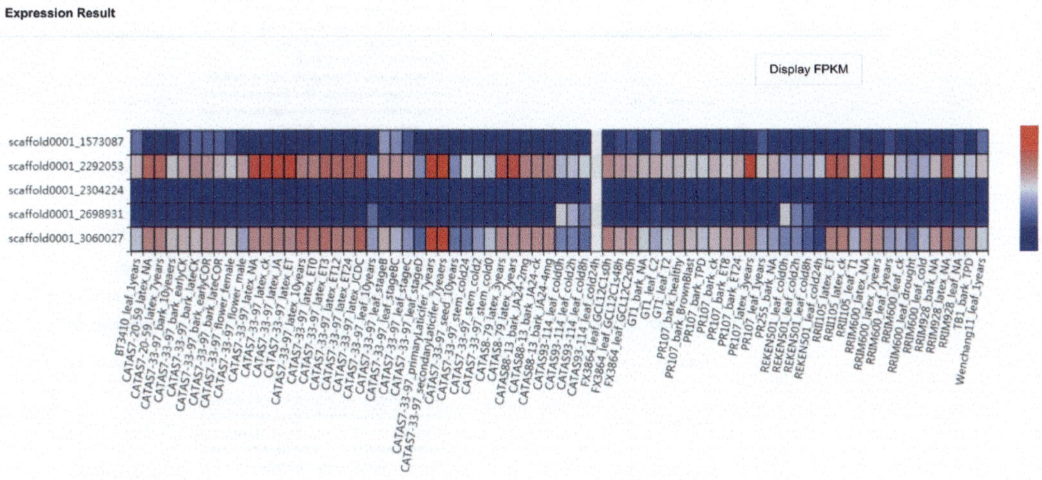

Fig. 9.5 A heatmap showing gene expression search results. Each row represents the expression data of a gene across many samples, and each column shows the expression in a particular sample

The WGCNA co-expression network is also searchable using gene ID. The top 20 adjacent nodes would be displayed in the results page (Fig. 9.6). The detailed information of target node genes would be shown in a table located below the network. The user can either view the gene page through the links in the table, or view the network by clicking on the node genes.

The blast and blat programs have been incorporated to provide homology searches against genome, CDS, and protein sequences. These tools require nucleotide or protein sequences to align with the sequences deposited

Fig. 9.6 A page showing the co-expression network search result. The network on the upper part of the page shows the top 20 nearby nodes, while the table below it shows detailed information of the nodes

Gene network

Total : 21

Gene : scaffold0001_403263

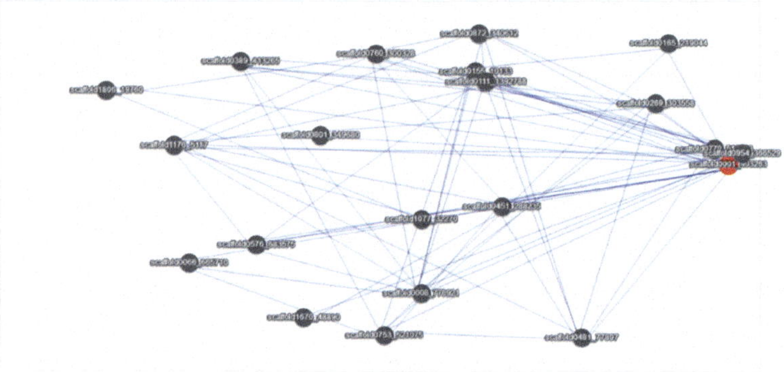

Target gene	Description	P-values
scaffold0165_219044		0.387198633615088
scaffold0753_521075	protein binding; F-box protein PP2-A15; Phloem protein 2	0.362419449622731
scaffold0066_665710	calmodulin binding; integral component of membrane; calcium-transporting ATPase activity; calcium ion transmembrane transport; membrane; nucleotide binding; ATP binding; metal ion binding; Calcium-transporting ATPase 4, plasma membrane-type; E1-E2 ATPase; haloacid dehalogenase-like hydrolase; Ca2+-ATPase N terminal autoinhibitory domain; Cation transporting ATPase, C-terminus; Cation transporter/ATPase, N-terminus	0.41612638217271
scaffold0389_413265	IST1 homolog; Regulator of Vps4 activity in the MVB pathway	0.368086664941461
scaffold0779_519203	transmembrane transport; membrane; drug transmembrane transport; antiporter activity; drug transmembrane transporter activity; Protein TRANSPARENT TESTA 12; MatE	0.602070438030851
scaffold0155_10133	catalytic activity; carbohydrate binding; carbohydrate metabolic process; Rhamnogalacturonate lyase; Rhamnogalacturonate lyase family; Polysaccharide lyase family 4, domain II	0.378702524206816
scaffold1176_5117	zinc ion binding; sequence-specific DNA binding ; DNA binding transcription factor activity; regulation of transcription, DNA-templated; GATA transcription factor 9 ; GATA zinc finger	0.444676919400072
scaffold0872_340612	ADP binding; protein binding; Putative disease resistance RPP13-like protein 1 ; NB-ARC domain	0.352853404641219
scaffold0954_366529		0.424794611633043
scaffold0481_77897	KIP1-like protein	0.433422126411193
scaffold0269_303558		0.419198884986866
scaffold1806_19760	protein binding; ADP binding; Putative disease resistance protein RGA3; Leucine rich repeat; NB-ARC domain	0.352488061474712
scaffold0801_349680	protein phosphorylation; protein kinase activity; transferase activity, transferring phosphorus-containing groups; ATP binding; Probable inactive receptor kinase RLK902; Protein tyrosine kinase	0.349120145544767
scaffold0576_683575	Beta-carotene isomerase D27, chloroplastic; Domain of unknown function (DUF4033)	0.392029440352796
scaffold0111_1392788	protein binding; F-box protein At5g49610 ; F-box domain	0.359582870706392
scaffold0760_300328		0.440431996183902
scaffold0008_778921	hydrolase activity, hydrolyzing O-glycosyl compounds; carbohydrate metabolic process; Glucan endo-1,3-beta-D-glucosidase; Glycosyl hydrolases family 17; X8 domain	0.393546829362885
scaffold1670_48890	nucleic acid binding; RNA-DNA hybrid ribonuclease activity; Putative ribonuclease H protein At1g65750	0.373386980332042
scaffold0451_288235	Agamous-like MADS-box protein AGL15	0.492728777065286
scaffold1077_32270		0.389971700988634

in HeveaDB, and alignments with the highest scores are returned.

Users can retrieve data on rubber germplasm (Wickham clones and wild Brazilian genotypes) by searching with the germplasm names. There are two entries in the main menu, the user can go to http://hevea.catas.cn/search/v1/toGermplasms for Wickham germplasm and to http://hevea.catas.cn/search/v1/toPhenotype for wild Brazilian germplasm. Fuzzy searching is allowed in this search engine.

9.4 General Architecture

HeveaDB 1.0 is hosted by CATAS, Hainan and is currently running on a Linux (CentOS release 6.6) virtual cluster platform. Eight-core processors with 32 GB memory resources were allocated for the database. The software environment was set up as follows: Apache Tomcat 7.0.64, Java openjdk-1.8.0.161, Perl 5.10.1, Apache HTTP server 2.2.15, MySQL 5.6.38. Genome information may be browsed in GBrowse 2.54 and JBrowse 1.12.3. Gene expression is visualized by ECharts 3.2.2, and the co-expression network is drawn with Cytoscape.js 2.7.2. BLAST 2.2.26 and BLAT 3.5 are used to align with the genome. Other pages are generated as static web pages.

9.5 Prospects of Future Development

Rapid advances in sequencing technologies have accelerated the analysis of genomes of non-model plant species (Cheng et al. 2018b), including *H. brasiliensis*. The emergence of huge amounts of sequence datasets requires a comprehensive hub for the deposition, utilization and analysis of data by the rubber research community. HeveaDB is a data hub with functions to store, share, distribute, and re-utilize the genomic, transcriptomic, and phenotypic data of the *Hevea* rubber tree. Using the Reyan 7-33-97 genome as the reference, HeveaDB provides an interface for visualization of the genome, annotated genes, gene expression, co-expression networks, and markers. The database content has been made more accessible through the search engines and downloadable information. However, HeveaDB is still at a preliminary level of development when compared with databases of model organisms. In order for HeveaDB to be even more accessible and functional for genome analysis, the database needs to be continually improved. In terms of new tools, additional functions need to be developed so as to facilitate post-genomic data utilization and analyses online. Three categories of functions have been identified for development in the near future: map drawing, sequence retrieval, and other functions, as shown in Table 9.5. In terms of sequence content, there is a need for continual efforts in capturing all available ESTs, full-length cDNAs and transcriptomes in order to ensure a comprehensive repository. With the current momentum of genomic applica-

Table 9.5 Three categories of functions to be provided in future versions of HeveaDB

1. Map draw
Clustering heatmap drawing
GO/KEGG enrichment and clustering heatmap
GO KEGG enrichment
Dot plot map
Histogram map
2. Sequence retrieval
CDS
cDNA
Peptide
Gene
Annotation
TSS up/down stream sequences
Optional position sequences
3. Other functions
Mean value calculation of gene expression corresponding to transcriptome replicates
DEG_analysis from transcriptomes
Correlation analysis
PCA
Contig/motif prediction
Transcriptional factor annotation and prediction

tions in improving rubber tree productivity, rapid generation of new sequence data is expected for both genomes and transcriptomes in the near future; these will be updated periodically to HeveaDB as they become available to the rubber research community.

Acknowledgements The HeveaDB data hub project was organized under the IRRDB Molecular Biology and Physiology Specialist Group and was financially supported by the Central Public-interest Scientific Institution Basal Research Fund for Innovative Research Team Program (NO. 17CXTD-28) of CATAS. The author thanks Prof. Jiannan Zhou from CATAS for his coordination between members of IRRDB. Also thanks to Dr. Keng-See Chow from the Malaysian Rubber Board, Dr. Yongjun Fang and Dr. Zewei An from CATAS for their insightful suggestions for the construction of HeveaDB.

References

Chao J, Chen Y, Wu S, Tian W-M (2015) Comparative transcriptome analysis of latex from rubber tree clone CATAS8-79 and PR107 reveals new cues for the regulation of latex regeneration and duration of latex flow. BMC Plant Biol 15. http://www.ncbi.nlm.nih.gov/pmc/articles/PMC4410575/

Cheng H, Gao J, Cai H, Zhu J, Huang H (2016) Gain-of-function in *Arabidopsis* (GAINA) for identifying functional genes in *Hevea brasiliensis*. SpringerPlus 5:1853

Cheng H, Chen X, Fang J, An Z, Hu Y, Huang H (2018a) Comparative transcriptome analysis reveals an early gene expression profile that contributes to cold resistance in *Hevea brasiliensis* (the Para rubber tree). Tree Physiol 38:1409–1423

Cheng H, Wang Y, Sun M (2018b) Comparison of gene expression profiles in nonmodel Eukaryotic organisms with RNA-seq. In: Wang Y, Sun M (eds) Transcriptome data analysis. Springer, New York, NY, pp 3–16. http://link.springer.com/10.1007/978-1-4939-7710-9_1

Chow K-S, Wan K-L, Isa MNM, Bahari A, Tan S-H, Harikrishna K, Yeang H-Y (2007) Insights into rubber biosynthesis from transcriptome analysis of *Hevea brasiliensis* latex. J Exp Bot 58:2429–2440

Fang Y, Mei H, Zhou B, Xiao X, Yang M, Huang Y, Long X, Hu S, Tang C (2016) De novo transcriptome analysis reveals distinct defense mechanisms by young and mature leaves of *Hevea brasiliensis* (Para Rubber Tree). Sci Rep 6:33151

Lau N-S, Makita Y, Kawashima M, Taylor TD, Kondo S, Othman AS, Shu-Chien AC, Matsui M (2016) The rubber tree genome shows expansion of gene family associated with rubber biosynthesis. Sci Rep 6. http://www.nature.com/articles/srep28594. Last accessed 19 March 2018

Li D, Deng Z, Qin B, Liu X, Men Z (2012) De novo assembly and characterization of bark transcriptome using Illumina sequencing and development of EST-SSR markers in rubber tree (*Hevea brasiliensis* Muell. Arg.). BMC Genom 13:192

Liu J-P, Xia Z-Q, Tian X-Y, Li Y-J (2015) Transcriptome sequencing and analysis of rubber tree (*Hevea brasiliensis* Muell.) to discover putative genes associated with tapping panel dryness (TPD). BMC Genom 16:398

Makita Y, Kawashima M, Lau NS, Othman AS, Matsui M (2018) Construction of Pará rubber tree genome and multi-transcriptome database accelerates rubber researches. BMC Genom 19:922

Pootakham W, Ruang-Areerate P, Jomchai N, Sonthirod C, Sangsrakru D, Yoocha T, Theerawattanasuk K, Nirapathpongporn K, Romruensukharom P, Tragoonrung S, Tangphatsornruang S (2015) Construction of a high-density integrated genetic linkage map of rubber tree (*Hevea brasiliensis*) using genotyping-by-sequencing (GBS). Front Plant Sci 6:367

Pootakham W, Sonthirod C, Naktang C, Ruang-Areerate P, Yoocha T, Sangsrakru D, Theerawattanasuk K, Rattanawong R, Lekawipat N, Tangphatsornruang S (2017) De novo hybrid assembly of the rubber tree genome reveals evidence of paleotetraploidy in *Hevea* species. Sci Rep 7:41457

Rahman AYA, Usharraj AO, Misra BB, Thottathil GP, Jayasekaran K, Feng Y, Hou S, Ong SY, Ng FL, Lee LS, Tan HS, Sakaff MKLM, Teh BS, Khoo BF, Badai SS, Aziz NA, Yuryev A, Knudsen B, Dionne-Laporte A, Mchunu NP, Yu Q, Langston BJ, Freitas TAK, Young AG, Chen R, Wang L, Najimudin N, Saito JA, Alam M (2013) Draft genome sequence of the rubber tree *Hevea brasiliensis*. BMC Genom 14:75

Tang C, Yang M, Fang Y, Luo Y, Gao S, Xiao X, An Z, Zhou B, Zhang B, Tan X, Yeang H-Y, Qin Y, Yang J, Lin Q, Mei H, Montoro P, Long X, Qi J, Hua Y, He Z, Sun M, Li W, Zeng X, Cheng H, Liu Y, Yang J, Tian W, Zhuang N, Zeng R, Li D, He P, Li Z, Zou Z, Li S, Li C, Wang J, Wei D, Lai C-Q, Luo W, Yu J, Hu S, Huang H (2016) The rubber tree genome reveals new insights into rubber production and species adaptation. Nat Plants 2:16073

Triwitayakorn K, Chatkulkawin P, Kanjanawattanawong S, Sraphet S, Yoocha T, Sangsrakru D, Chanprasert J, Ngamphiw C, Jomchai N, Therawattanasuk K, Tangphatsornruang S (2011) Transcriptome sequencing of *Hevea brasiliensis* for development of microsatellite markers and construction of a genetic linkage map. DNA Res 18:471–482

Xia Z, Xu H, Zhai J, Li D, Luo H, He C, Huang X (2011) RNA-Seq analysis and de novo transcriptome assembly of *Hevea brasiliensis*. Plant Mol Biol 77:299–308

Zhen X, Hu D (1994) Catalogue of rubber germplasm resources in China, 1st edn. China Agricultural Press

10 New Developments in Rubber Particle Biogenesis of Rubber-Producing Species

Muhammad Akbar Abdul Ghaffar and Katrina Cornish

Abstract

Natural rubber (NR) is a polymer critical to modern economies. However, the entire global supply is produced by one tropical tree species, *Hevea brasiliensis* (para rubber tree), derived from perhaps 20 plants collected over a century ago. This species is grown as clones making it even more genetically uniform. Efforts to diversify rubber production have led to ongoing development of alternative natural rubber-producing species. *Taraxacum kok-saghyz* (*T. kok-saghyz*) is an herbaceous plant of the *Asteraceae* family native to Kazakhstan. Similar to those in *H. brasiliensis* tree trunks, *T. kok-saghyz* roots produce rubber latex in multi-nucleate pipe-like laticifers. Fundamental knowledge of *T. kok-saghyz* anatomy, laticifer development and rubber biosynthesis is needed to accelerate its domestication. The origin of rubber particles may be particularly relevant to rubber yield. Microscopy methods were used to analyze rubber particle ontogeny in detail. Rubber particle ontogeny occurred before laticifers, apparently within the secretory pathway of the endoplasmic reticulum–Golgi vesicular complex. This is the first report of the origin of rubber particles in *T. kok-saghyz* seedlings. Unexpectedly, rubber particle development bifurcated in roots of mature plants, and laticifer plastids became a second rubber particle ontological site.

10.1 Introduction

Hevea brasiliensis (Müll., Arg.) is the sole rubber-producing crop plant of economic importance and is mostly cultivated in Southeast Asia. Currently, China is the largest consumer of natural rubber, followed by the United States and the European Union. The United States is the biggest rubber goods importer with 991,000 tons in 2018 and 1,002,000 tons in 2019 (IRSG, 2019). The dependence on a single clonally propagated species largely grown in one region puts the global supply at significant risk. Since rubber demand continues to rise, and *H. brasiliensis* acreage is likely near its maximum, alternative rubber crops suited to mechanized agriculture and temperate zones are needed. Synthetic rubber production cannot make up for natural rubber shortfalls because many manufacturing industries require the heat dispersion,

M. A. A. Ghaffar
Latex Harvesting Technologies and Physiology Unit, Production Development Division, Malaysian Rubber Board, 47000, Sungai Buloh, Selangor, Malaysia

K. Cornish (✉)
Departments of Horticulture and Crop Science, and Food, Agricultural and Biological Engineering, Ohio Agricultural Research and Development Center, College of Food, Agricultural and Environmental Science, The Ohio State University, 1680 Madison Avenue, Wooster, OH, USA
e-mail: cornish.19@osu.edu

M. Matsui and K.-S. Chow (eds.), *The Rubber Tree Genome*, Compendium of Plant Genomes,
https://doi.org/10.1007/978-3-030-42258-5_10

flexibility, durability, elasticity and resilience that only natural rubber can provide.

The *H. brasiliensis* rubber supply relies heavily on Southeast Asian smallholders. Rubber production may be limited by poor management practices, climate change and disease. Alternative rubber crops providing geographical and genetic diversity in the global supply are likely to affect the global rubber trade by simultaneously calming price volatility and providing supply security. Guayule (*Parthenium argentatum*, Gray) and rubber root dandelion (*Taraxacum kok-saghyz,* Rodin) (sometimes called Russian dandelion, even though it is not from Russia) are the leading natural rubber alternatives. Both species make high-quality rubber and can be harvested annually. Guayule is under development in semi-arid regions, including the southwestern USA and the European Mediterranean region, whereas *T. kok-saghyz* is the crop of choice for wetter regions with snowy winters (Buranov and Elmuradov 2010; Cornish 2017; Cornish et al. 2013). *T. kok-saghyz* rubber is very similar in quality to that from *H. brasiliensis* and may prove to be a good model species for research into rubber particle ontogeny, biochemistry and genetic investigation of yield limiting factors.

10.2 Laticifer Cells and Rubber Production

Rubber-producing species are found in the eudicotyledonous branch of the angiosperms, and rubber is not produced in monocotyledonous plants, gymnosperms or lower plant classes (Bonner and Galston 1947), although a few fungi do produce rubber (Ohya et al. 1998). Latex-producing plants are quite common ($\sim$12,500 species) (Esau 1965), although only about 20% of them make rubber (Bonner 1991). Also, many laticiferous plants make latex without making rubber in that latex (Esau 1965), such as *Ficus callosa* (proteins) and *Brosiraum galactodendron* (waxes) (Bonner and Galston 1947) and *Carica papaya* (crude proteins) (Macalood et al. 2013).

Latex is more common in tropical regions than temperature regions, occurring in 14% and 6% of species, respectively (Agrawal and Konno 2009). A single genus may include multiple latex-producing tropical species, but their temperature zone cousins often do not make latex. Latex production and rubber-containing latex are thought to have evolved convergently to provide an important adaptive function to selection pressures that may not exist today, which may explain why this ability is scattered among the Eudicot genera and species.

Latex is produced in specialized vessels called laticifers (pipe-like multi-nucleate cell systems). *H. brasiliensis* is the best-known example of laticifers producing rubber particles in the latex (de Faÿ and Jacob 1989). Many other rubber-producing plants such as *Ficus elastica* and *T. kok-saghyz* also develop laticifers (Artschwager and McGuire 1943; Metcalfe 1967; Esau 1965). These occur in at least 20 genera (Esau 1965; Hagel et al. 2008). Two main types of laticifers have evolved, namely non-articulated or articulated laticifers (Mahlberg 1993; Hagel et al. 2008). Non-articulated laticifers develop from single cells: the nucleus divides without triggering cell division and the cell becomes multinucleate. Repeated nuclear divisions lead to cell expansion as the laticifer forms and grows during plant growth. Articulated laticifers are derived from a series of cells, which align and then unite into a single multi-nucleate vessel as the intervening walls dissolve (Esau 1977). However, rubber particles also can be made in unspecialized cells. The best-known example of this cell type is in *P. argentatum* which produces its rubber particles in parenchyma cells in the bark cortex (Bonner and Galston 1947).

Laticifers were originally thought to serve as food conduits or storage compartments due to their high sugar content and the myriad of other metabolites (Esau 1965). Many isoforms of sucrose transporters are present in laticifers (Dusotoit-Coucaud et al. 2009). *Hevea brasiliensis* latex contains proteins, carbohydrates, fats, sterols, antioxidants, and ascorbic acid amongst other compounds (Jacob et al. 1993). Other products in latex may be cytotoxic and it was suggested that laticifers evolved to sequester such compounds away from vascular tissue (Hagel et al. 2008). It was also thought that the rubber particles produced by laticifers were part of the excretory system since the terpenoid compounds it contains, including rubber and resin, were thought to be non-functional by-products of cellular metabolism (Esau 1965). Terpenes often remain in the cells where they are formed. The laticifers' location in the bark, adjacent to the vascular system, supports the idea that they may serve as a sink of catabolizable carbohydrate (Jacob et al. 1993). Also, laticifers can readily absorb water from surrounding tissues, and may be involved in the regulation of water balance in the plant and perhaps transport oxygen (Esau 1965).

The role of latex in herbivory or pathogen protection appears to be generally accepted (Esau 1965; Dussourd and Eisner 1987; Lewinsohn 1991), as latex bleeds from the plant after wounding (de Faÿ and Jacob 1989). Many proteins produced in latices are clearly related to defense, such as pathogenesis-related polypeptides and chitinases (Hagel et al. 2008). This trait appears to have independently evolved several times possibly to reduce herbivory (Agrawal and Konno 2009). The chemical composition of latex varies (Metcalfe 1967) perhaps due to species-specific biosynthesis, trafficking and compartmentalization of biochemicals and proteins in their laticifers. This may be why some latex-producing plants produce rubber while others do not. It is not known if laticifers evolved before plants developed the ability to synthesize rubber. Within latex the aqueous emulsion is formed by suspended small, membrane-bound particles with hydrophobic interiors, which may contain rubber, other smaller polyisoprenes and terpenes (de Faÿ and Jacob 1989). In the study reported here, latex is the sub-fraction of rubber in individual particles suspended in aqueous cytosol. Latex does not include particles which have coagulated during the life of the plant.

Rubber biosynthesis is mediated by a rubber transferase complex bound to the rubber particle uni-lamella membrane (Cherian et al. 2019; Yamashita et al. 2016). New rubber molecules are initiated by binding of an allylic pyrophosphate, usually cytoplasmic farnesyl pyrophosphate, and then rubber is polymerized from the hydrophilic monomeric substrate, isopentenyl pyrophosphate, which is produced by both the plastidic methylerythritol-4-phosphate pathway and the cytosolic mevalonate pathway (Hopkins and Hüner 2008). Rubber molecules are compartmentalized into the rubber particle interior as they grow until an uncharacterized termination event releases the polymer from the enzyme complex (Cherian et al. 2019; Cornish 1993; Men et al. 2018). There are considerable commonalities in rubber particle structure, rubber biosynthesis and the rubber transferase complex among different species.

However, the ontogeny of rubber particles is not yet well understood. Most studies of rubber storage vessels in *H. brasiliensis* and *P. argentatum* have ignored rubber particle ontogeny. Histological studies in *H. brasiliensis* studied plants aged from weeks to months (Hebant 1981) to 10-years-old (Gomez 1983). Laticifers also have been histologically characterized by their age (Dickenson 1969). The youngest laticifers are located in the active differentiating regions next to the cambium or in apical meristems, whereas the oldest laticifers are associated with secondary phloem. However, rubber particle ontogeny was not reported. The anatomy and morphology of *P. argentatum* has been described but not rubber particle ontogeny (Artschwager 1943). A de novo site of rubber particles was suggested, in an anatomical study of plants aged 2.5, 8 and 36 months (Backhaus and Walsh 1983). Similarly, although morphological and anatomical structures, including root laticifers, have been characterized in *T. kok-saghyz,* no laticifer or rubber particle ontogeny was described (Artschwager and McGuire 1943). An endoplasmic reticulum (ER) origin for some latex and rubber particle proteins was indicated in a heterologous expression study (Brown et al. 2017).

In our preliminary studies of *T. kok-saghyz,* laticifer initiation and early development occurred in plants younger than those previously studied and so rubber particles also may first develop in very young plants. Also, seedlings often have different secondary compounds than found in older plants (Barton 2007), and this different composition may affect ontological processes. Seedling growth of higher plants is accompanied by rough ER cisternae proliferation which is tied to the accumulation of protein bodies in parenchyma cells as seen in different legumes (Harris and Chrispeels 1980). Rubber particle ontogeny may follow a similar secretory protein pathway process, including the endoplasmic reticulum and Golgi apparatus, due to their function as entry ports and as routes that deliver proteins to multiple distinct compartments (Vitale et al. 1993). An expression study in *Nicotiana benthamiana* showed that four *H. brasiliensis* latex proteins are associated with the endoplasmic reticulum, namely, rubber elongation factor (REF) which has soluble and rubber particle membrane associated and bound forms, small rubber particle protein (SRPP), a soluble *cis*-prenyltransferase (CPT), and a bridging protein (HRBP or CBP) (Brown et al. 2017), supporting a secretory protein pathway origin of rubber particles. The similarity of *T. kok-saghyz* rubber and latex associated proteins to those in *H. brasiliensis* rubber and latex may reflect a similar particle ontogeny. A better understanding of rubber particle ontogeny, accumulation and storage, may give insights into why rubber is being produced at all, and lead to new approaches to control rubber deficiency problems like tapping panel dryness (Krishnakumar et al. 2001).

10.3 Novel Ontogenies Produce *Taraxacum kok-saghyz* Rubber Particles

Glycosylated proteins have been identified in membranes of rubber particles purified from different species supporting an origin in the rough ER (Cornish 2001a), but the detailed process of particle ontogeny is not known. Thus, rubber particle ontogeny was investigated in both young *T. kok-saghyz* plants (6 days to 2 months old) and more mature plants. The mature plants included those producing flowers after 60 to 70 days of growth, and in plants from 4 to 12 months old. Transmission electron microscopy (TEM) was used to visualize the subcellular site of rubber particle origination and its relationship with vesicular trafficking pathways. Harvested seedling roots were hand sectioned at the hypocotyl and fixed in 2% paraformaldehyde, 3% glutaraldehyde and 0.1 M potassium phosphate buffer (PB) overnight. Samples were post-fixed in 1% uranyl acetate and 1% osmium tetroxide for 1 h and then rinsed 3x with double-distilled water. Fixed root samples were dehydrated using a series of ethanol concentrations (25, 50, 70 and 90%), then infiltrated with a series of propylene oxide and EM Bed-812 resin mixtures, as follows: samples were put into a EM Bed-812 resin and propylene oxide mixture (at 1:2 ratio) for 1 h, followed by an equal ratio (1:1) for 2 h and ending with a 2 EM Bed-812 resin:1 propylene oxide mixture for another 2 h. Lastly, sections were embedded in 100% EM Bed-812 resin and dried overnight at 60 °C. The resin blocks were sliced (70 nm thick) using a LEICA EM-UC6 Ultra Microtome (LEICA, Vienna, Austria). A 3% aqueous uranyl acetate and Reynold's lead citrate were used to stain the sectioned tissues. Thirty samples of each developmental stage were investigated by TEM (Hitachi H-7500, Tokyo, Japan) and images collected with an Optronics QuantiFire-Model S99835 digital camera. Rubber particle diameter was measured using ImageJ (US National Institute of Health, Bethesda, MD) and averages ($n = 30 \pm$ s.e) calculated.

Our findings are based solely on microscopy observations. The rubber particles that we observed were identified based on their dense layer and osmiophilic nature. We also took into consideration other organelles near the rubber particles (under the microscope) that we could relate to the rubber particle ontogeny process. The vesicles observed in micrographs of *T. kok-saghyz* roots were similar to vesicles reported in tobacco BY-2 cells (Ritzenthaler et al. 2002), *Arabidopsis thaliana* (Langhans et al. 2008) and *Scherffelia dubia* (Donohoe et al. 2013). The secretory pathway between the ER and the Golgi has bidirectional movement (anterograde and retrograde) of membrane traffic, which is assisted by coat protein complexes (COP) that ensure direction and fidelity (Brandizzi and Barlowe 2013). The retrograde route is mediated by protein complex I (COPI) from the Golgi, whilst protein complex II (COPII) operates within the anterograde pathway starting from the ER (Brandizzi and Barlowe 2013). Two types of plant COPI proteins, COPIa and COPIb, have been distinguished (Staehelin and Kang 2008). COPIa exclusively buds from *cis*-Golgi cisternae. COPIa was lightly stained, confined to the ER-*cis*-Golgi interface, and co-localized with COPII vesicles. In contrast, COPIb mostly buds from the *medial-*, *trans-* and early *trans*-Golgi network (TGN) cisternae and has darkly staining luminal content like the Golgi cisternae (Staehelin and Kang 2008).

These apparent COPII vesiculation events in *T. kok-saghyz* roots (Fig. 10.1, symbol 1, the COPII route or the anterograde pathway; Fig. 10.2a, b) indicate that vesicles came from ER *en route* to the *cis*-Golgi (Golgi entry point). These vesicles were electron dense, certainly denser than vesicles budded from the TGN or medial-Golgi (Fig. 10.2b), indicating their likely synthesis of electron dense rubber. They then appeared to merge with the *cis*-Golgi (Fig. 10.2c, d). The micrographs strongly suggest that the *T. kok-saghyz* rubber particles originated in the endoplasmic reticulum–Golgi vesicular complex. This pattern of cytoplasmic rubber particle formation was observed in seedlings and in mature roots, and we denote it "the cytoplasmic pathway". The mechanism is in line with known ER functions during anterograde transport of newly synthesized proteins and lipids initiated at the ER (Vitale et al. 1993; Brandizzi and Barlowe 2013) before being trafficked to other cellular locations (Vitale et al. 1993), and the cytoplasmic pathway of rubber particle formation described above is consistent with this.

Cytoplasmic pathway-produced rubber particles were visible in seedlings 6 days after germination, and the micrographs indicated that rubber particles were only made this way in young plants aged from 6 days to 2 months. In mature plants, cytoplasmic rubber particle ontogeny began either with the formation of COPIb-like secretory vesicles by the *trans-* and *medial-*Golgi (Fig. 10.1, symbol 5; Fig. 10.3a, b) which later developed into small rubber particles, or small rubber particles were directly budded from the ER cisternae (Fig. 10.1, symbol 10; Fig. 10.3c, d). Free small rubber particles accumulated adjacent to the tonoplast (Fig. 10.4a). These small particles then either moved directly into the vacuole (Fig. 10.1, symbol 2; Fig. 10.5a), or first merged in the cytosol to form larger particles before being translocated into the laticifer vacuole (Fig. 10.1, symbol 3; Fig. 10.5b).

Small rubber particles were produced in mature roots exactly as in younger roots except that only those merged into larger particles in the cytosol were able to translocate into the vacuoles. As in the young roots, the small particles accumulated near the tonoplast (Fig. 10.1, symbol 10; Fig. 10.4a) before merging into larger particles. Cups consisting of dense cytoplasm formed at the cytoplasm tonoplast interface (Fig. 10.5e), surrounded by ER (Fig. 10.5f). The rubber particles contained within these cups increased in size maintaining a spherical shape as they grew. The tonoplast then cleaved (Fig. 10.4a, d), releasing the mature rubber particles into the vacuole (Figs. 10.4d, g and 10.5e) and then reformed (Fig. 10.4j). In roots of all ages, many particles not formed in a cytoplasmic cup remained tethered to the tonoplast after translocation into the vacuole. Such tethered particles become a site of small particle aggregation in young roots (Fig. 10.1, symbol 4; Fig. 10.5d), and merged into large irregular particles in mature roots (Fig. 10.1; symbol 12; Fig. 10.5g, h).

Unexpectedly, the micrographs of mature roots revealed a second route of rubber particle formation distinct from the cytoplasmic pathway in which rubber particles were formed in vesicular plastids, in a process we coin "the vesicular pathway". The *T. kok-saghyz* vesicular plastids seem similar to those that make lipoprotein particles in the laticifers of opium poppies, *Papaver somniferum* (Bajpai et al. 2000) and *Papaver bracteatum* (Nessler and Mahlberg 1979).

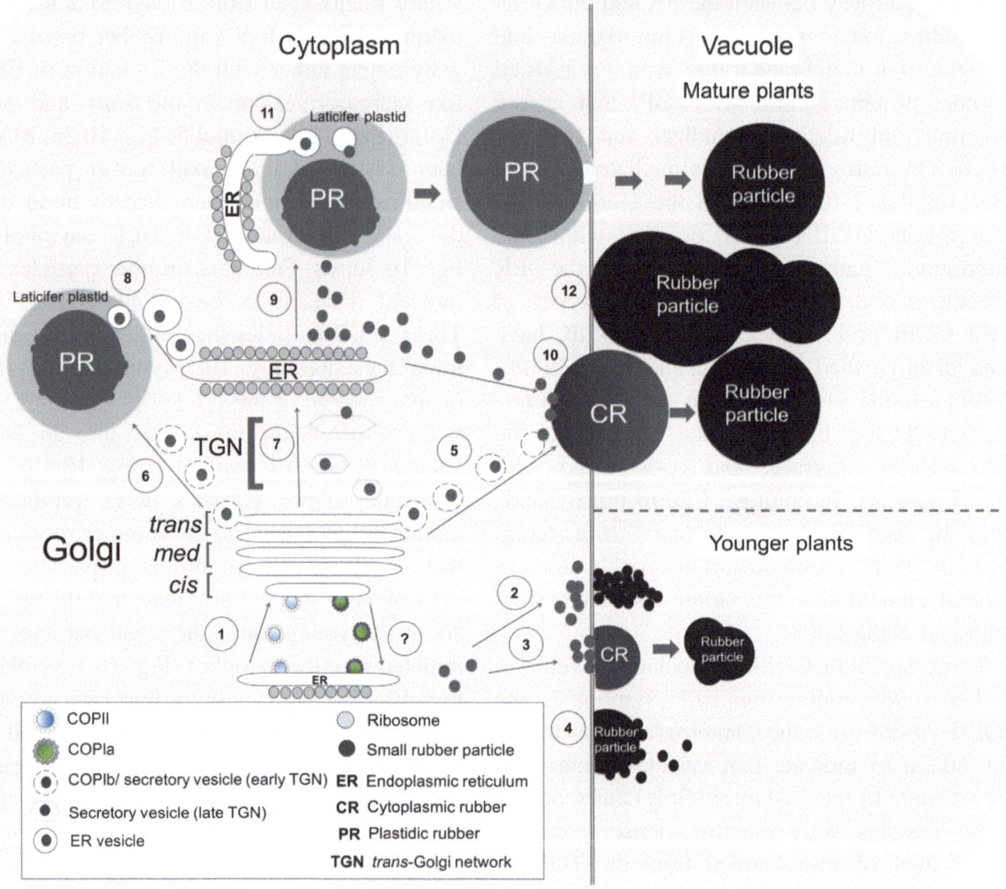

Fig. 10.1 Model of rubber particle ontogeny indicating the formation of two types of rubber particles, the plastidic and cytoplasmic rubber in *Taraxacum kok-saghyz*. Rubber particle formation in roots of young plants is indicated by symbols 1 to 4, whereas rubber particle formation in roots of mature plant is indicated by symbols 1 to 12. The detailed explanation is described in the accompanying text. (?) symbol: the COPIa (the retrograde pathway from the *cis*-Golgi to the ER) was not observed. COPIa and COPII may be distinguished based on the location where they are observed, with COPIa being found near to the Golgi and COPII at the ER

Mature rubber particles from both *T. kok-saghyz* synthetic routes were translocated into the vacuoles (Figs. 10.3c, d, and 10.8q), a common storage compartment for mature rubber particles in most species (Goss 1991). The two disparate *T. kok-saghyz* routes of rubber particle ontogeny and development imply that distinct vesicular sorting pathways and location-specific secretory vesicles produce the plastidic and cytoplasmic rubber particles.

Rubber particles are made in mature roots by the vesicular pathway as follows. Vesicles (secretory and clathrin-coated vesicles) from the *medial-* and *trans*-TGN (Fig. 10.1, symbol 7; Fig. 10.2b) moved to the ER cisternae in a retrograde manner. Vesicles then migrated to the laticifer plastids releasing small rubber particles into the plastid lumen (Fig. 10.1, symbol 8; Fig. 10.6 c). TGN-producing vesicles appeared to migrate to the ER (Fig. 10.1, symbol 7; Fig. 10.7a, b) and

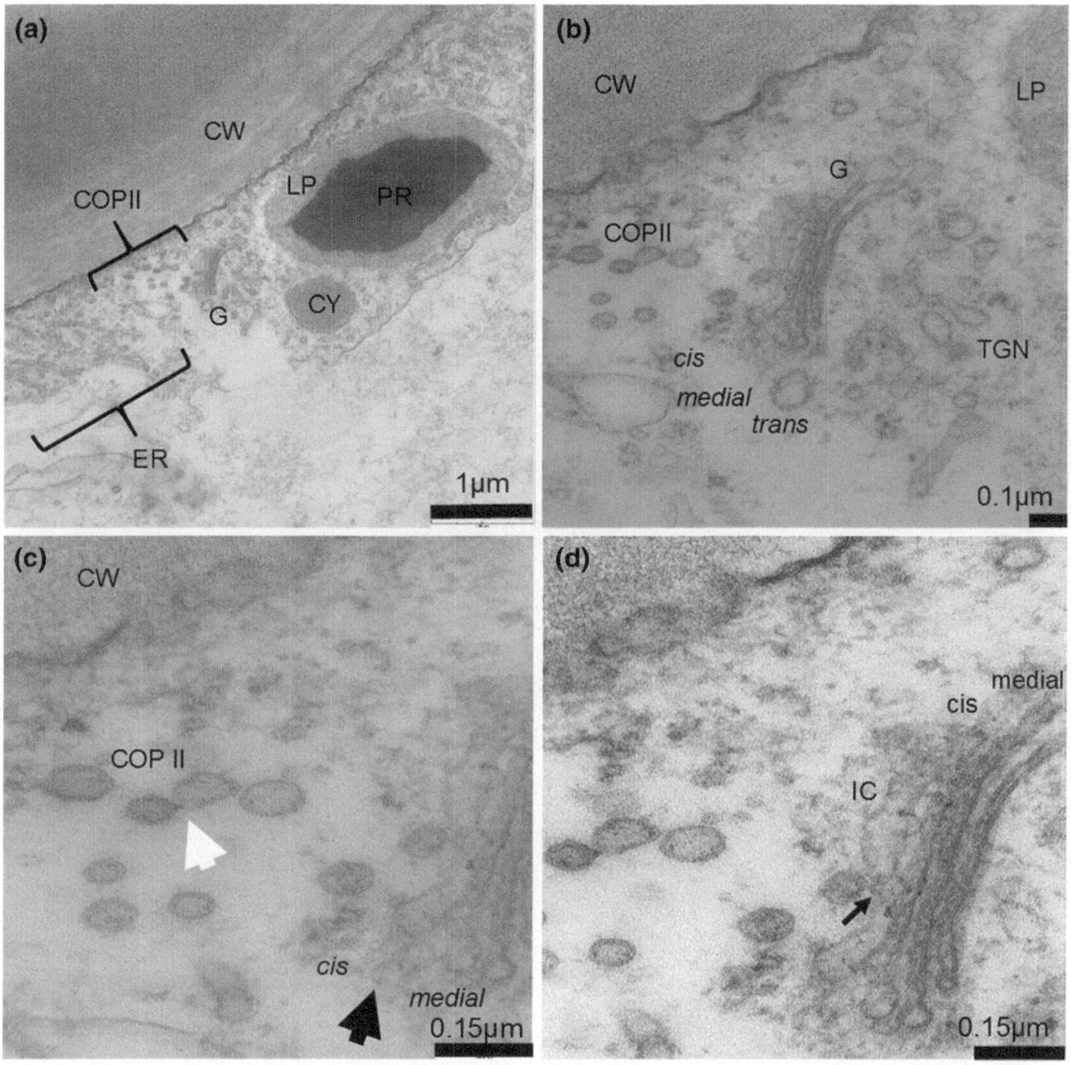

Fig. 10.2 Longitudinal section of mature root of *Taraxacum kok-saghyz*: **a** Overall view of the laticifer cell which shows parts of ER group, COPII group, Golgi and the laticifer plastid (LP) where the small rubber particles merge to become irregular rubber particles; **b**, **c**, **d** Enlargement of (a) which depicts the COPII entering the *cis* part of the Golgi; **c** The *white arrow* indicates the fusion of the vesicles, whereas the *black arrow* indicates the merging of COPII with the *cis*-Golgi; and **d** The *black arrow* indicates the developed IC. Key: Cell wall (CW); crystal (CY); vesicles or putatively coat protein complex II (COPII); Golgi body (G); intermediate compartment (IC); laticifer plastid (LP); plastidic rubber (PR); endoplasmic reticulum (ER); *trans*-Golgi network (TGN)

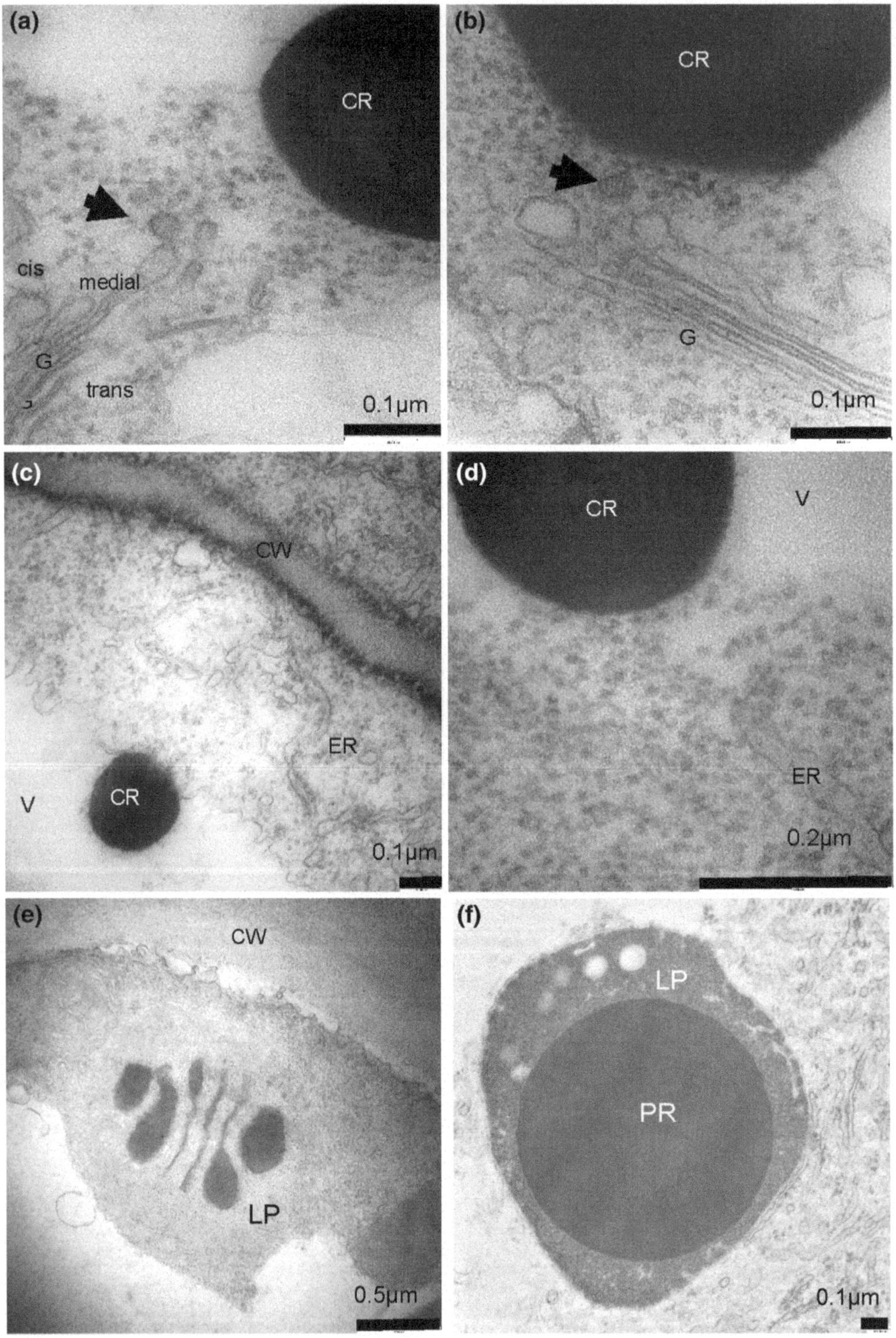

Fig. 10.3 Cytoplasmic (**a**, **b**, **c**, **d**) and plastidic (**e**, **f**) rubber particles developed from the small rubber particles in *Taraxacum kok-saghyz* roots. **a**, **b** Arrows indicate COPIb captured nearby cytoplasmic rubber; **c**, **d** Small rubber particles produced from the ER cisternae and vesicle form cytoplasmic rubber; **e** Longitudinal section of laticifer plastid showing layers formed by the small rubber particles; **f** Longitudinal section of fully developed plastidic rubber (PR) in the laticifer plastid (LP). Key: Cell wall (CW); cytoplasmic rubber (CR); endoplasmic reticulum (ER); Golgi body (G); laticifer plastid (LP); plastidic rubber (PR); vacuole (V)

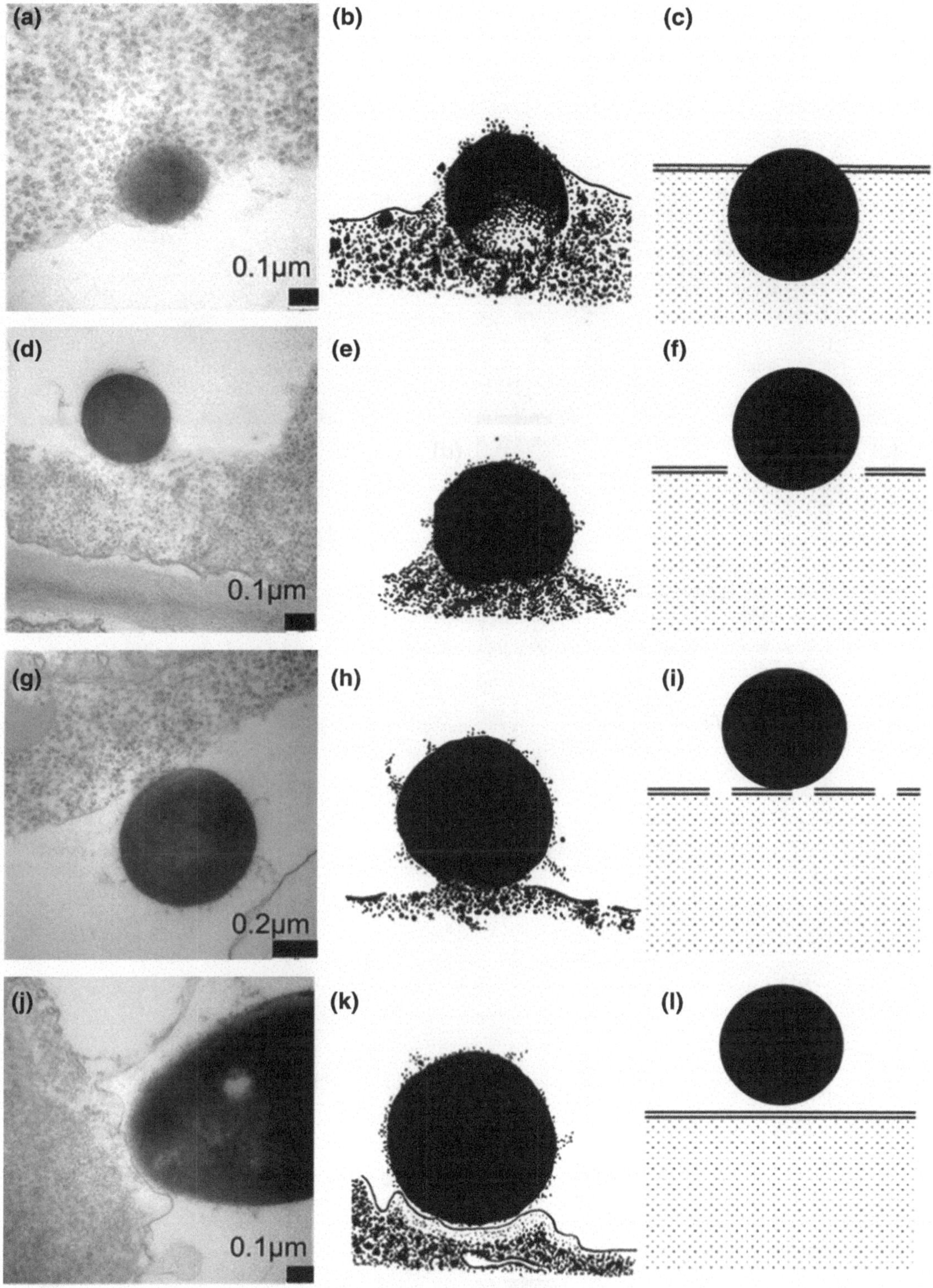

Fig. 10.4 The formation of cytoplasmic rubber particles (CR) showing micrographic images (**a**, **d**, **g**, **j**), illustrations (**b**, **e**, **h**, **k**), and diagrams (**c**, **f**, **i**, **l**) of the process in which small rubber particles accumulate near the tonoplast, spherical-shaped CR traverse into the vacuole, the tonoplast cleaves and then reforms after CR is compartmentalized within the vacuole

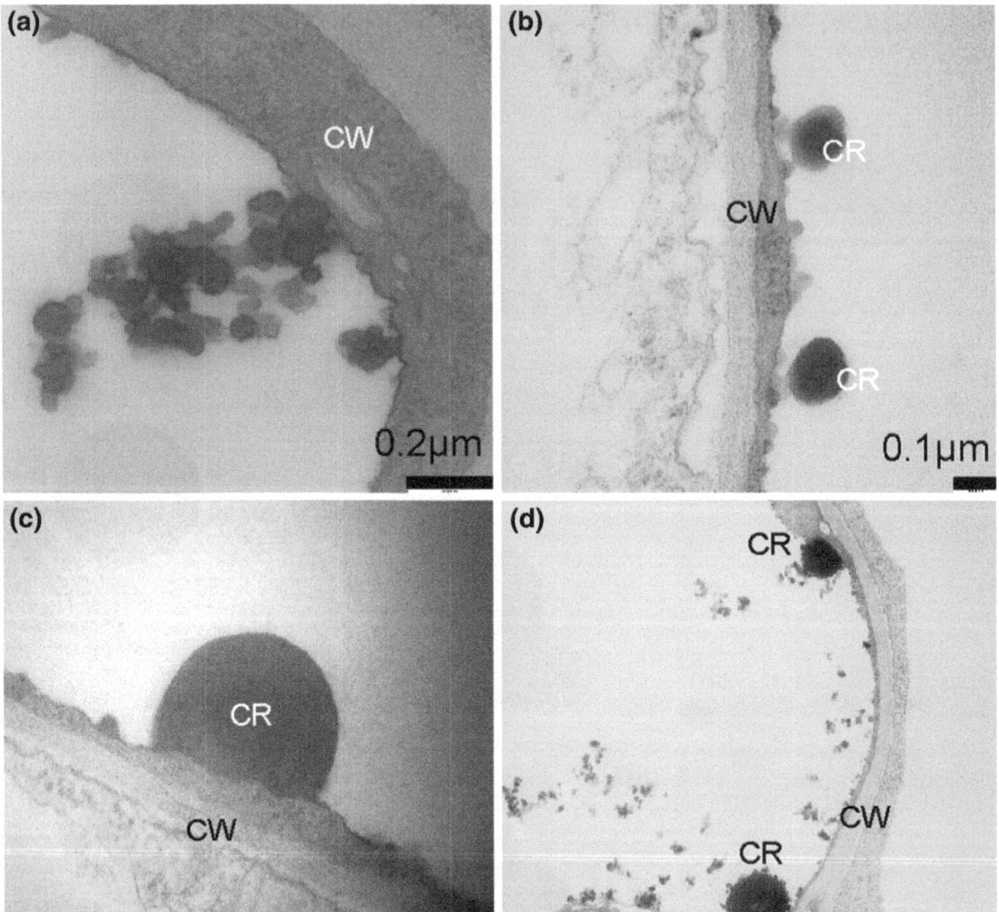

Fig. 10.5 **a**, **b**, **c**, **d** The translocation of CR in the seedlings with **a** direct translocation of small rubber particles into the vacuole; **b** Irregular rubber particles formed by small rubber particles translocated in the vacuole; **c** CR tethered to the tonoplast; **d** CR attracts small particles as they aggregate. **e**, **f**, **g**, **h** The translocation of CR in the mature plant with; **e** Longitudinal section of cup at the border of the cytoplasm and tonoplast as well as the extrusion process of CR; **f** The cups surrounded with ER; **g**, **h** Tethered CR form large irregular particles. Key: Cell wall (CW); cytoplasmic rubber (CR)

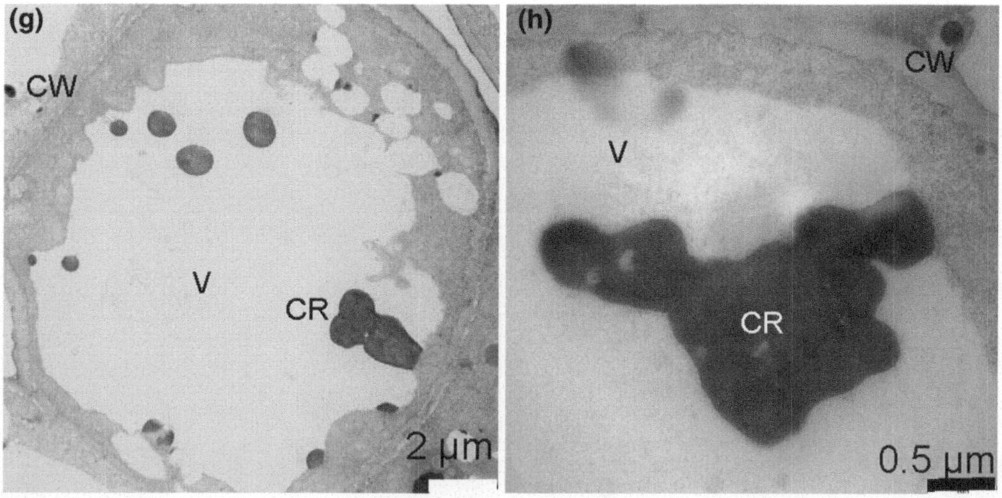

Fig. 10.5 (continued)

to laticifer plastids (Fig. 10.7a). The ER cisternae also appeared to be directly connected to the laticifer plastids (Fig. 10.1, symbol 11; Fig. 10.6 b, c (black arrow)). Small rubber particles were apparently translocated directly into the plastid (Fig. 10.1, symbol 9; Fig. 10.7d) as well as into vacuoles by the ER cisternae.

In the plastid, irregularly shaped larger particles were formed as the juvenile particles first formed layers (Fig. 10.3e) which then merged (Figs. 10.2a, 10.6c and 10.7d). These large particles co-existed with small particles (Fig. 10.8g). In contrast, some plastids produced a single large spherical rubber particle (Fig. 10.3f), which earlier had been thought to be protein crystals (Heinrich, 1967). These plastids migrated to the tonoplast (Fig. 10.8m) where the laticifer plastid membrane opened up to release the rubber particle (Fig. 10.8n) in a manner similar to the cytoplasmic cup used by the cytoplasmic pathway (Fig. 10.8q). After release, the tonoplast resealed (Fig. 10.8r and s) and the empty plastid dissipated into the cytoplasm (Fig. 10.8q, arrow).

Thus, vacuoles are a storage compartment for mature rubber particles but rubber biosynthesis and true particle growth both occur in the cytosol. Rubber biosynthesis is catalyzed by a rubber polymerizing complex embedded in the membrane of the particle (Cornish and Backhaus 1990; Cherian et al. 2019), and this cannot occur inside the vacuole. Vacuoles contain acid phosphatases, which would dephosphorylate any rubber substrates that might cross the tonoplast as well as the active end of growing rubber chains. Three putative phosphatase homologs of the *Lycopersicon esculentum* [AAG40473] and *Arabidopsis thaliana* putative phosphatases [NP 173, 213 and AAM63155] were found in the *T. kok-saghyz* genomic database at the Ohio State University, and 10 acid phosphatase genes are known in the common dandelion, *T. officinale* (Kashin et al. 2005).

The biochemical requirements for rubber biosynthesis and the basic architecture of rubber particles have been conserved among rubber-producing species, even those without laticifers (Cornish et al. 1993; Cornish 2001a, b). Thus, the mechanisms of rubber particle ontogeny may prove to be broadly applicable among rubber-producing species. In addition, our *T. kok-saghyz* model of rubber particle ontogeny (Fig. 10.1) may inform

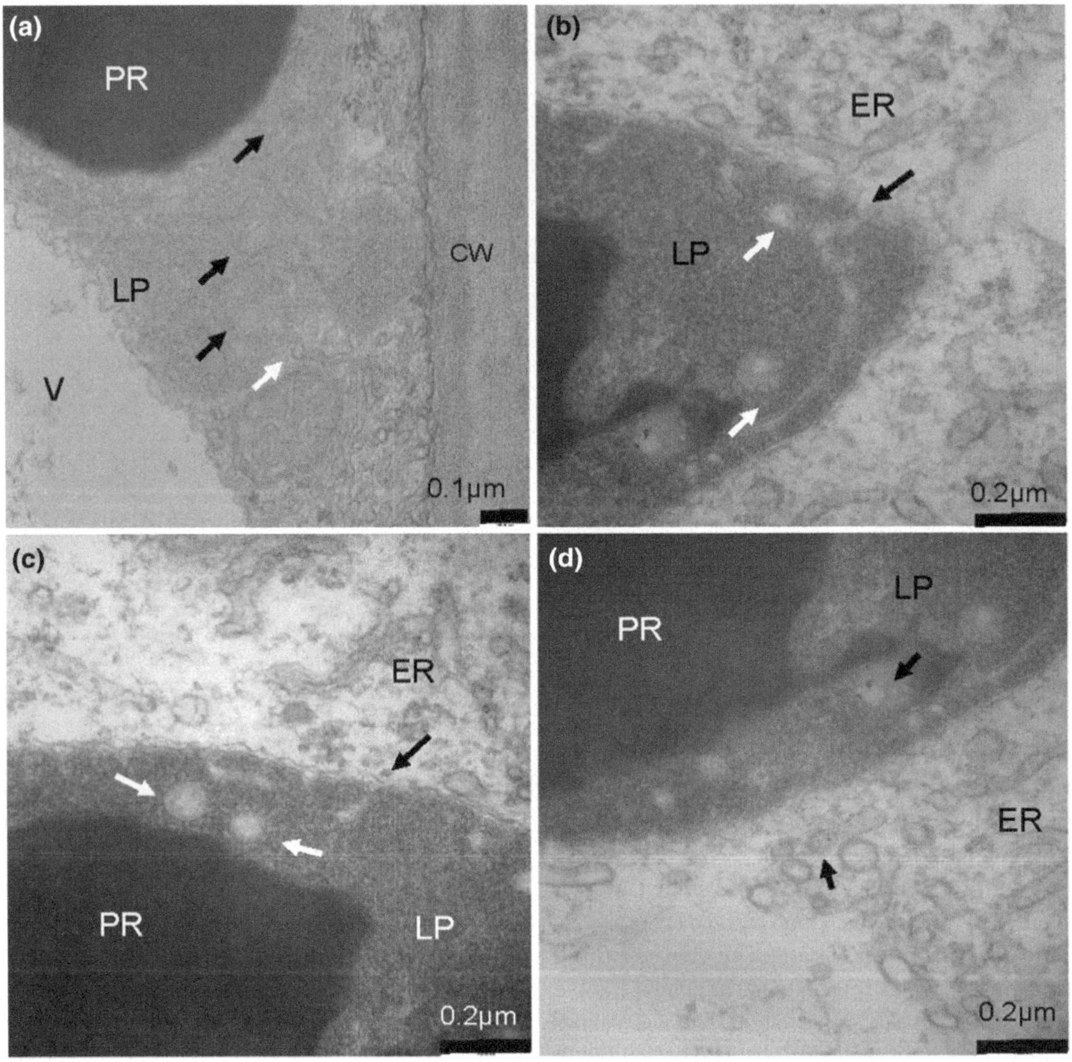

Fig. 10.6 Longitudinal section of *Taraxacum kok-saghyz roots* with endoplasmic reticulum (ER) and ER secretory vesicles connected to the plastid; **a** An ER (*white arrow*) connection with the laticifer plastid (LP) and the vesicles (V) (*black arrows*) can be seen crossing the plastid; **b** Invagination of ER (*black arrow*) that connects the outside ER with LP as ER vesicles are developed (*white arrows*) within the LP; **c** Similar observation as in **b**; **d** Arrows indicate vesicles containing small rubber particles that are found both inside and outside (cytosol) of the LP suggesting that the vesicle can also transverse into the LP. Key: Cell wall (CW); endoplasmic reticulum (ER); laticifer plastid (LP); plastidic rubber (PR); vacuole (V)

differences in rubber yield among genotypes, or as affected by cultural practices or environmental and edaphic factors. For example, cold-induced rubber biosynthesis may be solely mediated by the plastidic rubber particle ontogenetic pathway—in such a case, experiments on juvenile plants may not be relevant to the mature crop system. Similarly, in a vertical hydroponic farming system, with multiple harvests to maximize rubber yield, it will be important to know if roots regrown from mature plants are physiologically juvenile (one pathway) or mature (two pathways).

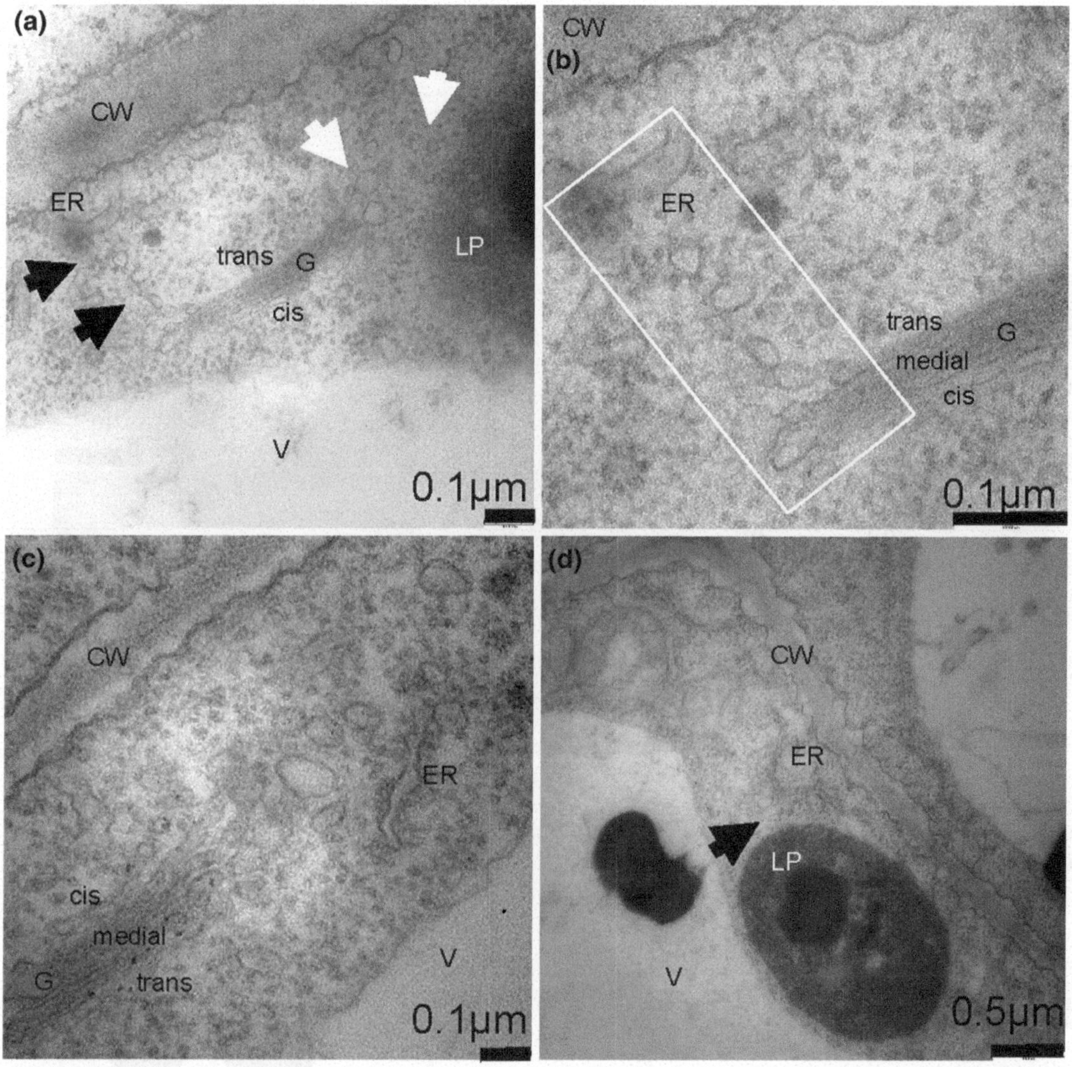

Fig. 10.7 **a** Golgi produced different vesicles for ER (*black arrows*) and laticifer plastid (*white arrows*); **b** Enlargement of (**a**) shows vesicles from *medial*- and *trans*-area of the Golgi were directed to the ER; **c** Another example of vesicles produced from *medial*, *trans*- and early TGN directed to the ER; **d** Arrows indicate small rubber particles being translocated into the laticifer plastid (LP). Key: Cell wall (CW); endoplasmic reticulum (ER); Golgi body (G); laticifer plastid (LP); vacuole (V)

10.4 Conclusion

This is the first time that rubber particle ontogeny has been anatomically detailed in a rubber-producing species. Two developmentally distinct subsets of rubber particles are formed which may differentially accumulate in response to internal or external factors. This regulation may be key to maximizing yield as *T. kok-saghyz* enters the commercial sector. These rubber particle synthetic pathways may be in common among laticiferous rubber-producing species.

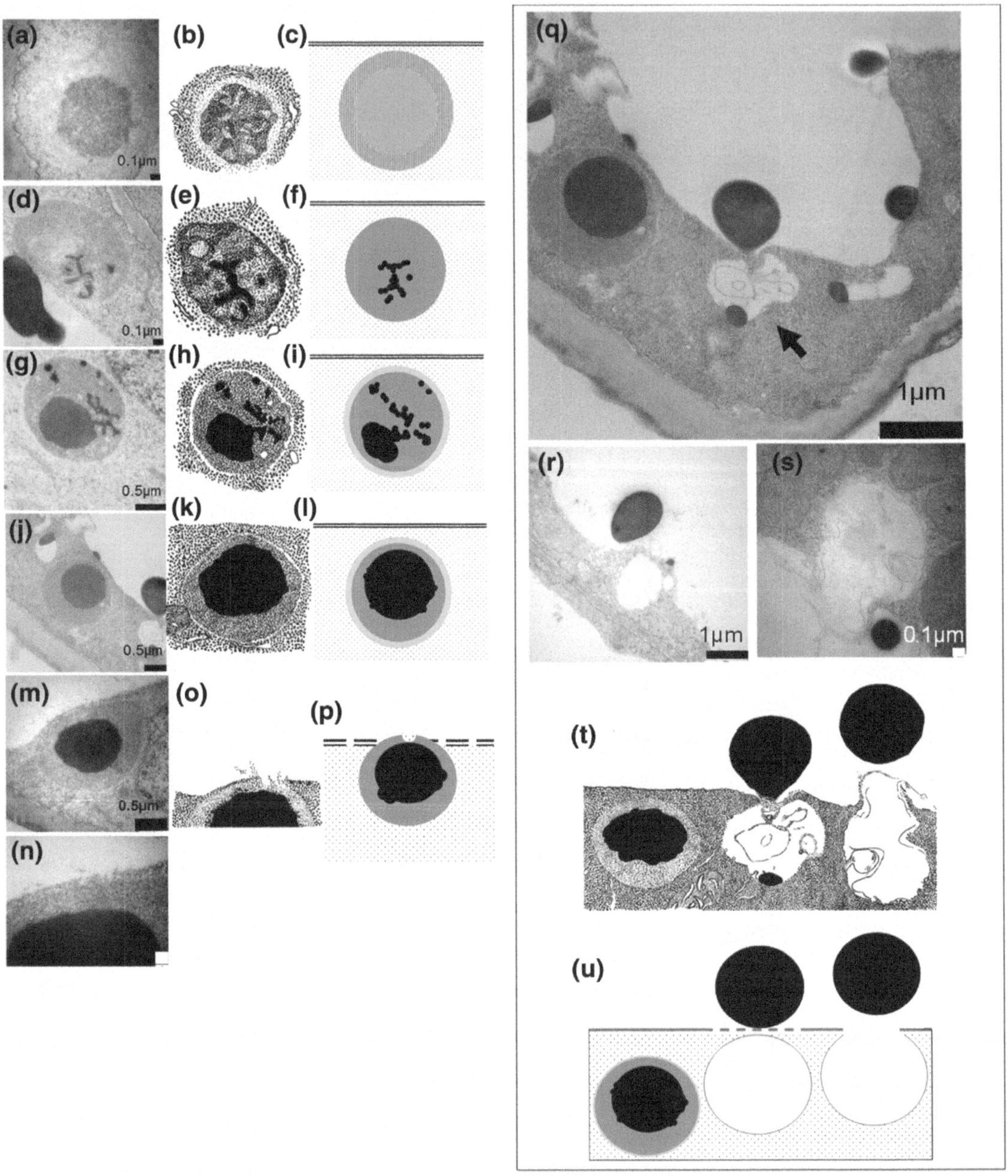

Fig. 10.8 The formation of plastidic rubber particles (PR) showing micrographic images **a**, **d**, **g**, **j**, **m**, **n**, **q**, **r**, **s**, illustrations **b**, **e**, **h**, **k**, **o**, **t**, and diagrams **c**, **f**, **i**, **l**, **p**, **u** of the process in which **a**, **b**, **c** Small rubber particles originate and; **d**, **e**, **f** Accumulate in laticifer plastids (LP); **g**, **h**, **i** The development of a new rubber particle body by agglomeration of small RP; **j**, **k**, **l** Formation of irregularly shaped PR; **m**, **n**, **o**, **p**, **q**, **r**, **s**, **t**, **u** Release of irregularly shaped PR into the vacuoles after which the empty plastid dissipates into the cytoplasm (**s**–**u**)

References

Agrawal AA, Konno K (2009) Latex: a model for understanding mechanisms ecology, and evolution of plant defense against herbivory. Annu Rev Ecol Evol Syst 40:311–331

Artschwager E (1943) Contribution to the morphology and anatomy of guayule (*Parthenium argentatum*) tech. Bull 842:1–33

Artschwager E, McGuire RC (1943) Contribution to the morphology and anatomy of the Russian dandelion (*Taraxacum kok-saghyz*). Tech Bull 843:1–24

Backhaus RA, Walsh S (1983) The ontogeny of rubber formation in guayule, *Parthenium argentatum* Gray. Bot Gaz 144:391–400

Bajpai S, Gupta AP, Gupta MM, Sharma S, Govil CM, Kumar S (2000) Inter-relation between descriptors and morphine yield in Asian germplasm of opium poppy *Papaver somniferum*. Gen Res Crop Evol. 47:315–322

Barton KE (2007) Early ontogenetic patterns in chemical defense in *Plantago* (Plantaginaceae): genetic variation and trade-offs. Am J Bot 94:56–66

Bonner J, Galston AW (1947) The physiology and biochemistry of rubber formation in plants. Bot Rev 13:543–596

Bonner J (1991) The history of rubber in guayule natural rubber: a technical publication with emphasis on recent findings. The University of Arizona, pp 1–16

Brandizzi F, Barlowe C (2013) Organization of the ER-Golgi interface for membrane traffic control. Nat Rev Mol Cell Biol. 6:382–392

Brown D, Feeney M, Ahmadi M, Lonoce C, Sajari R, Di Cola A, Frigerio L (2017) Subcellular localization and interactions among rubber particle proteins from *Hevea brasiliensis*. J Exp Bot 68:5045–5055

Buranov AU, Elmuradov BJ (2010) Extraction and characterization of latex and natural rubber from rubber-bearing plants. J Agric Food Chem 58:734–743

Cherian S, Ryu SB, Cornish K (2019) Natural rubber biosynthesis in plants, the rubber transferase complex, and metabolic engineering progress and prospects. Plant Biotechnol J 17(11):2041–2061

Cornish K (1993) The separate roles of plant *cis* and *trans* prenyl transferases in *cis*-1, 4-polyisoprene biosynthesis. Eur J Biochem 218:267–271

Cornish K (2001a) Similarities and differences in rubber biochemistry among plant species. Phytochem 57:1123–1134

Cornish K (2001b) Biochemistry of natural rubber, a vital raw material, emphasizing biosynthetic rate molecular weight and compartmentalization, in evolutionarily divergent plant species. Nat Prod Reps 18:1–8

Cornish K (2017) Alternative natural rubber crops: why should we care? Technol Innov 18:245–256

Cornish K, Backhaus RA (1990) Rubber transferase activity in rubber particles of guayule. Phytochem 29:3809–3813

Cornish K, Siler DJ, Grosjean OK, Goodman N (1993) Fundamental similarities in rubber particle architecture and function in three evolutionarily divergent plant species. J Nat Rub Ress 8:275–285

Cornish K, Bates GM, McNulty SK, Kopicky SE, Grewal S, Rossington J, Michel FC Jr, Walker S, Kleinhenz MD (2013) Buckeye gold storage: a study into rubber production in *Taraxacum kok-saghyz* with an emphasis on post-harvest storage. USA Tire Technol Int 10:36–38

de Faÿ E, Jacob JL (1989) Anatomical organization of the laticiferous system in the bark in physiology of rubber tree latex, pp 4–14

Dickenson PB (1969) Electron microscopical studies of latex vessel system of *Hevea brasiliensis*. J Rubb Res Inst Malaya 21:543–559

Donohoe BS, Kang BH, Gerl MJ, Gergeley ZR, McMichael CM, Bednarek SY, Staehelin LA (2013) *Cis*-Golgi cisternal assembly and biosynthetic activation occur sequentially in plants and algae. Traffic 14:551–567

Dusotoit-Coucaud A, Brunel N, Kongsawadworakul P, Viboonjun U, Lacointe A, Julien J, Chrestin H, Sakr S (2009) Sucrose importation into laticifers of *Hevea brasiliensis*, in relation to ethylene stimulation of latex production. Ann Bot 104:635–647

Dussourd DE, Eisner T (1987) Vein-cutting behavior: insect counterploy to the latex defense of plants. Science 237:898–901

Esau K (1965) Secretory structures in plant anatomy, 2nd edn. pp 308–335

Esau K (1977) Secretory structures in anatomy of seed plants, 2nd edn. pp 199–214

Gomez JB (1983) Anatomy of the laticiferous system in physiology of latex (rubber) production. Rubber Research Institute of Malaysia, Kuala Lumpur, pp 2–5

Goss R (1991) The morphology, anatomy and ultrastructure of guayule. In: Wayne J, Emily EW (eds) Guayule natural rubber: a technical publication with emphasis on recent findings, Guayule Administrative Management Committee and USDA Cooperative State Research Service. The University of Arizona, Arizona, pp 33–45

Hagel JM, Yeung EC, Facchini PJ (2008) Got milk? The secret life of laticifers. Trends Plant Sci 13:631–639

Harris N, Chrispeels MJ (1980) The endoplasmic reticulum of mung-bean cotyledons: quantitative morphology of cisternal and tubular ER during seedling growth. Planta 148:293–303

Hebant C (1981) Ontogénie des laticifères du système primaire de l'*Hevea brasiliensis*: Une Étude Ultrastructurale et Cytochimique. Can J Bot 59:974–985

Heinrich G (1967) Light and electron microscopic investigations on the milk tubes of *Taraxacum bicorne*. Flora 158:413–420

Hopkins WG, Hüner NPA (2008) Introduction to plant physiology, Chapter 27. pp 459–480

IRSG (2019) The world rubber industry outlook—review and prospects to 2028, pp 37–38

Jacob J, d'Auzac J, Prevôt J (1993) The composition of natural latex from *Hevea brasiliensis*. Clin Rev Allergy 11:325–337

Kashin A, Anfalov, VE, Demochko YA (2005) Studying allozyme variation in sexual and apomictic *Taraxacum* and *Pilosella* (Asteraceae) populations. Rus J Gen 41:144–154

Krishnakumar R, Cornish K, Jacob J (2001) Rubber biosynthesis in tapping panel dryness affected *Hevea* trees. J Rub Res 4:131–139

Langhans M, Marcote MJ, Pimpl P, Virgili-López G, Robinson DG, Aniento F (2008) In vivo trafficking and localization of p24 proteins in plant cells. Traffic 9:770–785

Lewinsohn TM (1991) The geographical distribution of plant latex. Chemoecol. 2:64–68

Macalood JS, Vicente HJ, Boniao RD, Gorospe JG, Roa EC (2013) Chemical analysis of *Carica papaya* L. crude latex. Am J Plant Sci 4:1941–1948

Mahlberg PG (1993) Laticifers: an historical perspective. Bot Rev 59:1–23

Men X, Wang F, Chen G, Zhang HB, Xian M (2018) Biosynthesis of natural rubber: current state and perspectives. Int J Mol Sci 20:E50. https://doi.org/10.3390/ijms20010050

Metcalfe CR (1967) Distribution of latex in plant kingdom. Econ Bot 21:115–127

Nessler CL, Mahlberg PG (1979) Plastids in laticifers of papaver. II. Enzyme cytochemistry of membrane-bound inclusions of laticifer plastids in *P. bracteatum* Lindl. (Papaveraceae). Am J Bot 66:274–279

Ohya N, Takizawa J, Kawahara S, Tanaka Y (1998) Molecular weight distribution of polyisoprene from *Lactarius volemus*. Phytochem 48:781–786

Ritzenthaler C, Laporte C, Gaire F, Dunoyer P, Schmitt C, Duval S, Piequet A, Loudes AM, Rohfritsch O, Stussi-Garaud C, Pfeiffer P (2002) Grapevine fanleaf virus replication occurs on endoplasmic reticulum-derived membranes. J Virol 76:8808–8819

Staehelin LA, Kang BH (2008) Nanoscale architecture of endoplasmic reticulum export sites and of Golgi membranes as determined by electron tomography. Plant Physiol 147:1454–1468

Vitale A, Ceriotti A, Denecke J (1993) The role of the endoplasmic reticulum in protein synthesis, modification and intracellular transport. J Exp Bot 44:1417–1444

Yamashita S, Yamaguchi H, Waki T, Aoki Y, Mizuno M, Yanbe F, Ishii T, Funaki A, Tozawa Y, Miyagi-Inoue Y, Fushihara K, Nakayama T Takahashi S (2016) Identification and reconstitution of the rubber biosynthetic machinery on rubber particles from *Hevea brasiliensis*. eLife 5:e19022

Perspectives and Ongoing Challenges

11

Katrina Cornish

Abstract

This new Rubber Tree Genome book includes the latest information on the status and application of modern molecular and genomic approaches and resources for natural rubber production, in multiple species, as well as detailed historical information on the origin and development of the current *Hevea brasiliensis* natural rubber industry. Furthermore, it includes new information on rubber particle ontogeny using the analogous species, *Taraxacum kok-saghyz*. In this chapter, I present some perspectives and challenges to our essential supplies of natural rubber, touching on the detailed information in this book and what I believe must be done in the future to ensure continued global rubber security. Also, I discuss some important research directions, rooted in the fundamental biology of rubber production, some of which are currently underdeveloped or largely ignored.

11.1 Rubber Security

Natural rubber production is a critical agricultural commodity essential to all sectors of manufacturing. Demand continues to increase but many challenges are faced by producers, trying to maintain current supplies and, once prices rise sufficiently, to increase supplies to meet increasing market demands. Disease, political disputes and global warming-induced climate change pose the greatest threats to rubber production. Based on the global spread of fatal tree diseases, it appears inevitable that deadly South American Leaf Blight (SALB, caused by *Microcyclus ulei*) will cross to Southeast Asia, where the industry is yet to introduce resistant trees at scale. Even before this occurs, the current unsustainably low price of natural rubber, which does not provide a living wage to rubber tappers in most producing countries, leads to labor shortages and threatens supplies. The industry is yet to develop effective mechanized production methods, and the current hand tapping latex collection system has changed little, since it was developed by Henry Ridley, Director of the Singapore Botanic Gardens, in 1889 (Wycherley 1958).

Price increase to sustainable levels is now essential because the global moratorium on further deforestation of tropical rain forests prevents planting on newly cleared forest lands in SALB-free regions. Thus, production cannot keep moving to areas with lower labor costs, and

K. Cornish (✉)
Departments of Horticulture and Crop Science, and Food, Agricultural and Biological Engineering, Ohio Agricultural Research and Development Center, College of Food Agricultural and Environmental Science, The Ohio State University, 1680 Madison Avenue, Wooster, OH 44691-4096, USA
e-mail: cornish.19@osu.edu

M. Matsui and K.-S. Chow (eds.), *The Rubber Tree Genome*, Compendium of Plant Genomes,
https://doi.org/10.1007/978-3-030-42258-5_11

importing cheap immigrant labor has limited the effectiveness when tappers have to live in the production area.

It is possible to increase production from existing plantations, but this is difficult to accomplish. For example, the application of ethephon can increase rubber yields (Abraham et al. 1968; Yew et al. 1998) but this practice has not yet been widely adapted because of the perceived risks associated with ethephon. Even though effective ethephon management practices are now well understood, earlier poor management led to over-application and induced tapping panel dryness (Sivakumaran and Pakianathan 1983). Smallholders have not forgotten this effect, and many remain reluctant to try again. Also, crop management in Indonesia could be significantly improved to bring its production levels up to those achieved in neighboring countries from identical rubber tree clones. However, this requires government investment and training of extension professionals.

Conventional plant breeding is always very slow in tree species; for example, in China, only five *H. brasiliensis* generations have been achieved in 100 years (see Chap. 2), partly because rubber yields from mature trees under regular tapping cycles are very difficult to predict from seedlings or virgin trees. Nonetheless, even though breeding is very time-consuming, the Malaysian Rubber Board's pedigree collection bred in phases since 1928 (Ong et al. 1994) and more recent collections of wild Brazilian germplasm (Adifaiz et al. 2017; Ong and Tan 1987) appear to contain some individual genotypes with outstanding potential. These collections provide an essential source of experimental material for researchers.

The pressure on the rubber supply, coupled with an increasing awareness of rubber supply line insecurity, and price volatility largely caused by futures traders, has led to considerable investment by industry (mostly tire companies) in the development of alternative rubber-producing species in temperate regions to increase the biodiversity of rubber production. The leading species are *T. kok-saghyz* and *Parthenium argentatum*, and with *H. brasiliensis*, these can allow rubber to be produced in most parts of the world, not just in specific tropical regions. However, issues of scale remain significant impediments to the alternative crops because small acreages and pilot processing plants cannot produce rubber at the same price as vast acreages of hand-tapped *H. brasiliensis* (Cornish 2017).

11.2 The Biology of Rubber Production

Biology is the foundation of rubber production, and it is the coordinated whole plant system that must be understood to maximize the efficiency of experiments, practices and genetic approaches designed to increase rubber yield.

11.2.1 A Consideration of Root Stocks

Commercial rubber trees are produced by grafting buds harvested from high yielding clonal trees onto seedling root stocks and, once sufficiently established, the original shoot is excised. However, it should not be forgotten that latex production is a product of biology and is strongly influenced by the physiology of the intact tree. It is likely that different performance among identical clones, and among disparate soils and environments, reflects the root stock variation which, despite a limited ancestry, is still genetically and physiologically heterogeneous.

Inbred lines of *H. brasiliensis* are not yet available and not likely to be achieved through conventional inbreeding of this tree species. However, reciprocal grafts could easily be done by obtaining seeds generated from clonal plants. Also, since *H. brasiliensis* is amenable to tissue culture, it is feasible to generate cloned plants from new germplasm, and hence cloned root stocks from different genotypes. Both methods could be very useful tools to investigate the influence of root stock genetics, physiology and biochemistry on latex yield in the above ground laticifers. For example, rubber yield and molecular weight are strongly affected by the concentration of monomeric (isopentenyl pyrophosphate, IPP) and initiating (farnesyl pyrophosphate, FPP) rubber polymer substrates in the laticifer but are even more sensitive to the concentration of the essential magnesium cation activator in the laticifer cytoplasm (da Costa et al. 2004, 2006; Scott et al. 2003). Magnesium is taken up from the soil in which the tree is grown and so is dependent upon the efficiency of the magnesium transmembrane transporters in the root cell plasmalemma and then the internal transporters in the vasculature and laticifers. Cytoplasmic magnesium concentration can be changed by differential expression of magnesium transporters, and also very probably by the concentration of soluble magnesium in the soil which, in turn, can be changed by fertilizers and chelators. I find it surprising that the effect of root stock and magnesium continue to be largely ignored by the rubber production industry when both are readily addressable research questions.

11.2.2 Molecular Resources, Approaches and the Biochemistry of Rubber Biosynthesis

Molecular resources have only recently been developed for rubber-producing species, and several genomes, transcriptomes and proteomes are now available for *H. brasiliensis* and other species (Chow et al. 2007; Dai et al. 2013; Lau et al. 2016; Lin et al. 2017; Luo et al. 2017; Mantello et al. 2014; Ponciano et al. 2012; Rahman et al. 2013; Stonebloom and Scheller 2019; Tang et al. 2016; Tong et al. 2017; Wahler et al., 2012; Wang et al., 2015). However, the *H. brasiliensis* genome is large and contains more than 70% repeat sequences (Tang et al. 2016). Additional resources include chloroplast and mitochondrial genome sequences (Shearman et al. 2014; Tangphatsornruang et al. 2011; Zhang et al. 2017). Chloroplasts and mitochondria are important producers of IPP and can export excess IPP into the cytoplasm where it supplements the IPP pool available for rubber polymerization. The high $K_m{}^{IPP-Mg}$ of all rubber transferases (RT-ases) characterized to date means that rubber is only made when IPP is accumulated to concentrations non-limiting to all the other IPP-requiring enzymes present in the cytosol (Cornish 2001a, b). New technologies, such as CRISPR/Cas9 gene editing have also been used in rubber-producing plants (Iaffaldano et al. 2016).

Rubber production requires complex developmental regulation (Chow et al. 2012). For example, an important yield determinant in laticiferous species is the number of laticifers per unit cross-sectional area. This was made very clear in the investigation of rubber particle ontogeny and laticifer development in *Taraxacum kok-saghyz* (Abdul Ghaffar and Cornish 2019). The genetics of laticifer ontogeny are not well understood (Sando et al. 2009) but, since rubber particles appear before laticifers in *T. kok-saghyz*, it seems possible that rubber particles may trigger laticifer development. The induction of more rubber particles may concomitantly lead to the induction of more laticifers. Similarly, the creation of rubber particles may be triggered by the expression of all or some of the subunits making up the RT-ase complex (Berthelot et al. 2014; Cherian et al. 2019; Collins-Silva et al. 2012; Cornish et al. 2018; Dai et al. 2017; Lakusta et al. 2019; Yamashita et al. 2016). Overexpression of a critical subunit might lead to expression of the other components or all these processes may be controlled by a key transcription factor. Transcription factors certainly operate in *H. brasiliensis* (Li et al. 2016). The available genomic resources now make it possible to address some of these questions.

The regulation of rubber biosynthesis is somewhat distinct from the machinery required for synthesis. By this I mean that although rubber obviously cannot be made in species without RT-ases, laticifers are not always used to make rubber particles (*Parthenium argentatum* is the best characterized non-laciferous rubber-producing species). Also, even when sufficient RT-ase complexes are present to support higher rubber synthetic rate, molecular studies, which alter rubber particle protein composition or substrate availability and lead to increased amounts of rubber in seedlings, have yet to maintain such gains in field crops (Dong et al. 2013; Hillebrand et al. 2012; Placido et al. 2019; Stolze et al. 2017). Clearly, feedback loops are acting in rubber plants which prevent a perceived over-production of rubber, presumably because of some adverse effect on primary metabolism, possibly due to substrate deficits or accumulation of toxic byproducts. Such feedback clearly changes under different environmental conditions, because rubber biosynthetic rates change with environmental factors (Benedict et al. 2008; Cornish and Backhaus 2003; Ji et al. 1993; Kreuzberger et al. 2016). Thus, a fundamental understanding of the genetic and biochemical regulation of rubber biosynthesis is required to inform effective molecular modification to predictably increase rubber yield.

In-depth molecular and metabolomic investigation of plants within individual species with naturally large differences in rubber content will certainly improve our fundamental understanding of rubber biosynthesis. In addition to the issues raised in the previous paragraph, these plants perhaps have even more value in the development of selectable molecular markers, quantitative trait loci and molecular maps—all of which may be used to inform selection and molecular approaches, and be applicable to seedlings without requiring initial phenotyping of mature plants (Luo et al. 2017; Mantello et al. 2014; Rosa et al. 2018; Shearman et al. 2015). The Malaysian collection may also be sufficiently diverse to permit meaningful genome-wide association mapping approaches. These germplasm sources would become even more powerful if homozygous lines could be developed from high and low rubber yielding parents. Haploid induction from germ cells, or by manipulation of *cen* genes (Britt and Kuppu 2016; Karimi-Ashtiyani et al. 2015; Ravi and Chan 2010) followed by chromosome doubling via colchicine poisoning (Luo et al. 2018), are two potential avenues to achieve this goal.

Interspecific approaches have already yielded considerable fundamental biochemical and genetic information about rubber biosynthesis and can expand on what may be provided from a single species, especially a tree such as *H. brasiliensis*. The rubber and rubber particle composition produced by *T. kok-saghyz* are very similar to those produced by *H. brasiliensis* (Cornish et al. 2015) indicating either strong conservation of the rubber biosynthetic machinery or strong convergent evolution. In all species investigated so far, the basic rubber particle structure has been highly conserved within the Eudicots, where all higher plant rubber-producing plants are classified (Metcalfe 1967). Rubber particles are always made in the cytosol, are often moved into vacuolar storage, have a uni-lamellar membrane (although the membrane lipid and protein composition may differ), have a membrane-bound RT-ase complex, use the same substrates and cofactors, and compartmentalize the rubber polymers into the particle interior (Cornish 2001a, b; Cornish et al. 1999; Siler et al. 1997; Singh et al. 2003). However, the details of rubber production differ among plants and these differences may be exploited to improve yield or quality in other species. This potential has become extremely important as additional rubber-producing species enter the agricultural arena and contribute their own newly developed genomic resources.

11.3 Conclusions

The information in this book provides a consolidated foundation of rubber genome resources currently available and some of the associated studies currently published, focused primarily on *H. brasiliensis*. However, comparative genomic studies are becoming increasingly informative in identifying key pathways and genes associated with or key to rubber biosynthesis and yield, and there are many fascinating research questions still awaiting elucidation. As alternative rubber-producing crops are introduced, it is important to maintain polymer quality so that the needs of the rubber manufacturing industry continue to be met. In addition, premium niche applications must be identified to permit these crops to be scaled up—they have a very long way to go before they can become a rubber commodity producer like *H. brasiliensis*. Nonetheless, biodiversity of the global natural rubber supply is essential to rubber security and interdisciplinary research on multiple rubber-producing species is needed to achieve this goal. Natural rubber is, and will continue to be, a major contributor to the circular bioeconomy and sustainable production is essential.

References

Abdul Ghaffar MA, Cornish K (2019) Histological study of laticifer development in *Taraxacum kok-saghyz* roots. J Rubb Sci, in press

Abraham PD, Wycherley PR, Pakianathan SW (1968) Stimulation of latex flow *in Hevea brasiliensis* by 4-amino-3, 5, 6-trichloropicolinic acid and 2-chloroethanephosphonic acid. J Rubb Res Inst Malaya 20(5):291–305

Adifaiz AF, Maiden NA, Shamin NA, Zarawi AG, Rafii MY (2017) Potential genotypes of the 1995 RRIM *Hevea* germplasm collection for future rubber breeding and selection programme. J Rubb Res 20 (4):242–260

Benedict CR, Greer PJ, Foster MA (2008) The physiological and biochemical responses of guayule to the low temperature of the Chihuahuan Desert in the biosynthesis of rubber. Ind Crops Prod 27:225–235

Berthelot K, Lecomte S, Estevez Y, Zhendre V, Henry S, Thevenot J, Dufourc EJ, Alves ID, Peruch F (2014) Rubber particle proteins, HbREF and HbSRPP, show different interactions with model membranes. Biochim Biophys Acta 1838:287–299

Britt AB, Kuppu S (2016) Cenh3: an emerging player in haploid induction technology. Front Plant Sci 7:357

Cherian S, Ryu SB, Cornish K (2019) Natural rubber biosynthesis in plants, the rubber transferase complex, and metabolic engineering progress and prospects. Plant Biotechnol J 17(11):2041–2061

Chow KS, Wan KL, Isa MN, Bahari A, Tan SH, Harikrishna K, Yeang HY (2007) Insights into rubber biosynthesis from transcriptome analysis of *Hevea brasiliensis* latex. J Exp Bot 58:2429–2440

Chow KS, Mohd-Noor M, Bahari A, Ghazali AK, Alias H, Zainuddin ZM, Hoh CC, Wan KI (2012) Metabolic routes affecting rubber biosynthesis in *Hevea brasiliensis* latex. J Exp Bot 63:1863–1871

Collins-Silva J, Nural AS, Skaggs A, Scott DJ, Hathwai U, Woolsey R, Schegg K, McMahan CM, Whalen M, Cornish K, Shintani DK (2012) Altered levels of the *Taraxacum kok-saghyz* (Russian dandelion) small rubber particle protein, TkSRPP3, result in qualitative and quantitative changes in rubber metabolism. Phytochem 79:46–56

Cornish K (2001a) Similarities and differences in rubber biochemistry among plant species. Phytochem 57:1123–1134

Cornish K (2001b) Biochemistry of natural rubber, a vital raw material, emphasizing biosynthetic rate, molecular weight and compartmentalization, in evolutionarily divergent plant species. Nat Prod Rep 18:182–189

Cornish K (2017) Alternative natural rubber crops: why should we care? Technol Innov 18:245–256

Cornish K, Backhaus RA (2003) Induction of rubber transferase activity in guayule (*Parthenium argentatum* Gray) by low temperatures. Ind Crops Prod 17:83–92

Cornish K, Wood DF, Windle JJ (1999) Rubber particles from four different species, examined by transmission electron microscopy and electron-paramagnetic-resonance spin labeling, are found to consist of a homogeneous rubber core enclosed by a contiguous, monolayer biomembrane. Planta 210:85–96

Cornish K, Xie W, Kostyal D, Shintani D, Hamilton RG (2015) Immunological analysis of the alternate natural rubber crop *Taraxacum kok-saghyz* indicates multiple proteins cross-reactive with *Hevea brasiliensis* latex allergens. J Biotechn Biomat 5:201–207

Cornish K, Scott DJ, Xie W, Mau CDJ, Zheng YF, Liu X-H, Prestwich GD (2018) Unusual subunits are directly involved in binding substrates for natural rubber biosynthesis in multiple plant species. Phytochem 156:55–72

da Costa BMT, Keasling JD, Cornish K (2004) Regulation of rubber biosynthetic rate and molecular weight in *Hevea brasiliensis* by metal cofactor. Biomacromol 6:279–289

da Costa BMT, Keasling JD, McMahan CM, Cornish K (2006) Magnesium ion regulation of in vitro rubber biosynthesis by *Parthenium argentatum* Gray. Phytochem 67:1621–1628

Dai L, Kang G, Li Y, Nie Z, Duan C, Zeng R (2013) In-depth proteome analysis of the rubber particle of *Hevea brasiliensis* (para rubber tree). Plant Mol Biol 82:155–168

Dai L, Nie Z, Kang G, Li L, Zheng R (2017) Identification and subcellular localization analysis of two rubber elongation factor isoforms on *Hevea brasiliensis* rubber particles. Plant Physiol Biochem 111:96–106

Dong N, Ponciano G, McMahan CM, Coffelt TA, Johnson L, Creelman RA, Whalen MA, Cornish K (2013) Overexpression of 3-hydroxy-3-methylglutaryl coenzyme A reductase in *Parthenium argentatum* (guayule). Ind Crops Prod 46:15–24

Hillebrand A, Post JJ, Wurbs D, Wahler D, Lenders M, Krzyzanek V, Prüfer D, Gronover CS (2012) Down-regulation of small rubber particle protein expression affects integrity of rubber particles and rubber content in *Taraxacum brevicorniculatum*. PLoS ONE 7: e41874

Iaffaldano B, Zhang Y, Cornish K (2016) CRISPR/Cas9 genome editing of rubber producing dandelion

Taraxacum kok-saghyz using *Agrobacterium rhizogenes* without selection. Ind Crops Prod 98:356–362

Ji W, Benedict CR, Foster MA (1993) Seasonal variation in rubber biosynthesis, 3- hydroxy-3-methylglutaryl-coenzyme A reductase, and rubber transferase activities in *Parthenium argentatum* in the Chihuahuan Desert. Plant Physiol 103:535–542

Karimi-Ashtiyani R, Ishii T, Niessen M, Stein N, Heckmass S, Gurushidze M, Banaei-Moghaddam AM, Fuchs J, Schubert V, Koch K, Weiss O, Demidov D, Schmidt, K, Kumlehn J, Houben A (2015) Point mutation impairs centromeric CENH3 loading and induces haploid plants. Proc Natl Acad Sci U S A 112:11211

Kreuzberger M, Hahn T, Zibek S, Schiemann J, Thiele K (2016) Seasonal pattern of biomass and rubber and inulin of wild Russian dandelion (*Taraxacum kok-saghyz* L. Rodin) under experimental field conditions. Eur J Agro 80:66–77

Lakusta AM, Kwon M, Kwon EG, Stonebloom S, Scheller HV, Ro D (2019) Molecular studies of the protein complexes involving *cis*-prenyltransferase in guayule *Parthenium argentatum*), an alternative rubber producing plant. Front Plant Sci 10:165

Lau NS, Makita Y, Kawashima M, Taylor TD, Kondo S, Othman AS, Shu-Chien AC, Matsui M (2016) The rubber tree genome shows expansion of gene family associated with rubber biosynthesis. Sci Rep 6:28594

Li HL, Wei LR, Guo D, Wang Y, Zhu JH, Chen XT, Peng SQ (2016) HbMADS4, a MADS-box transcription factor from *Hevea brasiliensis*, negatively regulates HbSRPP. Front Plant Sci 7:1709

Lin T, Xu X, Ruan J, Liu S, Wu S, Shao X, Wang X, Gan L, Qin B, Yushuang Yang Y, Cheng Z, Yang S, Zhang Z, Xiong G, Huang S, Yu H, Li J (2017) Genome analysis of *Taraxacum kok-saghyz* Rodin provides new insights into rubber biosynthesis. Natl Sci Rev 5:78–87

Luo Z, Iafaldano BJ, Zhuang X, Fresnedo-Ramırez J, Cornish K (2017) Analysis of the first *Taraxacum kok-saghyz* transcriptome reveals potential rubber yield related SNPs. Sci Rep 7:9939

Luo Z, Iaffaldano BJ, Cornish K (2018) Colchicine-induced polyploidy has the potential to improve rubber yield in *Taraxacum kok-saghyz*. Ind Crops Prod 112:75–78

Mantello CC, Cardosa-Silva CB, da Silva CC, de Souza LM, Scaloppi Junior EJ, Goncalves PS, Vicentini R, de Souza AP (2014) De novo assembly and transcriptome analysis of the rubber tree (*Hevea brasiliensis*) and SNP markers development for rubber biosynthesis pathways. PLoS ONE 9:e102665

Metcalfe CR (1967) Distribution of latex in the plant kingdom. Econ Bot 21:115–127

Ong SH, Tan H (1987) Utilization of *Hevea* genetic resources in the RRIM. Malays Appl Biol 16(1):145–155

Ong SH, Ramli O, Othman H, Masahuling B, Naimah I (1994) Rubber breeding, progress and strategies to meet future needs of the plantation industry. In: Proceedings of the International Planters Conference, 24–26, Kuala Lumpur, pp 53–70

Placido D, Dong N, Dong C, Cruz VMV, Dierig D, Cahoon RE, Kang B, Huynh T, Whalen M, Ponciano G, McMahan CM (2019) Downregulation of a CYP74 rubber particle protein increases natural rubber production in *Parthenium argentatum*. Front Plant Sci 10:760

Ponciano G, McMahan CM, Xie W, Lazo GR, Coffelt TA, Collins-Silva J, Nural-Taban A, Gollery M, Shintani DK, Whalen MC (2012) Transcriptome and gene expression analysis in cold-acclimated guayule (*Parthenium argentatum*) rubber-producing tissue. Phytochem 79:57–66

Rahman AYA, Usharraj A, Misra B, Thottathil G, Jayasekaran K, Feng Y, Hou S, Ong SY, Ng FL, Lee LS, Tan HS, Sakaff MKLM, Teh BS, Khoo B, Badai SS, Aziz NA, Yuryev A, Knudsen B, Dionne-Laporte A, Mchunu N, Yu Q, Langston B, Freitas TA, Young A, Chen R, Wang L, Najimudin N, Saito J, Alam M (2013) Draft genome sequence of the rubber tree *Hevea brasiliensis*. BMC Genomics 14:75

Ravi M, Chan SWL (2010) Haploid plants produced by centromere-mediated genome elimination. Nature 464:615

Rosa JRBF, Mantello CC, Garcia D, De Souza LM, Da Silva CC, Gazaffi R, Da Silva CC, Toledo-Silva G, Cubry P, Garcia AaF, De Souza AP, Le Guen V (2018) QTL detection for growth and latex production in a full-sib rubber tree population cultivated under suboptimal climate conditions. BMC Plant Biol 18:223

Sando T, Hayashi T, Takeda T, Akiyama Y, Nakazawa Y, Fukusaki E, Kobayashi A (2009) Histochemical study of detailed laticifer structure and rubber biosynthesis-related protein localization in *Hevea brasiliensis* using spectral confocal laser scanning microscopy. Planta 230:215–225

Scott DJ, da Costa BMT, Espy SC, Keasling JD, Cornish K (2003) Activation and inhibition of rubber transferases by metal cofactor and pyrophosphate substrate. Phytochem 64:121–132

Shearman JR, Sangsrakru D, Ruang-Areerate P, Sonthirod C, Uthaipaisanwong P, Yoocha T, Poopear S, Theerawattanasuk K, Tragoonrung S, Tangphatsornruang S (2014) Assembly and analysis of a male sterile rubber tree mitochondrial genome reveals DNA rearrangement events and a novel transcript. BMC Plant Biol 14:45

Shearman JR, Sangsrakru D, Jomchai N, Ruang-Areerate P, Sonthirod C, Naktang C, Theerawattanasuk K, Tragoonrung S, Tangphatsornruang S (2015) SNP identification from RNA sequencing and linkage map construction of rubber tree for anchoring the draft genome. PLoS ONE 10:e0121961

Siler DJ, Goodrich-Tanrikulu M, Cornish K, Stafford AE, McKeon TA (1997) Composition of rubber particles of *Hevea brasiliensis*, *Parthenium argentatum*, *Ficus elastica* and *Euphorbia lactiflua* indicates unconventional surface structure. Plant Physiol Biochem 35:881–889

Singh AP, Wi SG, Chung GC, Kim YS, Kang H (2003) The micromorphology and protein characterization of rubber particles in *Ficus carica*, *Ficus benghalensis* and *Hevea brasiliensis*. J Exp Bot 54:985–992

Sivakumaran S, Pakianathan SW (1983) Studies on tree dryness: A simple and rapid method of inducing dryness in *Hevea* trees. J Rubb Res Inst Malays 31 (2):88–101

Stolze A, Wanke A, van Deenen N, Geyer R, Prüfer D, Gronover CS (2017) Development of rubber-enriched dandelion varieties by metabolic engineering of the inulin pathway. Plant Biotechnol J 15:740–753

Stonebloom SH, Scheller HV (2019) Transcriptome analysis of rubber biosynthesis in guayule (*Parthenium argentatum* Gray). BMC Plant Biol 19:71

Tang C, Yang M, Fang Y, Luo Y, Gao S, Xiao X, An Z et al (2016) The rubber tree genome reveals new insights into rubber production and species adaptation. Nat Plants 2:16073

Tangphatsornruang S, Uthaipaisanwong P, Sangsrakru D, Chanprasert J, Yoocha T, Jomchai N, Tragoonrung S (2011) Characterization of the complete chloroplast genome of *Hevea brasiliensis* reveals genome rearrangement, RNA editing sites and phylogenetic relationships. Gene 475:104–112

Tong Z, Wang D, Sun Y, Yang Q, Meng X, Wang L, Feng W, Li L, Wurtele ES, Wany X (2017) Comparative proteomics of rubber latex revealed multiple protein species of REF/SRPP family respond diversely to ethylene stimulation among different rubber tree clones. Int J Mol Sci 18:958–972

Wahler D, Colby T, Kowalski NA, Harzen A, Wotzka SY, Hillebrand A, Fischer R et al (2012) Proteomic analysis of latex from the rubber-producing plant *Taraxacum brevicorniculatum*. Proteomics 12:901–905

Wang X, Wang D, Sun Y, Yang Q, Chang L, Wang L, Meng X, Huang Q, Jin X, Tong Z (2015) Comprehensive proteomics analysis of laticifer latex reveals new insights into ethylene stimulation of natural rubber production. Sci Rep 5:13778

Wycherley PR (1958) The Singapore Botanic Gardens and rubber in Malaya. Gar Bull Singap 17:175–186

Yamashita S, Yamaguchi H, Waki T, Aoki Y, Mizuno M, Yanbe F, Ishii T, Funaki A, Tozawa Y, Miyagi-Inoue Y, Fushihara K, Nakayama T, Takahashi S (2016) Identification and reconstitution of the rubber biosynthetic machinery on rubber particles from *Hevea brasiliensis*. eLife 5:e19022

Yew FK, Sivakumaran S, Chong DT (1998) RRIMFLOW system of exploitation: recent improvements and update on yield performance. In: Seminar of low intensity tapping system, Malaysian Rubber Board, pp 42–65

Zhang Y, Iaffaldano BJ, Zhuang X, Cardina J, Cornish K (2017) Chloroplast genome resources and molecular markers differentiate rubber dandelion species from weedy relatives. BMC Plant Biol 17:34–48

The manufacturer's authorised representative in the EU is Springer
Nature Customer Service Centre GmbH, Europaplatz 3, 69115 Heidelberg,
Germany. If you have any concerns regarding our products, please
contact ProductSafety@springernature.com

Printed and bound by CPI Group (UK) Ltd, Croydon, CR0 4YY
30/09/2026
02225646-0020